MODERN

TRIGONOMETRY

ABOUT THE AUTHOR

Kaj L. Nielsen received the degrees of B.S. from the University of Michigan, M.A. from Syracuse University, and Ph.D. from the University of Illinois. He has held teaching positions at Syracuse, Illinois, Brown, Butler, Purdue, and Louisiana State University. He has been associated with a number of industries and the U. S. Navy Department as a research engineer and scientist. He is presently Director of the Systems Analysis Division at Battelle Memorial Institute in Columbus, Ohio. Dr. Nielsen has written numerous articles based upon original research in mathematics and published in leading mathematical and engineering journals. He is the author or coauthor of *College Mathematics*, *Differential Equations*, *Logarithmic and Trigonometric Tables*, *Plane and Spherical Trigonometry*, and *Problems in Plane Geometry*, in the College Outline Series, and of *Mathematics for Practical Use*, in the Everyday Handbooks Series. Dr. Nielsen is listed in *American Men of Science*, *Who is Who in the Midwest*, *Who is Who in the Computer Field*, and *The World Directory of Mathematicians*.

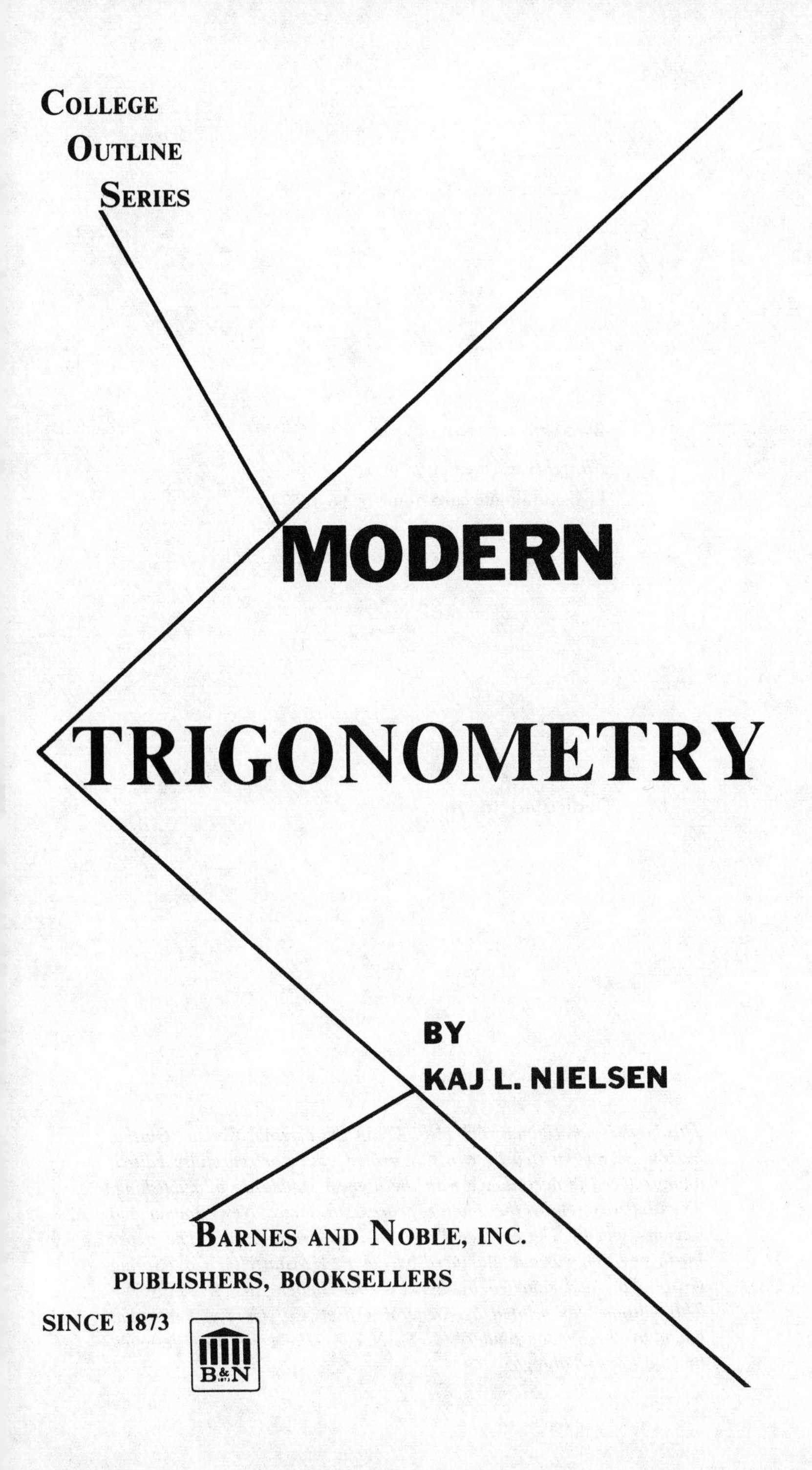

COLLEGE OUTLINE SERIES

MODERN TRIGONOMETRY

BY

KAJ L. NIELSEN

BARNES AND NOBLE, INC.
PUBLISHERS, BOOKSELLERS
SINCE 1873
B&N

L. C. catalogue card number: 66-18573

Dedicated to Jan

This book is an original work (No. 47) in the original College Outline Series. It was written by a distinguished educator, carefully edited, and produced in accordance with the highest standards of publishing. The text was set by the Photon process in Times New Roman and Techno Bold by The Science Press, Inc. (Ephrata, Pa.). The paper for this edition was manufactured by the P. H. Glatfelter Co. (Spring Grove, Pa.) and supplied by Perkins and Squier Co. (N.Y., N.Y.). This edition was printed by General Offset Co. (N.Y., N.Y.) and bound by the Sendor Bindery (N.Y., N.Y.). The cover was designed by Rod Lopez-Fabrega.

Preface

This book covers the fundamentals of a course in plane and spherical trigonometry. It may be read by those who are beginning the study of trigonometry; or it may be used as a supplement to any standard textbook or as a review for the reader who has already completed a course in trigonometry. The subject is presented in a manner recently referred to as the modern approach; the subject of trigonometry is viewed as that part of mathematical investigations which is performed by means of trigonometric functions, and the real number system is used for the argument of the trigonometric function.

The definitions of the functions are based on the formation of ordered pairs and start with the real numbers as the argument. This is followed by the specialization to the trigonometric functions of the general angle and finally to the restricted domain of acute angles for the numerical values of the trigonometric functions. Considerable space is devoted to the logical development of the functions, their inverses, properties, graphical representation; their relation to exponential, logarithmic, and hyperbolic functions and complex numbers; and their series representation.

Although this book is primarily concerned with the analytical rather than computational aspects of trigonometry,* several sections on the application of the trigonometric functions to plane and solid geometry are included. A small set of tables can be found at the back of the book, and reference is made to more extensive tables. The principles involved in using the tables are explained in the Appendix.

It is the author's belief that a thorough knowledge of any branch of mathematics cannot be obtained without solving problems. Consequently, a few typical exercises are included at

*For a presentation of computational trigonometry, see Kaj L. Nielsen and John H. VanLonkhuyzen, *Plane and Spherical Trigonometry* (New York: Barnes and Noble, 1954).

various places in the book. Answers are given for most of the exercises so that the reader may check his work. A set of examinations, with answers, is also furnished; it is hoped that it may aid the student in preparing for any examination or in testing his mastery of the subject.

Sprinkled throughout the book are sections of a more advanced or detailed nature. These are marked with a ⋆ and may be omitted without losing continuity.

The author gratefully acknowledges his indebtedness to his many students who, during his years of teaching this subject, have contributed many viewpoints which are incorporated in this presentation. The author is indebted to Joan Katkocin, Dr. Gladys Walterhouse, and other staff members of Barnes and Noble, Inc., who did their usual excellent job in arriving at the finished product.

K.L.N.

Columbus, Ohio

Table of Contents

5 TRIGONOMETRIC EQUATIONS

6 EXPONENTIAL AND LOGARITHMIC FUNCTIONS

7 GEOMETRIC APPLICATIONS

8 COMPLEX NUMBERS

9 HYPERBOLIC FUNCTIONS

10 SERIES

11 SPHERICAL TRIANGLES

Tabulated Bibliography of Standard Textbooks

This *College Outline* is keyed to standard textbooks in two ways.

1. If you are studying one of the following textbooks, consult the cross references here listed to find which pages of this *Outline* summarize the appropriate chapter of your text. (Roman numerals refer to the textbook chapters, Arabic figures to the corresponding pages of this *Outline*.) If your book includes Algebra, see also *College Outline Series* No. 105, *College Mathematics* by Kaj L. Nielsen.

2. If you are using this *Outline* as your basis for study and need a different treatment of a topic, consult the pages of any of the standard textbooks as indicated in the Quick Reference Table on pages xiii–xvi.

Brink, *Plane Trigonometry*, 3rd ed., 1959, Appleton-Century-Crofts.
I(1–19); II(6–9,20–22); III(22–30,117–122); IV(23–27,64–87); V(17,30–32); VI(36–44,88–95); VII(44–63,95–102); VIII(103–114); IX(117–122); X(122–140); XI(141–174).

Brixey and Andree, *Modern Trigonometry*, 1955, Holt, Rinehart and Winston.
I(1–19); II(6–9,20–29); III(103–114); IV(Appendix); V(20–23,64–87); VI(30–33, 36–63); VII(80–83); VIII(141–151); IX(none).

Cameron, *Algebra and Trigonometry*, rev. ed., 1965, Holt, Rinehart and Winston.
I(1–19); II–VI(none); VII(165–169); VIII(103–114); IX(20–35); X(36–87); XI(141–151); XII–XIII(none).

Fisher and Ziebur, *Integrated Algebra and Trigonometry*, 1958, Prentice-Hall.
I(none); II(3–9,14–18); III(103–114); IV(20–35); V(141–151); VI–VIII(none); IX(160–174); X(17,30–34,88–102).

Fort, *Trigonometry and the Elementary Transcendental Functions*, 1963, Macmillan.
I(3–9); II(1–19); III(6–9); IV(20–30); V(117–122); VI(36–43); VII(46–61); VIII(36–39,64–87); IX(none); X(160–174); XI(64–87,169–174); XII(103–114); XIII(Appendix); XIV(17,30–34); XV(88–102); XVI(122–134); XVII(Appendix); XVIII(141–151).

Fuller, *Plane Trigonometry*, 2nd ed., 1959, McGraw-Hill.
I(1–14); II(20–30); III(Appendix); IV(117–122); V(36–43); VI(44–61); VII(64–87); VIII(102–114); IX(122–134); X(122–134); XI(17,30–34,88–102); XII(141–151).

Goodman, *Plane Trigonometry*, 1959, Wiley.
I(20–30,117–122); II(103–114,Appendix); III(Appendix); IV(20–30); V(36–43); VI(122–140); VII(44–61); VIII(6–9,11–13); IX(64–87); X(88–102); XI(7,30–34); XII(122–134); XIII(134–140); XIV(141–151).

Hall and Kattsoff, *Modern Trigonometry*, 1961, Wiley.
I(none); II(1–19); III(6–9); IV(20–34); V(29,Appendix); VI(Appendix); VII(64–87); VIII(103–114); IX(117–122); X(36–43); XI(88–102); XII(122–134); XIII(122–134); XIV(44–61); XV(141–151).

Hall and Kattsoff, *Unified Algebra and Trigonometry*, 1962, Wiley.
I(1–19); II(none); III(3–6); IV(6–9,13,20–29,38–39); V(21–35); VI–IX(none); X(103–114,Appendix); XI(117–140); XII(44–61); XIII(141–151); XIV(160–174); XV(none); XVI(64–87); XVII(165–169).

Hart, *College Algebra and Trigonometry*, 1959, Heath.
I–II(none); III(3–19); IV–VI(none); VII(160–174); VIII(6–9,20–61,64–87); IX(103–114,Appendix); X(117–122); XI(6–9,20–22); XII(36–61,88–102); XIII(none); XIV(141–151); XV–XVI(none); XVII(1–3); XVIII–XX(none); XXI(17,30–34); XXII(122–140).

Hart, *Modern Plane Trigonometry*, 1961, Heath.
I(1–19); II(20–30); III(117–122); IV(103–114); V(117–122), VI(20–30,64–87); VII(36–43); VIII(44–61); IX(122–140); X(17,30–34); XI(141–151).

Hartley, *Trigonometry, Plane and Spherical*, rev. and enl. 1964, Odyssey.
I(103–114,Appendix); II(3–9); III(21–30); IV(36–44,88–102); V(64–87); VI(117–122); VII(44–61); VIII(122–140); IX(throughout the book); X(175–195); XI(195–202); XII(202–208).

Heineman, *Plane Trigonometry with Tables*, 3rd ed., 1964, McGraw-Hill.
I(1–30); II(20–30,117–122); III(36–43); IV(24–27); V(6–9,11–13,21–22); VI(64–87); VII(43–61); VIII(88–102); IX(70–87); X(103–114); XI(117–122); XII(122–140); XIII(17,30–34); XIV(141–151).

Hillman and Alexanderson, *Algebra and Trigonometry*, 1963, Allyn and Bacon.
I–V(none); VI(141–151); VII(6–9,20–61); VIII(17,30–34,64–87,103–114); IX(115–140); X(none).

Mancill, *Modern Analytical Trigonometry*, 1960, Dodd, Mead.
I–II(1–19); III(20–30,117–122); IV(36–43,88–102); V(122–140); VI(44–61); VII(64–87); VIII(17,30–34); IX(141–151); X(103–114); XI(122–140,Appendix); XII(9,80–84).

Miller and Green, *Algebra and Trigonometry*, 1962, Prentice-Hall.
I(1–6); II(none); III(141–151); IV(14–18); V(103–114); VI(20–30); VII(36–63,142); VIII(30–33); IX(88–102); X–XIII(none); XIV(115–140); XV(160–174); XVI(none).

Miller and Walsh, *Elementary and Advanced Trigonometry*, 1962, Harper.
I(1–19); II(20–30); III(6–9); IV(44–61); V(17,30–34); VI(36–43); VII(88–102); VIII(115–134); IX(134–140); X(none); XI(141–151); XII(none); XIII(160–174); XIV(152–159); XV–XVI(none).

Moore, *Modern Algebra with Trigonometry*, 1964, Macmillan.
I–IV(none); V(103–114,Appendix); VI(1–30); VII(20–102); VIII(141–151); IX(none); X(160–169).

Niles, *Algebra and Trigonometry*, 1965, Wiley.
I(1–3); II(none); III(3–19); IV(103–114); V(20–87); VI(115–140); VII(7,30–34, 88–102); VIII(141–151); IX–X(none); XI(160–174).

Niles, *Plane Trigonometry*, 1959, Wiley.
I(1–19); II(20–30); III(117–122); IV(20–30); V(36–43); VI(64–87); VII(44–61); VIII(103–114,Appendix); IX(122–140); X(7,30–34); XI(88–102); XII(141–151).

Palmer, Leigh, and Kimball, *Plane and Spherical Trigonometry*, 5th ed., 1950, McGraw-Hill.
I(1–19); II(20–30); III(36–43); IV(117–122); V(20–30); VI(64–87); VII(134–140); VIII(44–61); IX(122–140); X(88–102); XI(141–159); XII(175–208); XIII(103–114, Appendix).

Rees and Sparks, *Algebra and Trigonometry*, 1962, McGraw-Hill.
I–IV(none); V(3–19); VI(none); VII(6–9,20–30); VIII(36–43); IX(25–29,

Appendix); X(6–9,11–14); XI(64–87); XII–XIV(none); XV(44–61); XVI(141–151); XVII–XVIII(none); XIX(103–114,Appendix); XX(117–122); XXI(122–140); XXII(160–165); XXIII(none); XXIV(165–167); XXV(88–102); XXVI(17, 30–34); XXVII–XXVIII(none).

Rickey and Cole, *Plane Trigonometry*, rev. ed., 1964, Holt, Rinehart and Winston.
I(1–19); II(20–30); III(6–9,20–22); IV(64–87); V(117–122,Appendix); VI(36–61); VII(17,30–34,88–102); VIII(122–134); IX(141–151).

Rosenbach, Whitman, and Moskovitz, *Essentials of Plane Trigonometry*, 2nd ed., 1961, Ginn.
I(1–30); II(Appendix); III(117–122); IV(39–61); V(122–140); VI(17,30–34); VII(64–87); VIII(103–114,Appendix); IX(141–151); X(175–208).

Rutledge and Pond, *Modern Trigonometry*, 2nd ed., 1961, Prentice-Hall.
I(1–18); II(105–109,Appendix); III(20–30,36–44); IV(115–134); V(44–63); VI(123–127); VII(64–86); VIII(30–35,88–102); IX(141–151).

Sharp, *Elements of Plane Trigonometry*, 1958, Prentice-Hall.
I(none); II(1–19); III(6–9,20–22); IV(22–30); V(64–87); VI(36–43); VII(44–61); VIII(17,30–34); IX(117–140); X(70–87); XI(141–151).

Sparks and Rees, *Plane Trigonometry*, 5th ed., 1965, Prentice-Hall.
I(1–30); II(117–122); III(36–43); IV(25–29); V(6–9,20–22); VI(64–87); VII(44–61); VIII(103–114); IX(Appendix); X(88–102); XI(17,30–34); XII(141–151); XIII(9,80–84).

Spitzbart and Bardell, *College Algebra and Plane Trigonometry*, 1964, Addison-Wesley.
I(none); II(3–9); III(20–30); IV(none); V(11–14,117–122); VI(none); VII(64–87); VIII(none); IX(17,30–34,103–114); X(36–61); XI(165–169); XII–XIII(none); XIV(122–134); XV(141–151); XVI–XVII(none).

Spitzbart and Bardell, *Introductory Algebra and Trigonometry*, 1962, Addison-Wesley.
I–IV(none); V(1–19); VI(20–30); VII–VIII(none); IX(13–14,117–122); X(64–87); XI(17,30–34,103–114); XII(36–61); XIII–XIV(none); XV(122–140); XVI(141–151).

Spitzbart and Bardell, *Plane Trigonometry*, 1964, Addison-Wesley.
I(1–10,21–35); II(22,118–122); III(64–87); IV(36–63); V(103–114); VI(117, 122–137); VII(141–151).

Vance, *Modern Algebra and Trigonometry*, 1962, Addison-Wesley.
I(1–19); II–III(none); IV(3–9); V(14–18); VI(20–61); VII–IX(none); X(17, 30–34); XI–XII(none); XIII(103–114); XIV(64–87); XV(115–140); XVI(141–151).

QUICK REFERENCE TABLE TO STANDARD TEXTBOOKS

All Figures Refer to Pages.

Chapter in this Outline	Topic	Brink	Brixey, Andree	Cameron	Fisher, Ziebur	Fort	Fuller	Goodman	Hall, Katsoff (Trig)
1	Fundamental Concepts	1, 13	1, 31, 42	1	1	1, 9, 19	1	1, 122	1, 7, 25
2	Trigonometric Functions	22, 31, 84	31, 81, 112	143, 150	107	33, 68, 140	14, 168	11, 59, 150	42
3	Properties of Functions	59, 91, 107	91	181, 219	131	50, 57, 106	57, 70	73, 103	134, 171
4	Graphs	68	81, 117	209	127, 146	69	94	130	78
5	Equations	101, 120	97	225		148	163	143	145
6	Exponential and Logarithmic Functions	129, 206	42, 62	144	75	119	108	40	101
7	Geometric Applications	42, 156, 166	69	190	177	43, 76, 153	34, 128	20, 55, 89 159	118, 153
8	Complex Numbers	198	128	242	199	172	177	174	186
9	Hyperbolic Functions								
10	Series	207				83			
11	Spherical Triangles								
Appendix	Use of Tables	35, 159	42, 69	117	119	134	24	40	59, 101

QUICK REFERENCE TABLE TO STANDARD TEXTBOOKS

All Figures Refer to Pages.

Chapter in this Outline	Topic	Hall, Katsoff (Algebra)	Hart (College)	Hart (Trig)	Hartley	Heineman	Hillman, Alexanderson	Mancill	Miller, Green
1	Fundamental Concepts	1, 26	1, 58, 303	1, 218	31	1, 65	1–155	1–51	1, 28, 69
2	Trigonometric Functions	76, 112	136, 211, 361	29, 118, 181	41	7, 18, 182	191, 264	57, 153, 244	100, 151
3	Properties of Functions	90, 319	226	133, 144	83, 129	40, 87	201	84, 118	121
4	Graphs	382	164	20, 58, 125 182	101	76, 115	250	138	131
5	Equations	208	226	34, 137, 154	91	48, 108	281	94	161
6	Exponential and Logarithmic Functions	276	172	80	3	125, 206	274	186	84
7	Geometric Applications	297	193, 370	68, 105, 158	117, 150	149, 160, 210	304	101, 207	261
8	Complex Numbers	334	255	193	178	192	159	167	56, 144
9	Hyperbolic Functions								
10	Series	347	122						273
11	Spherical Triangles				229, 242, 260				
Appendix	Use of Tables	129, 285	172	80	3	125	304	186	91, 113

QUICK REFERENCE TABLE TO STANDARD TEXTBOOKS

All Figures Refer to Pages.

Chapter in this Outline	Topics	Miller, Walsh	Moore	Niles (Algebra)	Niles (Trig)	Palmer, Leigh, Kimball	Rees, Sparks	Rickey, Cole	Rosenbach, Whitman, Moskovitz
1	Fundamental Concepts	1, 49	1, 71	1, 54	1	1	1–82	1	
2	Trigonometric Functions	21, 85, 250	122, 154	110, 234, 199	18, 65, 160	16, 67	99	21, 46	1, 110
3	Properties of Functions	67, 106	137, 174	143, 169	80, 108	35, 109	110, 117, 186	95	56
4	Graphs	41, 76	159	151	92	76	130	56	121
5	Equations	117	184	234	169	160	308	123	74, 117
6	Exponential and Logarithmic Functions	161	93	89	128	223	238	84	132
7	Geometric Applications	138, 161	143	182, 208	44, 139	49, 92, 134	256,267	80, 143	39, 82
8	Complex Numbers	224	190	248	177	169	199	163	148
9	Hyperbolic Functions	263							
10	Series	239	234						
11	Spherical Triangles					191			164
Appendix	Use of Tables	120	112	89	128	223	117	84	26

QUICK REFERENCE TABLE TO STANDARD TEXTBOOKS

All Figures Refer to Pages.

Chapter in this Outline	Topics	Rutledge, Pond	Sharp	Sparks, Rees	Spitzbart, Bardell (College)	Spitzbart, Bardell (Int.)	Spitzbart, Bardell (Plane)	Vance
1	Fundamental Concepts	1	1–13	1	1	1–24	1	1, 87
2	Trigonometric Functions	44	21, 104	12, 57, 175	39, 156	106, 168, 201	1, 19	99, 227
3	Properties of Functions	70, 128	72, 90	54, 65, 95	189	233	82	117
4	Graphs	161	53	82, 175, 201	118	192	53	107, 280
5	Equations	75, 184	82	166	215	253		
6	Exponential and Logarithmic Functions	18	174	117	165	201	115	257
7	Geometric Applications	84, 148	122	38, 132	84, 276	168, 296	30, 137	280, 293
8	Complex Numbers	202	155	188	295	318	159	322
9	Hyperbolic Functions							
10	Series							
11	Spherical Triangles							
Appendix	Use of Tables	30	35, 174	27	169	209	26	

MODERN TRIGONOMETRY

GREEK ALPHABET

Letters		Names	Letters		Names	Letters		Names
Α	α	Alpha	Ι	ι	Iota	Ρ	ρ	Rho
Β	β	Beta	Κ	κ	Kappa	Σ	σ	Sigma
Γ	γ	Gamma	Λ	λ	Lambda	Τ	τ	Tau
Δ	δ	Delta	Μ	μ	Mu	Υ	υ	Upsilon
Ε	ϵ	Epsilon	Ν	ν	Nu	Φ	ϕ	Phi
Ζ	ζ	Zeta	Ξ	ξ	Xi	Χ	χ	Chi
Η	η	Eta	Ο	o	Omicron	Ψ	ψ	Psi
Θ	θ	Theta	Π	π	Pi	Ω	ω	Omega

SYMBOLS

$=$, is equal to
$\equiv$, is identical to
$\neq$, is not equal to
$\doteq$, is approximately equal to
$<$, is less than
$>$, is greater than
$\leq$, is less than or equal to
$\geq$, is greater than or equal to
ln, natural logarithm
log, common logarithm
$n!$, n factorial
$\overline{AP}$, line segment

$\therefore$, therefore
$\cong$, is congruent to
$\ldots$, and so on
P_1, P sub 1
$n \to \infty$, n approaches infinity
$\sqrt{n}$, square root of n
$\angle ABC$, angle with vertex at B
$\triangle ABC$, triangle ABC
Δx, increment of x
$f(x)$, value of function at x
Π, product of
Σ, sum of
ϕ, null set
$\subset$, is a subset of

chapter 1

Fundamental Concepts

1. INTRODUCTION

The general subject of mathematics is the sequential development of a logical structure. Beginning with arithmetic and proceeding through the topics of algebra, geometry, trigonometry, analytic geometry, the calculus, and higher mathematics, the student will discover that each topic forms a prerequisite for those that follow. The subject of trigonometry fits into the structure of mathematics after one has mastered the fundamentals of algebra and geometry. Although we shall review some of these topics, it will be assumed that the reader has a knowledge of the basic concepts.*

The subject of trigonometry began with the solution of triangles. However, the growth of man's mathematical and scientific knowledge has expanded the subject so that today the solution of triangles forms a very small, but still important, part of trigonometry. In this book we shall develop the subject along more analytical lines. If the reader is interested in computational trigonometry, we suggest *Plane and Spherical Trigonometry* by K. L. Nielsen and J. H. Vanlonkhuyzen, Barnes and Noble, 1954.

2. SETS

The term set is a basic concept in mathematics. The reader should have encountered this term in one of his courses of study, but we shall briefly review the idea of a set and establish uniform symbolism for its description. In general, a **set** is a collection of objects with some common property. The objects are called **elements** of the set. There are two common ways to describe a set.

*For a review see the first chapter of Kaj L. Nielsen, *College Mathematics* (New York: Barnes and Noble, 1958), and Marcus Horblit and Kaj L. Nielsen, *Plane Geometry Problems* (New York: Barnes and Noble, 1947).

One may simply enumerate all the elements of the set and enclose them within braces.

$$A = \{2,4,6,8\}$$

This describes A as a set consisting of four elements—the numbers 2, 4, 6, and 8. Or, one may select a general set from which all the elements of a particular set are to be chosen. This general set is called a **universe**. A particular set is then defined by naming the property that is characteristic of all the elements of this particular set. The definition must be carefully stated so that we always have a well-defined set. Using the example shown above, let the universe be the set of all positive even integers, and define the particular set as the positive even integers that are less than 10. This definition gives us the set $A = \{2,4,6,8\}$ consisting of the four numbers listed. Another symbolic description of a set is

$$A = \{x \mid x < 10\}$$

which is read "the set of all x such that x is less than 10." The "$\mid$" is a shorthand symbol for "such that," and the part after this vertical bar is the definition of the set. To use this notation, we must first know the universe or include a description of it in the part after the vertical bar. Thus if we state that the universe is the set of all positive even integers, we have

$$A = \{x \mid x < 10\} = \{2,4,6,8\}$$

Or, we could write

$$\begin{aligned} A &= \{x \mid x \text{ is a positive even integer less than } 10\} \\ &= \{2,4,6,8\} \end{aligned}$$

The set which contains no elements is called the empty or **null** set and is denoted by ϕ.

A set A is called a **subset** of a set B if every element of A is also an element of B. This condition is expressed symbolically by $A \subset B$. To illustrate this concept consider the examples.

If $A = \{1,3\}$ and $B = \{1,3,5\}$, then $A \subset B$.
If $A = \{1,3\}$ and $B = \{1,2,5\}$, then $A \not\subset B$.
The set of even integers is a subset of the set of integers.

The concept of a subset can be used to define what is meant by two sets being equal.

$$A = B \text{ if } A \subset B \text{ and } B \subset A$$

We shall use the concept of a set throughout this book and will explain the symbolic logic as each new symbol is introduced.

3. NUMBER SYSTEM

It will be assumed that the reader is familiar with the real number system:

positive integers	1, 2, 3, 4, 5, ...
zero integer	0
negative integers	$-1, -2, -3, -4, -5, \ldots$
fractions	$\frac{1}{2}, \frac{2}{3}, \frac{8}{7}, \frac{112}{234}, \ldots$
decimal fractions	.15, .01, .0025, 1.34, ...

and the following definitions:

a **rational number** *is one which can be expressed as the quotient of two integers;*

an **irrational number** *is one which cannot be expressed exactly as the quotient of two integers.*

The reader should also be familiar with the fact that the real numbers can be represented on a straight line as shown in Figure 1, where P_1 represents a fraction between 1 and 2.

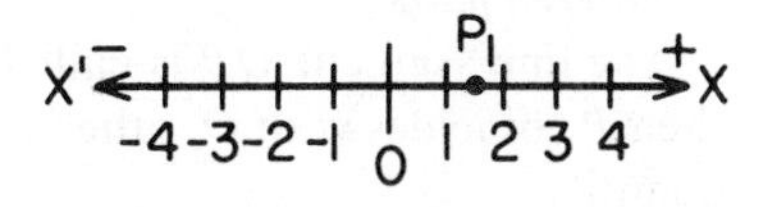

FIGURE 1

The comparison of numbers as to relative size, usually called the **trichotomy law**,

$$a < b \qquad a = b \qquad a > b$$

is a part of elementary algebra; we will not discuss it further here.

We also have the complex number system, composed of numbers of the form

$$a + bi$$

where i is the number such that $i^2 = -1$. In this book we shall concern ourselves mostly with the real numbers. However, in Chapter 8 we shall discuss the fundamental operations of the complex numbers and represent them in both rectangular and trigonometric form.

4. RECTANGULAR COORDINATES

Let us consider two number scales as pictured in Figure 1, and let them intersect each other at right angles at the point O (see

Figure 2). These two lines form a **rectangular coordinate system** in which the two lines, called the **axes**, are denoted by XX' and YY'. The point O is called the **origin**.

Let P be any point in the plane. The **horizontal coordinate** of P is the directed perpendicular distance x from YY' to P. It is said to be *positive* when P is to the *right* of the y-axis and *negative* when P is to the *left* of YY'; it is also called the **abscissa**. The **vertical coordinate** of P is the directed perpendicular distance y from XX' to P. It is said to be *positive* when P is *above* the x-axis and *negative* when P is *below* XX'; it is also called the **ordinate**.

The abscissa and ordinate together are called the **coordinates** of P and are denoted by the pair (x,y). An **ordered pair** is a set which consists of two elements, one of which has been designated as the first. This is usually indicated by the notation (x,y), which also specifies that x has been ordered to be the first element and y, the second. Thus the coordinates of a point P are an ordered pair, and the coordinates of all the points in the plane form a *set of ordered pairs*.

The line segment OP is called the radius vector, r; if r is zero then P coincides with O, otherwise r is considered to be a positive quantity.

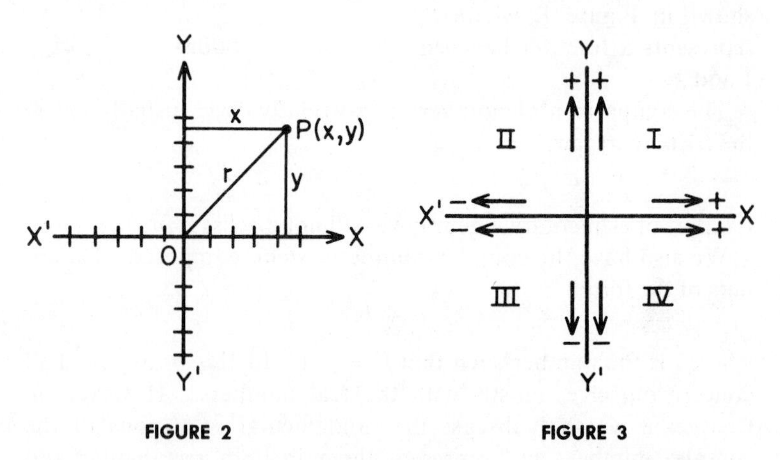

FIGURE 2 FIGURE 3

The coordinate axes divide the plane into four **quadrants**, which are numbered counterclockwise I, II, III, and IV. The point P may be located on this plane by simply knowing the values of its coordinates x and y. In the first quadrant both x and y are positive; in the second quadrant x is negative, y, positive; in the third quadrant $(-,-)$; and in the fourth quadrant $(+,-)$(see Figure 3).

To **plot** a point, $P(x,y)$, is to locate its position from the values of x and y.

ILLUSTRATION. Plot the points $P_1(2,1)$, $P_2(-3,2)$, $P_3(-2,-4)$, $P_4(3,-3)$.
SOLUTION. See Figure 4.

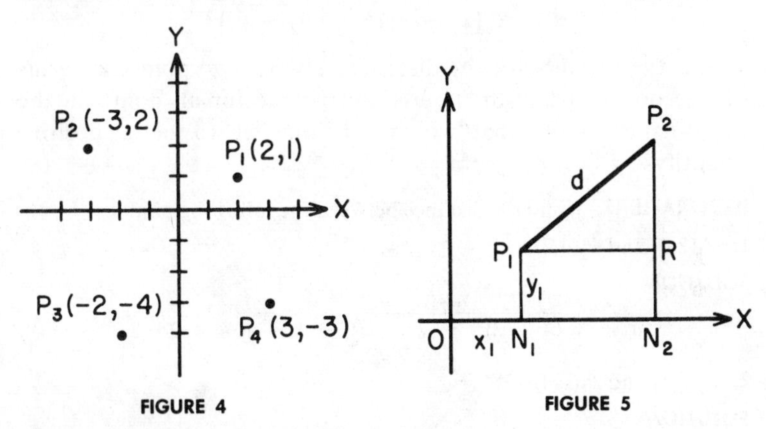

FIGURE 4 FIGURE 5

The rectangular coordinate system which we have defined is also called a **Cartesian coordinate system.**

The directed distance from one point to another may be either positive, negative, or zero; but the *length* of a line segment is defined so that it is never negative. The concept of the **absolute value** of a number is used to indicate a nonnegative quantity; it is denoted by two vertical lines and defined by

$$|a| = a, \text{ if } a \text{ is positive or zero and}$$
$$|a| = -a, \text{ if } a \text{ is negative}$$

Thus we see that the absolute value of a number always results in a *nonnegative quantity*.

The distance between two points in a plane may be found from their coordinates. Let the two points be $P_1(x_1,y_1)$ and $P_2(x_2,y_2)$. Plot these points on a coordinate system; see Figure 5. Form the right triangle P_1RP_2 and apply the Theorem of Pythagoras* to obtain

$$d^2 = \overline{P_1R}^2 + \overline{RP_2}^2$$

*The square on the hypotenuse of a right triangle is equal to the sum of the squares on the other two sides.

We note that

$$P_1R = N_1N_2 = ON_2 - ON_1 = x_2 - x_1 \quad \text{and}$$
$$RP_2 = N_2P_2 - N_2R = y_2 - y_1$$

Substituting these values into the formula on the bottom of page 5, we have, after taking the square root of both sides,

$$d = \sqrt{(x_2 - x_1)^2 + (y_2 - y_1)^2}$$

This is the formula for the distance between two points in terms of the coordinates of the points and is true for all points in the (x,y)-plane. Note that we are defining it to be a positive quantity.

ILLUSTRATIONS. Find the distance between the following points.

1. $P_1(2,4)$ and $P_2(4,2)$.

SOLUTION

$$d = \sqrt{(4 - 2)^2 + (2 - 4)^2} = \sqrt{4 + 4} = 2\sqrt{2}$$

2. $P_1(2,5)$ and $P_2(-4,-3)$.

SOLUTION

$$d = \sqrt{(-4 - 2)^2 + (-3 - 5)^2} = \sqrt{36 + 64} = 10$$

3. $P_1(3,-2)$ and $P_2(5,-2)$.

SOLUTION

$$d = \sqrt{(5 - 3)^2 + (-2 + 2)^2} = \sqrt{4 + 0} = 2$$

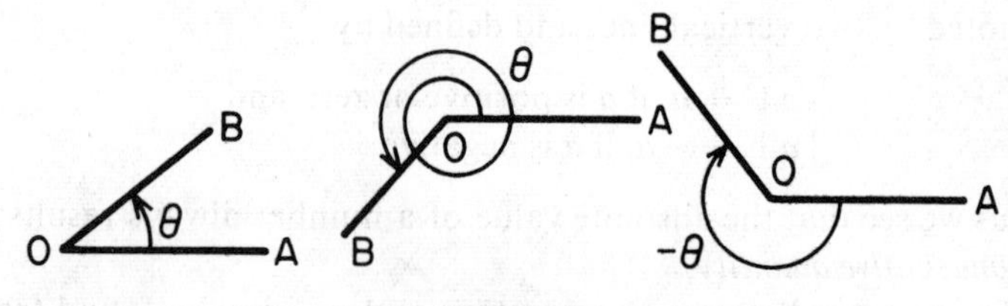

FIGURE 6

5. ANGLES

The reader should be familiar with the definition of an angle arising from the intersection of two lines. We shall discuss the angle in a more general sense. Let any line segment coinciding with OA (Figure 6) be revolved about one of its end points, say O, until it takes the position OB. The *revolution of the line has gen-*

erated the angle AOB. The line may be revolved about O in a clockwise or counterclockwise direction, and there may be no limit to the number of times it is revolved. This defines a general angle; we shall speak of it as being *positive* if the line is moved in a counterclockwise direction and as *negative*, if moved in a clockwise direction. The line OA is said to be the **initial side** and OB, the **terminal side**. The point O is called the **vertex** of the angle.

If we take a general angle and place it so that the vertex is at the origin of the rectangular coordinate system and its initial side coincides with the positive x-axis, the angle is said to be in its **standard position**. There are many ways to denote an angle: $\angle AOB$, angle at O, θ, etc. In this book we shall use Greek letters to denote angles, and shall make no distinction between the angle itself and the number that measures the angle.

The general angle may be measured in the usual sexagesimal measure or in the circular (radian) measure.

A. Sexagesimal Measure

The **degree** $= \dfrac{1}{360}$ of one revolution $= 1^\circ$.

The **minute** $= \dfrac{1}{60}$ of one degree $= 1'$.

The **second** $= \dfrac{1}{60}$ of one minute $= 1''$.

B. Radian Measure

A **radian** is defined to be the central angle subtended by an arc of a circle equal in length to the radius of the circle.

C. Conversion Units

The above systems of measurement are related by the following equivalents.

$$\textbf{1 degree} = \frac{\pi}{180} \textbf{ radians}$$

$$\textbf{1 radian} = \frac{180}{\pi} \textbf{ degrees}$$

To change from degrees to radians, multiply the number of degrees by $\pi/180$; *to change from radians to degrees, multiply the number of radians by* $180/\pi$. In decimal form we have the approximations

$$\mathbf{1^\circ = 0.017453 \text{ radians} \qquad 1 \text{ radian} = 57.2958^\circ}$$

ILLUSTRATIONS

1. Change $\pi/3$ radians to degrees.

SOLUTION

$$\frac{\pi}{3}\left(\frac{180}{\pi}\right) = 60^\circ$$

2. Change 45° to radian measure.

SOLUTION

$$45\left(\frac{\pi}{180}\right) = \frac{\pi}{4} \text{ radians}$$

The student should develop an agility for converting angles from sexagesimal measure to radian measure and vice versa. We exhibit the following table of frequently occurring angles for study and reference.

Degrees	30	45	60	90	120	135	150	180
Radians	$\frac{\pi}{6}$	$\frac{\pi}{4}$	$\frac{\pi}{3}$	$\frac{\pi}{2}$	$\frac{2\pi}{3}$	$\frac{3\pi}{4}$	$\frac{5\pi}{6}$	π
Degrees	210	225	240	270	300	315	330	360
Radians	$\frac{7\pi}{6}$	$\frac{5\pi}{4}$	$\frac{4\pi}{3}$	$\frac{3\pi}{2}$	$\frac{5\pi}{3}$	$\frac{7\pi}{4}$	$\frac{11\pi}{6}$	2π

We have the following terminology for angles.

Condition	Name
$\alpha = 90^\circ$	right angle
$\alpha < 90^\circ$	acute angle
$90^\circ < \alpha < 180^\circ$	obtuse angle
$\alpha = 180^\circ$	straight angle
$\alpha + \beta = 90^\circ$	complementary angles
$\alpha + \beta = 180^\circ$	supplementary angles

The geometric representation of an angle by an initial side and a terminal side is actually a representation of an infinite number of angles, since one or more revolutions added to the angle would result in the same geometric picture. Furthermore, the picture also represents the angle measured in the negative direction and complete revolutions added to this negative angle. Thus the angle shown in Figure 7 (page 9) is also a geometric representation of

$$120^\circ + 360^\circ = 480^\circ$$

$$\frac{2\pi}{3} + 4\pi = \frac{14\pi}{3}$$

$$120^\circ - 360^\circ = -240^\circ, \text{ etc.}$$

This fact can be expressed by using the symbolic representation for a set, where it is understood that n is an integer; i.e., $n = 0, \pm 1, \pm 2, \ldots$.

$$\{120^\circ + n\,360^\circ\} = \left\{\frac{2\pi}{3} + 2n\pi\right\}$$

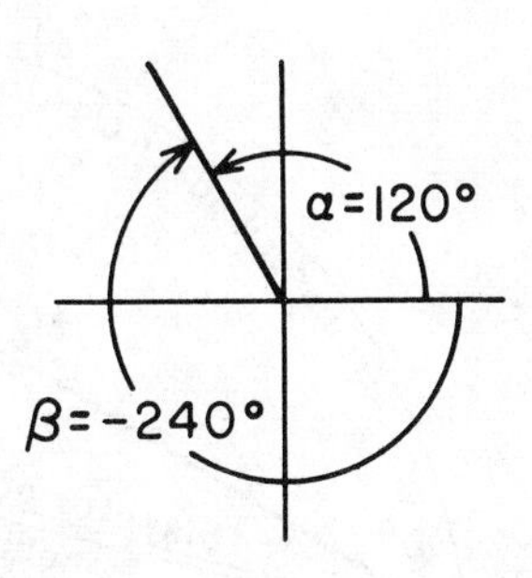

FIGURE 7

Thus the infinitely many angles shown in any geometric illustration can be expressed as a set of infinitely many elements $\{\alpha + 2n\pi\}$. Note that since n can be negative, we automatically include the negatively measured angles. In this notation, the angle α which is contained in the interval $0 \leq \alpha < 2\pi$ is often referred to as the "base" angle or "fundamental" angle of the set. Set notation is also used to express many angles with different terminal sides.

ILLUSTRATION. Tabulate the angles $\theta = \left\{\frac{\pi}{3} + \frac{n\pi}{2}\right\}$ for $n = 0, \pm 1, \pm 2, \pm 3, \pm 4$.

SOLUTION

n	0	± 1	± 2	± 3	± 4
θ	$\frac{\pi}{3}$	$\frac{5\pi}{6}, -\frac{\pi}{6}$	$\frac{4\pi}{3}, -\frac{2\pi}{3}$	$\frac{11\pi}{6}, -\frac{7\pi}{6}$	$\frac{7\pi}{3}, -\frac{5\pi}{3}$

6. TRIANGLES

A figure enclosing a plane surface with three straight lines is called a **triangle**. It consists of three sides and three vertices, thus forming three interior angles. The important elements of a triangle with which the reader should be familiar are shown in Figure 8 (page 10).

Triangles may be classified according to their sides.

I. If the three sides are of different length, the triangle is called a **scalene triangle.**

II. If two sides are equal, it is called an **isosceles triangle**.

III. If the three sides are equal, it is called an **equilateral triangle**.

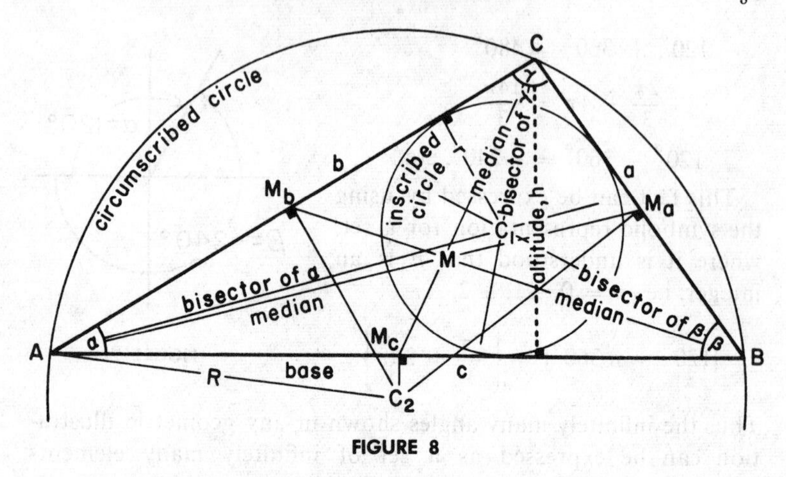

FIGURE 8

Triangles may also be classified according to their angles.

I. If the three angles are acute angles, the triangle is called an **acute triangle**.

II. If one angle is an obtuse angle, it is called an **obtuse triangle**.

III. If one angle is a right angle, it is called a **right triangle**.

Additional terminology for a right triangle is shown in Figure 9.

We now state four important properties of triangles.

I. The sum of the three angles of any triangle is 180°.

II. The acute angles of a right triangle are complementary.

III. The square on the hypotenuse of a right triangle is equal to the sum of the squares on the other two sides (Pythagorean Theorem).

IV. The area of a triangle is equal to $\frac{1}{2}$ the base times the altitude.

These properties lead to the formulas

$$\alpha + \beta + \gamma = 180^\circ \qquad A = \tfrac{1}{2} bh$$

In a right triangle with $\gamma = 90^\circ$,

$$\alpha + \beta = 90^\circ \qquad c^2 = a^2 + b^2$$

$$c = \sqrt{a^2 + b^2} \qquad a = \sqrt{c^2 - b^2} \qquad b = \sqrt{c^2 - a^2}$$

The equilateral triangle has the additional property that

$$\alpha = \beta = \gamma = 60^\circ \qquad h = \tfrac{1}{2} s\sqrt{3} \qquad A = \tfrac{1}{4} s^2\sqrt{3}$$

where s is the length of each side.

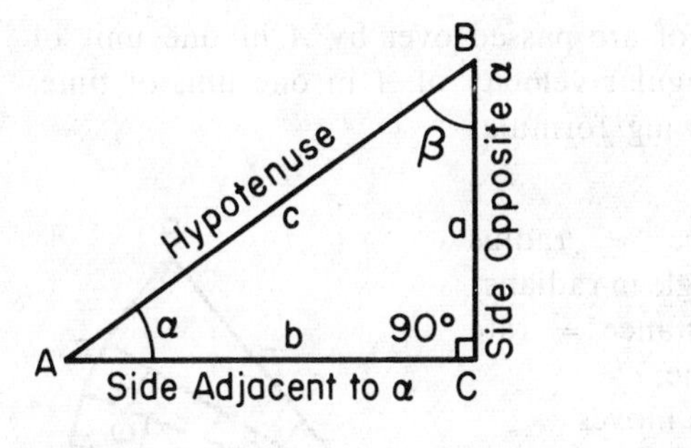

FIGURE 9

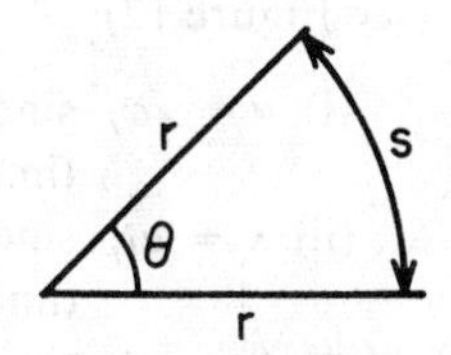

FIGURE 11

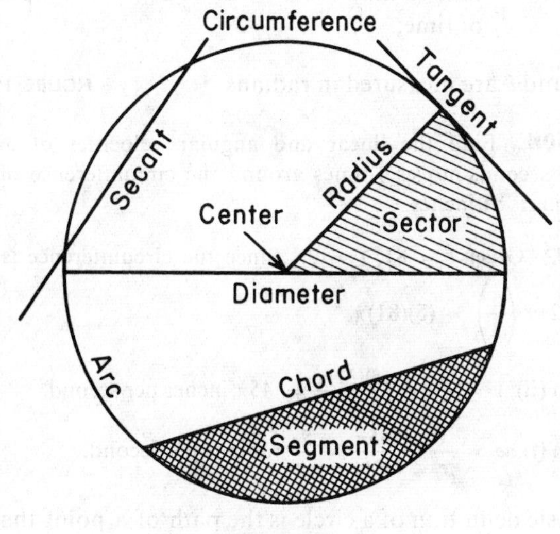

FIGURE 10

7. CIRCLES

The elements of a circle are shown in Figure 10. We recall two important formulas

$$\text{circumference, } \mathbf{C} = 2\pi r \qquad \text{area, } \mathbf{A} = \pi r^2$$

If we use radian measure for the central angle θ, we have two more useful formulas (see Figure 11).

$$\text{length of an arc, } \mathbf{s} = r\theta \qquad \text{area of a sector, } \mathbf{A} = \tfrac{1}{2} r^2\theta$$

If an object A is moving with uniform velocity on the circumference of a circle with center at O and radius r, then the linear velocity v is the length of arc passed over by A in one unit of time t. Let ω be the angular velocity of A in one unit of time. Then we have the following formulas (see Figure 12):

(i) $\mathbf{v} = r\omega$, since arc = radius times angle in radians;

(ii) $\mathbf{s} = \mathbf{v}t$, since distance = rate times time;

(iii) $\theta = \omega t$, since OA moves through ω radians per unit of time for t units of time;

where ω and θ are measured in radians.

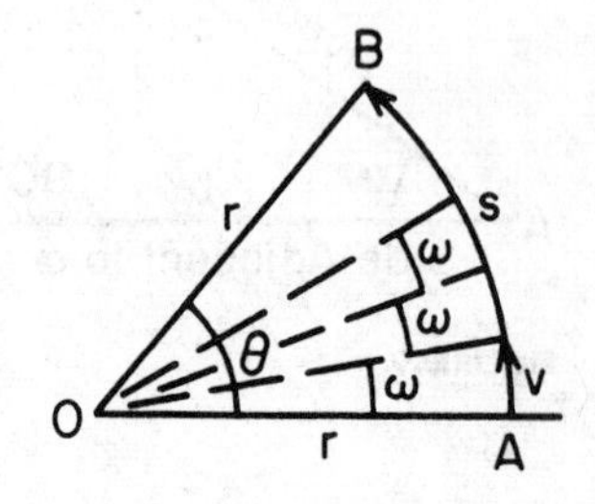

FIGURE 12

ILLUSTRATION. Find the linear and angular velocities of an object which in 9 seconds moves $\frac{5}{2}$ times around the circumference of a circle whose radius is 81 inches.

SOLUTION. Given $r = 81$, $t = 9$. Since the circumference is $2\pi r$, we have $s = 2\pi r\left(\frac{5}{2}\right) = (5)(81)\pi$.

From (ii), $v = \dfrac{s}{t} = \dfrac{(5)(81)\pi}{9} = 45\pi$ inches per second.

From (i), $\omega = \dfrac{v}{r} = \dfrac{45\pi}{81} = \dfrac{5\pi}{9}$ radians per second.

The basic definition of a circle is the path of a point that moves at a constant distance (called the radius) from a fixed point (called the center). Let us place the center at the origin of a rectangular coordinate system and let the moving point have coordinates (x,y). The distance from this point to the center is given by the distance formula.

$$\sqrt{(x - 0)^2 + (y - 0)^2} = \sqrt{x^2 + y^2}$$

Set this equal to the constant r and square both sides to obtain the relation

$$x^2 + y^2 = r^2$$

which represents the equation of a circle with its center at the origin. See Figure 13.

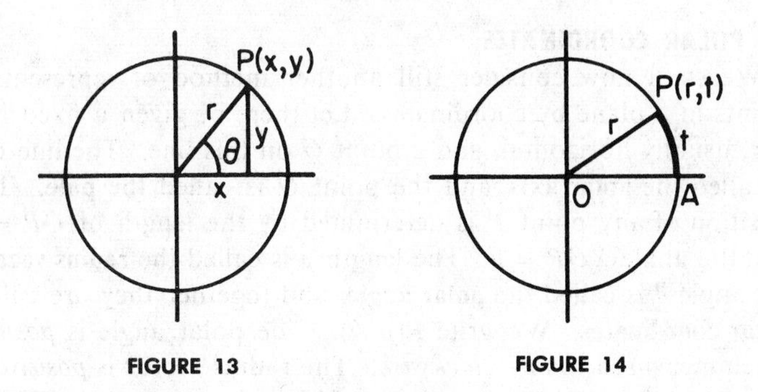

FIGURE 13 FIGURE 14

8. COORDINATE SYSTEMS ON THE CIRCLE

Consider a circle with its center at the origin and let A be the point of intersection of the circle and the positive x-axis (Figure 14). Start at the point A and measure a distance of t units along the circumference of the circle. The value t is *positive* if we measure the distance in a *counterclockwise* direction and *negative*, if in a *clockwise* direction. We can now identify a point P by the two values r and t, thus locating P by the **circular coordinates** (r,t). The circular coordinates differ from the rectangular coordinates in that for the same point P we have more than one pair of circular coordinates whereas the rectangular coordinates are unique. Thus the coordinates $(r,5)$ and $(r, 5 + 2\pi r)$ locate the same point. In fact, a point has an infinite number of circular coordinates $(r, t \pm 2n\pi r)$. Note that in the circular coordinates, the second coordinate t is a pure number representing a distance along the circumference of a circle and *is* **not** *necessarily the measure of an angle.*

A very special set of circular coordinates is the one in which the radius of the circle is one, i.e., $r = 1$; and $(1,t)$ form the coordinates of a point on the circumference of this circle. This circle is called the **unit circle** and has the important property that the distance t is equal to the radian measure of the angle formed by the x-axis and the line from the origin to P. This can be seen from the formula (see Section 7)

$$s = r\theta \ (\theta \text{ in radians})$$

since for $r = 1$ we have $s = \theta$. Thus, *on the unit circle, the circular coordinate* t *equals the radian measure of* θ.

9. POLAR COORDINATES

We shall now consider still another method of representing points in a plane by coordinates. Let there be given a fixed line OX, usually horizontal, and a point O on this line. The line OX is called the **polar axis**, and the point O is called the **pole**. The position of any point P is determined by the length of $OP = \rho$ and the angle $XOP = \theta$. The length ρ is called the **radius vector**; the angle θ is called the **polar angle**; and together they are called **polar coordinates**. We write $P(\rho, \theta)$. The polar angle is *positive* when measured *counterclockwise*. The radius vector is *positive* if it is measured on the *terminal side of* θ. A *negative* radius vector is measured on the terminal side of θ *produced through O*. Figure 15 shows three points.

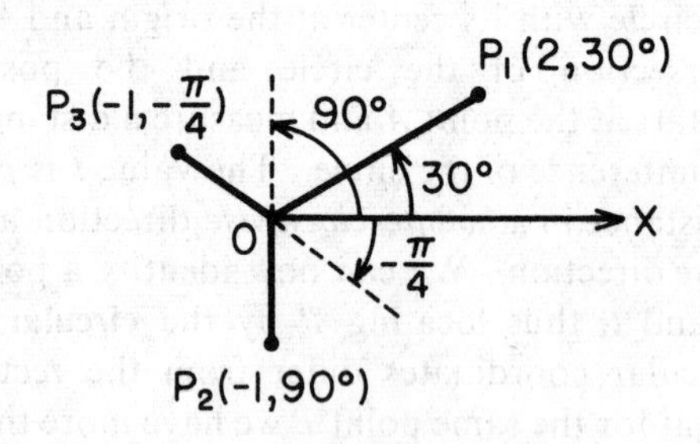

FIGURE 15

Since an angle in a plane can have more than one designation ($\theta = \theta + 2n\pi$), every point has an infinite number of polar coordinates. In order to be more definite, the angle θ is restricted to $0 \leq \theta \leq 360^\circ$ unless it is specifically stated otherwise. To ease the plotting of points, special graph paper, called *polar coordinate paper*, is available. Figure 16 shows some points plotted on this paper. The concentric circles about the origin measure units of ρ, and the angles are indicated.

We now have three methods of designating the coordinates of a point in the plane, namely $P(x,y)$, $P(r,t)$, and $P(\rho,\theta)$. Each of these representations is important in trigonometry. We shall discuss the relations among them in subsequent chapters.

10. FUNCTIONS

One of the most basic concepts in mathematics is the concept of a function.

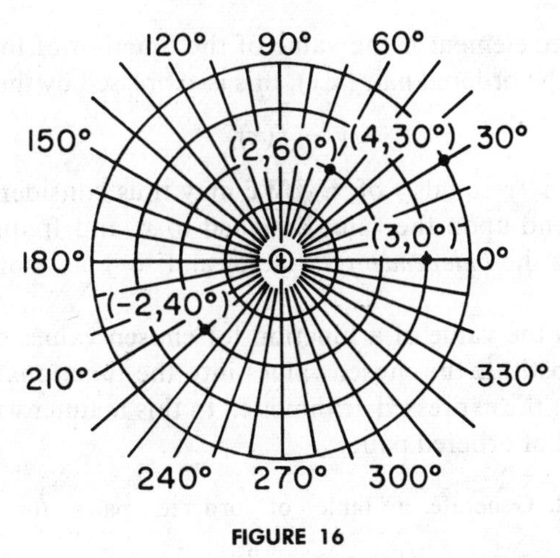

FIGURE 16

DEFINITION. *A* **function** *is a set of ordered pairs such that no two ordered pairs in the set have the same first element.*

The **domain** of definition of the function is the set of all the first elements of the ordered pairs, and the **range** of the function is the set of all the second elements of the ordered pairs. We have used the notation (x,y) to designate an ordered pair, and in this notation the definition of a function specifies that for each x there is one and only one y.

When considering a specific function, there is a *law of formation* to generate the set of ordered pairs. This law of formation is frequently expressed as a formula or an equation which establishes a *relation* among the symbols. For example, the formula for the area of a circle, $A = \pi r^2$, establishes a relation among the symbols A, r, and π. In this relation the symbol π stands for a specific value which does not change during the discussion. When a symbol represents a fixed number, it is called a **constant**. The symbols A and r will vary depending upon the size of the circle. A symbol which, throughout a discussion, may assume different values is called a **variable**.

The above formula will give us the area of a circle if we specify the radius. Thus the two variables A and r are so related that if we assign a value to r, then A is determined. In other words, we can generate the set of ordered pairs (r,A) from the given equation. We then say that *A is a function of r*.

The second element is the value of the function of the first element. For the ordered pair (x,y), this is expressed by the notation

$$y = f(x)$$

which is read "y equals f of x." We may thus consider the value of y to depend upon the value assigned to x, and frequently will refer to x as the *independent* variable, and to y as the *dependent* variable.

To obtain the value of a function for chosen values of the first element, substitute the chosen values into the functional equation and perform the expressed arithmetic. In this manner we can generate a table of ordered pairs.

ILLUSTRATION. Generate a table of ordered pairs for

$$f(x) = x^2 - 2x + 5$$

SOLUTION. Assign the following arbitrary values (others can, of course, be assigned) to x: $-2, -1, 0, 1, 2, 3$. Then

$$\begin{aligned} f(-2) &= (-2)^2 - 2(-2) + 5 = 13 \\ f(-1) &= (-1)^2 - 2(-1) + 5 = 8 \\ f(0) &= (0)^2 - 2(0) + 5 = 5 \\ f(1) &= (1)^2 - 2(1) + 5 = 4 \\ f(2) &= (2)^2 - 2(2) + 5 = 5 \\ f(3) &= (3)^2 - 2(3) + 5 = 8 \end{aligned}$$

and we have the following table of ordered pairs.

x	-2	-1	0	1	2	3
$f(x)$	13	8	5	4	5	8

In the above example, the values of $f(x)$ could be calculated for all values of x in the real number system; we have only exhibited a few of them in the table. We can then say that the function is defined for all real numbers or that the domain of definition is the real number system.

Let us consider the equation

$$x^2 + y^2 = 16$$

which defines a relation between x and y. We know that the geometric picture in the (x,y)-plane is a circle. We could generate a set of ordered pairs by giving values to x(or y) and calculating the corresponding values of y (or x), but we note that for each value of x, there are two values of y; e.g., $x = 0$, $y = 4$ and $x = 0, y = -4$. This relation does not define a function according to our definition; however, we can write the relation as two functions.

$$f_1 = y = \sqrt{16 - x^2} \qquad f_2 = y = -\sqrt{16 - x^2}$$

Let us now suppose that we wish to consider only real values; i.e., values in the real number system. This will limit the domain of definition, for now x cannot be greater than 4 nor less than -4. We can express this fact in two ways.

$$\text{domain: } -4 \leq x \leq 4 \quad \text{or} \quad [-4,4]$$

Each expression tells us to take all the values of x in the real number system between -4 and 4 including the end points of this interval.

We notice that as x takes on the values in this interval, the values of y in the function f_1 are all positive and do not exceed 4; and in f_2, they are all negative and not less than -4. This gives the range for each function which can be expressed by

$$\text{range of } f_1\text{: } 0 \leq y \leq 4 \quad \text{range of } f_2\text{: } -4 \leq y \leq 0$$

Consider a function f with domain $a \leq x \leq b$ and range $c \leq y \leq d$. If we interchange the role of the domain and the range and find a relation between x and y which yields a unique second element for each first element of the ordered pair, we have the *inverse function of f*, which we denote by f^{-1}. To find the inverse of a function, interchange x and y in the given relation, and then solve for y in terms of x. The domain of f becomes the range of f^{-1} and the range of f becomes the domain of f^{-1}.

ILLUSTRATIONS

1. Find the inverse of $y = 2x + 4$ defined for $-1 \leq x \leq 3$.

SOLUTION. Interchange x and y in the given relation.

$$x = 2y + 4$$

Solve for y to obtain

$$y = \tfrac{1}{2}(x - 4) = f^{-1}$$

Domain of f: $-1 \leq x \leq 3$; range of f: $2 \leq y \leq 10$.
Domain of f^{-1}: $2 \leq x \leq 10$; range of f^{-1}: $-1 \leq x \leq 3$.

2. Find the inverse of $y = \sqrt{25 - x^2}$ defined over the domain $-5 \le x \le 0$.

SOLUTION. Interchange x and y in the given equation.

$$x = \sqrt{25 - y^2}$$

To solve for y we must square both sides to yield

$$x^2 = 25 - y^2 \quad \text{or} \quad y^2 = 25 - x^2$$

so that $y = \pm\sqrt{25 - x^2}$. In order to obtain the inverse function, we need to choose one of the two signs, $\pm$. This is done by using the domain of definition of f; i.e.,

$$\text{domain of } f\text{: } -5 \le x \le 0 \qquad \text{range of } f\text{: } 0 \le y \le 5$$

so that for f^{-1}, we have

$$\text{domain of } f^{-1}\text{: } 0 \le x \le 5 \qquad \text{range of } f^{-1}\text{: } -5 \le y \le 0$$

Thus the inverse function is

$$f^{-1} = y = -\sqrt{25 - x^2}$$

The problem of finding the inverse function may necessitate some new definitions, as we shall see in the next chapter.

11. EXERCISE 1

1. Plot the points (2,5), (−1,3), (−2,−5), (2,−3), (0,4), (3,0), and (−2,0) in a rectangular coordinate system.
2. Find the distance between the points.
   a) $P_1(3,5)$ and $P_2(-1,2)$ b) $P_1(-3,-2)$ and $P_2(-1,-6)$
   c) $P_1(-1,5)$ and $P_2(1,-6)$ d) $P_1(-3,0)$ and $P_2(0,2)$
   e) $P_1(3,0)$ and $P_2(-2,0)$ f) $P_1(0,-2)$ and $P_2(0,4)$
3. Sketch the angles for $n = 0, 1, -1, 2, -2$.

   a) $\{0 + 2n\pi\}$ b) $\left\{\dfrac{\pi}{2} + n\pi\right\}$ c) $\left\{\dfrac{\pi}{3} + 2n\pi\right\}$

   d) $\left\{\pi + \dfrac{n\pi}{2}\right\}$ e) $\{30^\circ + n90^\circ\}$ f) $\{60^\circ + n60^\circ\}$
4. Find the area of the sector cut out of a circle with radius 2 units and a central angle measuring a) 1 radian, b) 1°, c) $\pi/3$ radians, d) 150°.
5. Find the distance traveled by a point on a tire during 4 revolutions if the outer diameter of the tire is 16 inches.
6. A pendulum 20 inches long vibrates $1^\circ 30'$ on each side of its mean position. Find the length of the arc through which it swings.

7. A wheel 12 feet in diameter makes 300 revolutions per minute. Find the linear velocity of a point on the rim in feet per second.
8. Plot the following points if the given coordinates are circular coordinates.
   a) $(1, \pi/3)$ b) $(2, \pi)$ c) $(3,1)$
   d) $(2, -3.14)$ e) $(1,1.57)$ f) $(1,1)$
9. Plot the following points if the given coordinates are polar coordinates.
   a) $(1, \pi/3)$ b) $(3, \pi)$ c) $(-2, \pi)$
   d) $(-1, 30^\circ)$ e) $(2, 180^\circ)$ f) $(-1, 90^\circ)$
10. Write the equation of a circle with its center at the origin and a radius equal to $\sqrt{2}$.
11. Given the function $y = \sqrt{4 - x^2}$, calculate a set of ordered pairs and plot this set on a rectangular coordinate system. Join the points with a smooth curve. Specify the domain and range of the function.
12. Given the function $y = 3x - 6$ defined for $0 \leq x \leq 3$, find the inverse function and specify its domain and range.
13. Change the following angles from degrees to radians: 30°, 90°, 150°, 270°, -57°, $13^\circ 10'$, 7°, -270°, 45°, $3'$.
14. Change the following angles from radians to degrees: $\frac{\pi}{2}$, $\frac{\pi}{3}$, $\frac{\pi}{7}$, 2, $\frac{3\pi}{4}$.
15. Find the coordinates of a point P on a circle of radius 3 units if P generates an angle of a) $\frac{\pi}{6}$ radians, b) $\frac{\pi}{4}$ radians, c) $\frac{\pi}{3}$ radians.
16. Simplify.

$$\{[(x + a)(x - a)m + a^2m]e\}$$
$$\times \left\{ \frac{\frac{r + x}{x} + \frac{x}{r - x}}{\frac{x}{r - x}} - \frac{4as}{3x[(e + y)^2 - (e - y)^2]} - \frac{4as}{3e[(y + x)^2 - (y - x)^2]} - \frac{4as}{3y[(e + x)^2 - (e - x)^2]} \right\} y$$

chapter 2

Trigonometric Functions

12. INTRODUCTION

In this chapter we shall discuss the trigonometric functions. There are six names given to these functions; to ease their notation, they are abbreviated as follows.

NAME	ABBREVIATION	NAME	ABBREVIATION
sine	sin	cotangent	cot
cosine	cos	secant	sec
tangent	tan	cosecant	csc

In order to discuss the functions we must express the independent variable, which can be a simple real number or a real number denoting the measure of an angle. We shall therefore write, for example, sin t or sin θ. The independent variable is also called the **argument**.

At one time there were four more functions called *versed sine* (vers), *coversed sine* (covers), *haversine* (hav), and *external secant* (ex sec). These were related to the trigonometric functions by the following definitions.

$$\text{vers } t = 1 - \cos t$$
$$\text{hav } t = \tfrac{1}{2}(1 - \cos t)$$
$$\text{covers } t = 1 - \sin t$$
$$\text{ex sec } t = \sec t - 1$$

However, these functions have disappeared from common usage with the possible exception of the haversine, which is used in the solution of spherical triangles.*

*See Kaj L. Nielsen and John H. Vanlonkhuyzen, *Plane and Spherical Trigonometry* (New York: Barnes and Noble, 1954), p. 118.

13. DEFINITIONS

We shall give three sets of definitions for the trigonometric functions. These definitions are *identical* over their common domain of definition.

A. Trigonometric Functions of a Real Number

Let t be any real number and T be the point on the unit circle with coordinates $(1,t)$. The point T also has the rectangular coordinates (x,y). See Figure 17. Then the trigonometric functions of the real number t are defined as follows.

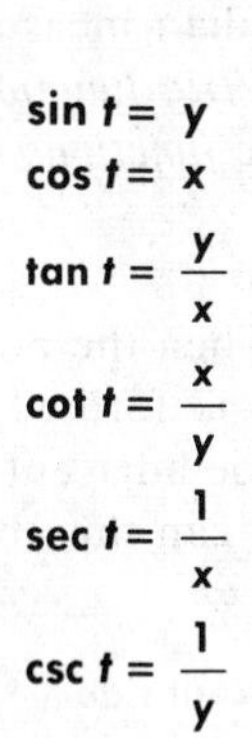

$$\sin t = y$$
$$\cos t = x$$
$$\tan t = \frac{y}{x}$$
$$\cot t = \frac{x}{y}$$
$$\sec t = \frac{1}{x}$$
$$\csc t = \frac{1}{y}$$

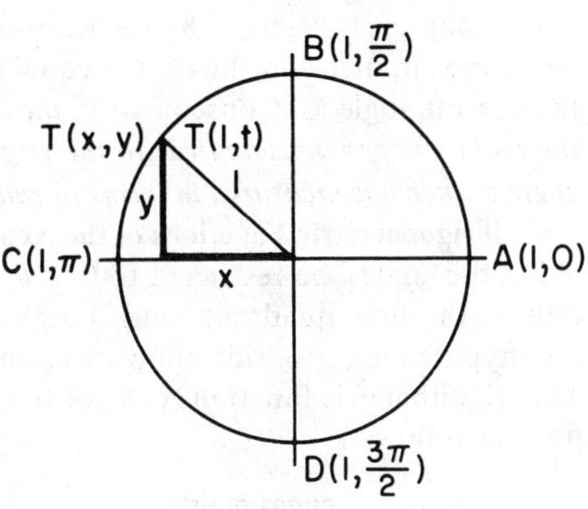

FIGURE 17

Since these definitions are given in terms of ratios, and division by zero is not permissible, it is necessary to limit the values of x and y for specific functions. Thus if $x = 0$, which happens when $t = \pi/2$ or $3\pi/2$, then $\tan t$ and $\sec t$ are undefined. If $y = 0$, then $\cot t$ and $\csc t$ are undefined.

B. Trigonometric Functions of a General Angle

Let θ be any angle placed in its standard position. Let P be any point (not the origin) having coordinates (x,y) and lying on the terminal side of θ. Let the radius vector of P be denoted by r. See Figure 18. Then the trigonometric functions of θ are defined as follows, on page 22.

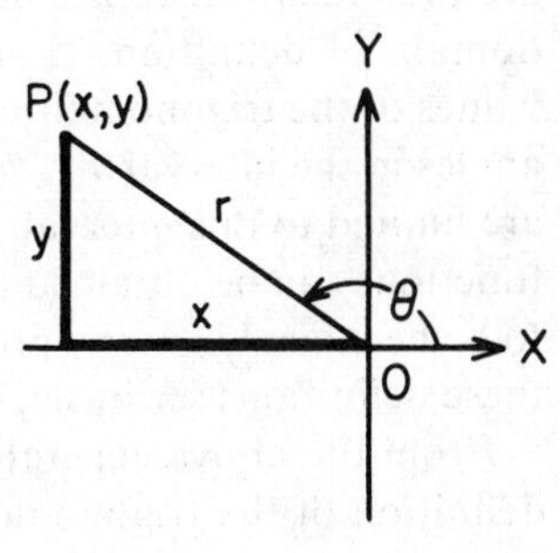

FIGURE 18

$$\sin\theta = \frac{\text{ordinate}}{\text{radius vector}} = \frac{y}{r} \qquad \cot\theta = \frac{\text{abscissa}}{\text{ordinate}} = \frac{x}{y}$$

$$\cos\theta = \frac{\text{abscissa}}{\text{radius vector}} = \frac{x}{r} \qquad \sec\theta = \frac{\text{radius vector}}{\text{abscissa}} = \frac{r}{x}$$

$$\tan\theta = \frac{\text{ordinate}}{\text{abscissa}} = \frac{y}{x} \qquad \csc\theta = \frac{\text{radius vector}}{\text{ordinate}} = \frac{r}{y}$$

We note that the same restrictions must be placed on x and y as those for the trigonometric functions of the real numbers.

In Chapter 1, Section 8, we demonstrated that on the unit circle the circular coordinate t is equal to the radian measure of the central angle θ. Consequently, *the trigonometric functions of the real number* t *are identical to the trigonometric functions of the angle* θ *when* θ *is measured in terms of radians.*

C. Trigonometric Functions of the Acute Angles

Let the angle θ be restricted to $0 < \theta < 90^\circ$. Then the point P falls in the first quadrant, and a right triangle is formed with r = hypotenuse, y = side opposite θ, and x = side adjacent to θ. The trigonometric functions of θ for $0 < \theta < 90^\circ$ can then be defined as follows.

$$\sin\theta = \frac{\text{opposite side}}{\text{hypotenuse}} \qquad \cot\theta = \frac{\text{adjacent side}}{\text{opposite side}}$$

$$\cos\theta = \frac{\text{adjacent side}}{\text{hypotenuse}} \qquad \sec\theta = \frac{\text{hypotenuse}}{\text{adjacent side}}$$

$$\tan\theta = \frac{\text{opposite side}}{\text{adjacent side}} \qquad \csc\theta = \frac{\text{hypotenuse}}{\text{opposite side}}$$

These last definitions for the trigonometric functions of an acute angle form a special case of the definitions given in B and are thus identical to the definitions A and B over this restricted domain of definition. Later we shall show that the numerical values of the trigonometric functions can be determined from the angles in the interval $0 \leq \theta \leq 90^\circ$ and, consequently, most tables are limited to this interval. The knowledge that the trigonometric functions can be obtained from the sides of a right triangle proves to be beneficial in many problems; the reader is urged to consider these definitions seriously.

From the above general definitions, we see that the domain of definition of the trigonometric functions can be taken to be the real number system with certain exceptions. These exceptions are also referred to as *inadmissible* values of the independent variable. Let R denote the set of real numbers and $n = 0, 1, 2, \ldots$, then the domain of definition for each function may be summarized as follows.

FUNCTION	DOMAIN	FUNCTION	DOMAIN
sin	R	cot	R except $\{\pm n\pi\}$
cos	R	sec	R except $\left\{\frac{\pi}{2} \pm n\pi\right\}$
tan	R except $\left\{\frac{\pi}{2} \pm n\pi\right\}$	csc	R except $\{\pm n\pi\}$

Since the trigonometric functions of real numbers are identical to the trigonometric functions of an angle (provided the angle is measured in radians) we may develop all the properties of the trigonometric functions by considering the independent variable to be an angle. The reader, however, should realize that these properties are also valid when the variable is a pure real number.

14. SIGNS OF THE TRIGONOMETRIC FUNCTIONS

Since the definitions of the trigonometric functions are given in terms of x and y and since these variables can be positive or negative depending on the location of P, we see that the trigonometric functions can be positive or negative depending on the size of the argument. The signs can be determined from the quadrant in which P falls, for this determines the signs of x and y. In checking the definitions remember that r is always positive. In the first quadrant, x and y are both positive and, therefore, all the functions are positive. In the second quadrant, x is negative and y is positive; thus, all the functions are negative except sin θ and csc θ. In the third quadrant, both x and y are negative, and in the fourth quadrant, x is positive and y is negative. After examining all the definitions for each quadrant, we can summarize the results in the following diagram.

We recommend that this diagram become a part of the reader's memory system for rapid recall.

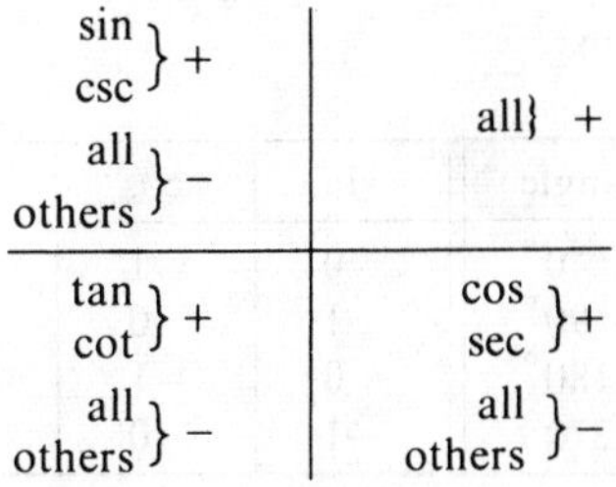

sin, csc } + all others } −	all } +
tan, cot } + all others } −	cos, sec } + all others } −

15. FUNCTIONS OF THE QUADRANTAL ANGLES

The functions of the quadrantal angles (0°, 90°, 180°, 270°) are determined directly from the definitions. These are also the trigonometric functions of the real numbers 0, $\frac{\pi}{2}$, π, and $\frac{3\pi}{2}$.

ILLUSTRATIONS

1. Find the functions of 0°.

SOLUTION. Place the angle 0° in its standard position. The initial side and terminal side will both coincide with the x-axis so that the coordinates of any point P on the terminal side will be $(x_1,0)$ and $r = \sqrt{x_1^2 + 0} = x_1$. Thus we have $x = x_1 = r$ and $y = 0$, and by substituting these values in the definitions, we obtain

$$\sin 0° = \frac{y}{r} = \frac{0}{r} = 0 \qquad \cot 0° = \frac{x}{y} = \frac{x_1}{0}\text{, undefined}$$

$$\cos 0° = \frac{x}{r} = \frac{x_1}{x_1} = 1 \qquad \sec 0° = \frac{r}{x} = \frac{x_1}{x_1} = 1$$

$$\tan 0° = \frac{y}{x} = \frac{0}{x_1} = 0 \qquad \csc 0° = \frac{r}{y} = \frac{x_1}{0}\text{, undefined}$$

2. Find the functions of 270°.

SOLUTION. Place 270° in its standard position. A point P on the terminal side will have $(0,-y_1)$ for its coordinates and $r = \sqrt{0 + y_1^2} = y_1$. Thus we have $x = 0$, $y = -y_1$, and $r = y_1$, and by the definitions,

$$\sin 270° = \frac{-y_1}{y_1} = -1 \qquad \cot 270° = \frac{0}{-y_1} = 0$$

$$\cos 270° = \frac{0}{y_1} = 0 \qquad \sec 270° = \frac{y_1}{0}\text{, undefined}$$

$$\tan 270° = \frac{-y_1}{0}\text{, undefined} \qquad \csc 270° = \frac{y_1}{-y_1} = -1$$

Checking all the quadrantal angles, we obtain the following table.

angle	sin	cos	tan	cot	sec	csc
0°	0	1	0	---	1	---
90°	1	0	---	0	---	1
180°	0	−1	0	---	−1	---
270°	−1	0	---	0	---	−1

16. FUNCTIONS OF NEGATIVE ARGUMENTS

To obtain the trigonometric functions of a negative argument, we shall employ the following theorem.

THEOREM. *If θ is any positive number, then*

$$\sin(-\theta) = -\sin\theta \qquad \cot(-\theta) = -\cot\theta$$
$$\cos(-\theta) = \cos\theta \qquad \sec(-\theta) = \sec\theta$$
$$\tan(-\theta) = -\tan\theta \qquad \csc(-\theta) = -\csc\theta$$

for all admissible values of θ.

PROOF. We shall develop the proof geometrically by considering θ to be an angle. Place the angle $-\theta$ in its standard position and construct the angle θ, numerically equal to $-\theta$, in its standard position. On the terminal sides of θ and $-\theta$ choose P_1 and P_2 so that $OP_1 = OP_2$. Then, since $|\theta| = |-\theta|$, $P_1P_2 \perp OX$ and is bisected by OX. Thus (see Figure 19);*

$$r_1 = r_2 \qquad x_1 = x_2 \qquad y_1 = -y_2$$

From the definition, we have

$$\sin(-\theta) = \frac{y_2}{r_2} = \frac{-y_1}{r_1} = -\sin\theta$$

$$\cos(-\theta) = \frac{x_2}{r_2} = \frac{x_1}{r_1} = \cos\theta$$

$$\tan(-\theta) = \frac{y_2}{x_2} = \frac{-y_1}{x_1} = -\tan\theta$$

etc.

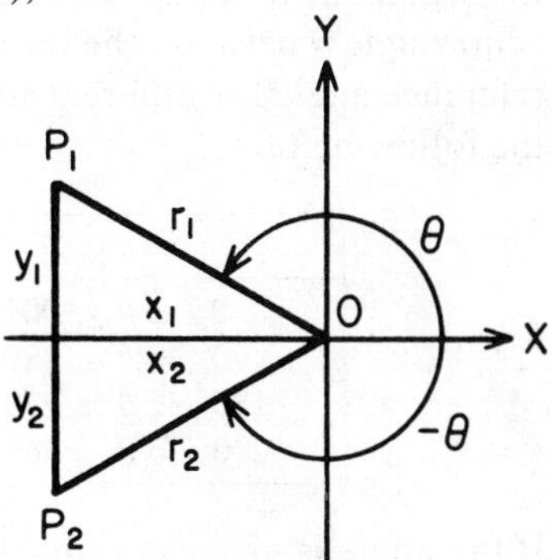

FIGURE 19

The above theorem may be more easily remembered by noting that the trigonometric functions of a negative argument and a positive argument are numerically equal. However, the trigonometric functions of a negative argument assume a negative value when defined in terms of a positive argument, with the exception of cosine and secant, which remain positive. We recommend the following procedure.

To find the function of a negative argument: first change to the function of the corresponding positive argument by using the above theorem, and then find the function of the positive argument.

*The reader who is familiar with the term *symmetry* may note that P_1 and P_2 are symmetrically placed with respect to the x-axis.

ILLUSTRATIONS

1. Find $\cos(-60^\circ)$.

SOLUTION. $\cos(-60^\circ) = \cos 60^\circ = \frac{1}{2}$.

2. Find $\tan\left(-\frac{\pi}{4}\right)$.

SOLUTION. $\tan\left(-\frac{\pi}{4}\right) = -\tan\left(\frac{\pi}{4}\right) = -1.$

17. REDUCTION TO $0 \leq t \leq \frac{\pi}{2}$

The trigonometric functions of any real number may be reduced to the trigonometric functions of a number in the interval $0 \leq t \leq \frac{\pi}{2}$. We shall again develop this concept by considering the argument to be an angle and define a **reference angle** as the acute angle α between the terminal side and the x-axis. Thus the reference angle for different intervals of θ can be found by using the following table.

θ	α
$0 < \theta < 90^\circ$	$\alpha = \theta$
$90^\circ < \theta < 180^\circ$	$\alpha = 180^\circ - \theta$
$180^\circ < \theta < 270^\circ$	$\alpha = \theta - 180^\circ$
$270^\circ < \theta < 360^\circ$	$\alpha = 360^\circ - \theta$

If the angle is given in radians, the equivalent radian measure is used in the above table; i.e., $180^\circ = \pi$, $360^\circ = 2\pi$.

THEOREM. *Any function of an angle θ, in any quadrant, is numerically equal to the same function of the reference angle for θ; i.e.,*

$$\textbf{(any function of } \theta) = \pm(\textbf{same function of } \alpha)$$

The "+" or "−" is determined by the quadrant in which the angle falls.

PROOF. Let θ be any angle in its standard position. On the terminal side of θ pick a point P and drop a perpendicular to the x-axis. See Figure 20. The reference angle α is an acute angle of the right triangle ORP, so we can apply the right triangle definitions to the functions of α; thus,

$$\sin\alpha = \frac{RP}{OP}, \qquad \cos\alpha = \frac{OR}{OP}, \text{ etc.}$$

Now $x = \pm OR$, $y = \pm RP$, and $r = OP$, the "+" or "−" depending upon the quadrant in which P falls. From the definitions in Section 13B, $\sin\theta = \dfrac{y}{r} = \pm\dfrac{RP}{OP} = \pm\sin\alpha$; similarly, for the other functions.

For $\theta > 2\pi$, we have the relation

$$\textbf{any function of } \theta = \textbf{same function of } (\theta - 2n\pi)$$

where n is any integer. The truth of this relation comes from the fact that θ and $\theta + 2n\pi$ have the same terminal side.

The property that any function of $\theta + 2n\pi$ is the same as the function of θ places the trigonometric functions in the class of periodic functions. Any function which repeats its values at a regular interval is said to be **periodic**, and the length of the interval is called its **period**. We shall encounter this property again in Chapter 4.

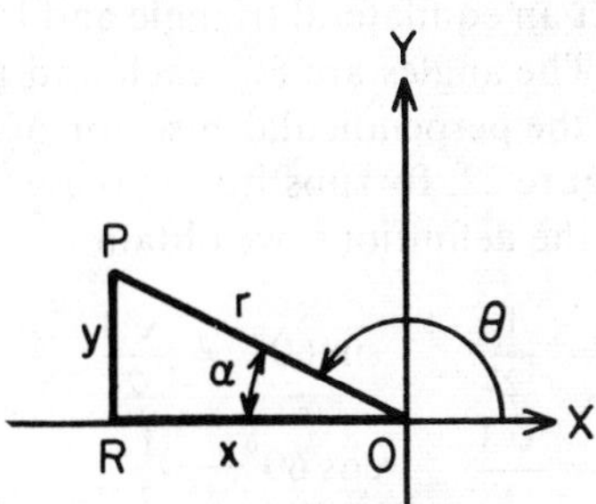

FIGURE 20

18. FUNCTIONS OF $\frac{\pi}{6}, \frac{\pi}{4}, \frac{\pi}{3}$ or 30°, 45°, 60°

The trigonometric functions of 30°, 45°, and 60° can be expressed in terms of exact numbers and, consequently, are used in many illustrations and in the plotting of functions involving the trigonometric ratios. These values can be derived by considering the coordinates on a unit circle. However, they can be generated just as easily by using the reader's knowledge of plane geometry and the special definitions in terms of the sides of a right triangle. Since the functions of the angles 30°, 45°, and 60° are identical to the functions of the real numbers $\frac{\pi}{6}$, $\frac{\pi}{4}$, and $\frac{\pi}{3}$, these values hold for the real numbers as well as the angles.

We begin by considering an isosceles right triangle ABC and letting the length of the equal legs be 1 unit each. The acute angles are each 45° and, by the Pythagorean Theorem, the hypotenuse is $\sqrt{2}$. See Figure 21. If these values are substituted into the definitions of the trigonometric functions, we obtain

FIGURE 21

$$\sin 45^\circ = \frac{1}{\sqrt{2}} = \frac{1}{2}\sqrt{2}$$

$$\cos 45^\circ = \frac{1}{\sqrt{2}} = \frac{1}{2}\sqrt{2}$$

$$\tan 45^\circ = \frac{1}{1} = 1 \qquad \sec 45^\circ = \frac{\sqrt{2}}{1} = \sqrt{2}$$

$$\cot 45^\circ = \frac{1}{1} = 1 \qquad \csc 45^\circ = \frac{\sqrt{2}}{1} = \sqrt{2}$$

Construct an equilateral triangle and let the length of each side be 2 units. The angles are 60° each and the bisector of any angle will also be the perpendicular bisector of the opposite side. Thus we have Figure 22. By substituting these values into the definitions, we obtain

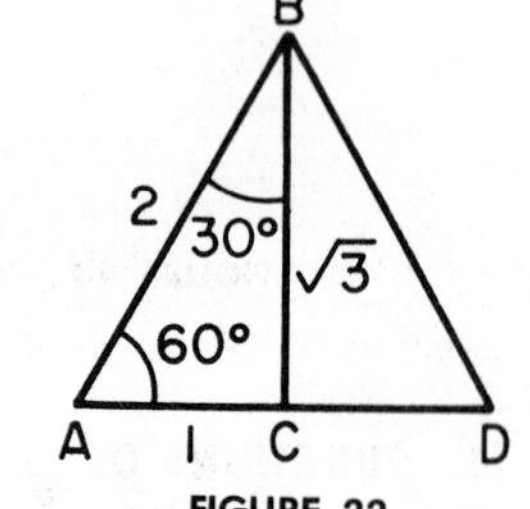

FIGURE 22

$$\sin 30^\circ = \frac{1}{2} \qquad \sin 60^\circ = \frac{\sqrt{3}}{2}$$

$$\cos 30^\circ = \frac{\sqrt{3}}{2} \qquad \cos 60^\circ = \frac{1}{2}$$

etc.

All the values are in the following table.

angle	sin	cos	tan	cot	sec	csc
30°	$\frac{1}{2}$	$\frac{1}{2}\sqrt{3}$	$\frac{1}{3}\sqrt{3}$	$\sqrt{3}$	$\frac{2}{3}\sqrt{3}$	2
45°	$\frac{1}{2}\sqrt{2}$	$\frac{1}{2}\sqrt{2}$	1	1	$\sqrt{2}$	$\sqrt{2}$
60°	$\frac{1}{2}\sqrt{3}$	$\frac{1}{2}$	$\sqrt{3}$	$\frac{1}{3}\sqrt{3}$	2	$\frac{2}{3}\sqrt{3}$

The above table also gives the numerical values of the functions of the angles θ which result in reference angles of 30°, 45°, or 60°. Thus the table can be used to obtain values of the trigonometric functions of the following angles by simply assigning the appropriate ± sign.

120°	135°	150°	210°	225°	240°	300°	315°	330°
$\frac{2\pi}{3}$	$\frac{3\pi}{4}$	$\frac{5\pi}{6}$	$\frac{7\pi}{6}$	$\frac{5\pi}{4}$	$\frac{4\pi}{3}$	$\frac{5\pi}{3}$	$\frac{7\pi}{4}$	$\frac{11\pi}{6}$

ILLUSTRATIONS

1. Find tan 135°.

SOLUTION. $\tan 135^\circ = -\tan(180^\circ - 135^\circ) = -\tan 45^\circ = -1.$

2. Find sin 210°.

SOLUTION. $\sin 210^\circ = -\sin(210^\circ - 180^\circ) = -\sin 30^\circ = -\frac{1}{2}.$

3. Find $\cos\frac{7\pi}{4}$.

SOLUTION. $\cos\frac{7\pi}{4} = \cos\left(2\pi - \frac{7\pi}{4}\right) = \cos\frac{\pi}{4} = \frac{\sqrt{2}}{2}.$

4. Find $\sec\left(-\frac{4\pi}{3}\right)$.

SOLUTION. $\sec\left(-\frac{4\pi}{3}\right) = \sec\left(\frac{4\pi}{3}\right) = -\sec\frac{\pi}{3} = -2.$

In Section 37, we shall derive the exact values of the trigonometric functions of $\frac{\pi}{5}$; thus giving us exact values of the functions of $\pi, \frac{\pi}{2}, \frac{\pi}{3}, \frac{\pi}{4}, \frac{\pi}{5}, \frac{\pi}{6}$, and, as we shall see, multiples of these numbers.

19. TABULAR VALUES

The trigonometric ratios which we have defined in previous sections are all unique numbers. Consequently, they can be calculated once and for all and their values placed in tables for chosen values of the argument. In general they will not be rational numbers so that the tables will exhibit the values by approximating decimal numbers. There are many tables which present the trigonometric functions for various intervals of the argument and to various degrees of accuracy, such as 4-place, 5-place, 7-place, 15-place, or other special tables.* The argument may be in degrees or radians. We only need to consider it in the

*For a 5-place table, see Kaj L. Nielsen, *Logarithmic and Trigonometric Tables* (2nd ed.; New York: Barnes and Noble, 1961).

interval $0 \leq \theta \leq \frac{\pi}{2}$. If only one type of table is available, it is a simple matter to change the argument by using the relations of Section 17. A brief description of how to use mathematical tables is given in the Appendix.

20. INVERSE TRIGONOMETRIC FUNCTIONS

We discussed the concept of inverse functions in Section 10. Since the trigonometric functions satisfy the definitions of a general function in their domain of definition, it is of interest to see if we can find the inverse of these functions. Let us recall that given a function f, we can find an inverse function f^{-1} by interchanging the independent and dependent variables and then solving for the new dependent variable in terms of the new independent variable. If the resulting relation satisfies the definition of a function, we have found the inverse function. Consider the function y given by

$$y = \sin x$$

and interchange x and y to obtain

$$x = \sin y$$

In order to solve this last equation for y, we need to introduce new terminology and symbols;

$$y = \arcsin x$$

is an expression for the number whose sine is x. For the trigonometric functions of angles, the expression $\arcsin x$ represents an *angle* (i.e., $\arcsin x = \theta$) whose sine is x (i.e., $\sin \theta = x$). The expressions $\arcsin x$ and $\sin y$ are **inverse relations**. Since for every number x in the domain $-1 \leq x \leq 1$ there are many angles whose sine is x, the relation does not satisfy the strict definition of a function.

ILLUSTRATION. Find $\arcsin \frac{1}{2}$.

SOLUTION. Let $y = \arcsin \frac{1}{2}$, then $\sin y = \frac{1}{2}$ and $y = \frac{\pi}{6}$ or $\frac{5\pi}{6}$. However, y can also have the values $\frac{\pi}{6} + 2n\pi$, where n is any integer. The general solution can be written

$$y = \arcsin \frac{1}{2} = \left\{ \frac{\pi}{6} + 2n\pi, \frac{5\pi}{6} + 2n\pi \right\}$$

In order to obtain the inverse function we restrict the range of the inverse relation and call this the *principal value.* In this book we shall use the following definition.

DEFINITION. *The* **principal value** *of arcsin x, arccsc x, arctan x, and arccot x is the smallest numerical value; of arccos x and arcsec x, it is the smallest positive value of the angle.*

The principal value is usually indicated by capitalizing the "A" in "arc." These values can be summarized.

$$-\frac{\pi}{2} \le \text{Arcsin}\, x \le \frac{\pi}{2} \qquad -\frac{\pi}{2} \le \text{Arccot}\, x \le \frac{\pi}{2}$$

$$0 \le \text{Arccos}\, x \le \pi \qquad 0 \le \text{Arcsec}\, x \le \pi$$

$$-\frac{\pi}{2} \le \text{Arctan}\, x \le \frac{\pi}{2} \qquad -\frac{\pi}{2} \le \text{Arccsc}\, x \le \frac{\pi}{2}$$

ILLUSTRATIONS

1. $\text{Arcsin}\,\frac{\sqrt{2}}{2} = 45°$. **2.** $\text{Arccot}\,(-\sqrt{3}) = -\frac{\pi}{6}$.

3. $\text{Arctan}\,(-1) = -45°$. **4.** $\text{Arcsec}\,(-\sqrt{2}) = 135°$.

5. $\text{Arccos}\,\frac{1}{2} = \frac{\pi}{3}$. **6.** $\text{Arccsc}\, 2 = \frac{\pi}{6}$.

NOTE. The inverse relations are also denoted by using the "-1" exponent; thus, arcsin x is written $\sin^{-1}x$ and read "the inverse sine of x." The principal value is denoted by capital S, thus $\text{Sin}^{-1}x$. In this notation we caution the reader against using -1 as an exponent; because $\sin^{-1}x \ne \frac{1}{\sin x}$.

Since the inverse trigonometric functions are numbers, we can find the trigonometric functions of the inverse. The values can frequently be found without finding the numerical value of the angle. This is accomplished by using the definitions of the trigonometric functions.

ILLUSTRATIONS

1. Find $\sin \arccos \frac{4}{5}$.

SOLUTION. Let $\theta = \arccos \frac{4}{5}$, then $\cos\theta = \frac{4}{5} = \frac{x}{r}$ and we can take $x = 4$ and $r = 5$, from which we obtain $y = \pm\sqrt{r^2 - x^2} = \pm 3$. Then $\sin \arccos \frac{4}{5} = \sin\theta = \frac{y}{r} = \pm\frac{3}{5}$.

2. Find $\cot \arctan(-\frac{1}{3})$.

SOLUTION. Let $\theta = \arctan\left(-\frac{1}{3}\right)$, then $\tan\,\theta = -\frac{1}{3} = \frac{y}{x}$. In this case, we can take $y = -1$ and $x = 3$, or $y = 1$ and $x = -3$. In either pairing we have $\cot \arctan\left(-\frac{1}{3}\right) = \cot\theta = \frac{x}{y} = -3$.

3. Find $\sin \operatorname{arccot} \sqrt{3}$.

SOLUTION. Let $\theta = \operatorname{arccot} \sqrt{3}$, then $\cot\theta = \sqrt{3} = \frac{x}{y}$. Let $x = \sqrt{3}$ and $y = 1$, then $r = 2$ and

$$\sin \operatorname{arccot} \sqrt{3} = \sin\theta = \frac{y}{r} = \frac{1}{2}$$

However, we could also have chosen $x = -\sqrt{3}$ and $y = -1$, still obtaining $r = 2$, from which

$$\sin \operatorname{arccot} \sqrt{3} = \sin\theta = \frac{y}{r} = -\frac{1}{2}$$

Both answers satisfy the statement of the problem. This can also be seen from the fact that if $\cot\theta = \sqrt{3}$, then θ can be equal to $30°$ and $210°$.

4. Find $\sin \operatorname{Arccot} \sqrt{3}$.

SOLUTION. This is the same problem as number 3 with the exception that now θ is limited to the principal value of $\operatorname{arccot} \sqrt{3}$ and must lie in the interval $-90° \le \theta \le 90°$. The only angle that satisfies this condition is $30°$ so that $\sin \operatorname{Arccot} \sqrt{3} = \frac{1}{2}$.

The inverse functions are an integral part of trigonometry; their properties will be discussed thoroughly in subsequent chapters.

SPECIAL NOTE. Throughout this chapter we have been writing the argument of the trigonometric functions in the form of numbers $\left(\text{for example } \frac{\pi}{2}\right)$ or angles measured in degrees (such as $30°$). This has been done to emphasize that the values of the trigonometric functions can be found by considering the argument to be an angle. However, care must be exercised when reading the argument. For example, $\sin 2$ and $\sin 2°$ are *not* the same. In the first case we wish to find the sine of the number 2 or its equivalent, the sine of 2 radians; in the second case we wish to find the sine of the angle whose measure is $2°$. The approximate values are

$$\sin 2 = 0.90930 \qquad \sin 2° = 0.03490$$

If the reader needs further clarification on this point, we recommend that he turn to the Appendix and study the use of the trigonometric tables.

21. EXERCISE 2. (For exercises requiring tables see Appendix.)

1. Complete the following table using exact numbers.

	sin	cos	tan	cot	sec	csc
30°						
210°						
45°						
120°						
150°						
300°						
225°						
330°						
135°						
60°						
315°						
240°						

2. State the quadrant in which θ lies under each of the following conditions.

a) $\sin\theta > 0, \cos\theta > 0$
b) $\sin\theta < 0, \tan\theta > 0$
c) $\sec\theta < 0, \sin\theta > 0$
d) $\cot\theta > 0, \sin\theta < 0$
e) $\csc\theta < 0, \tan\theta > 0$
f) $\cos\theta < 0, \tan\theta < 0$
g) $\sin\theta > 0, \sec\theta < 0$
h) $\cot\theta > 0, \csc\theta < 0$
i) $\cos\theta < 0, \sin\theta < 0$
j) $\tan\theta < 0, \cos\theta > 0$

3. Find the following values.

a) $\sin\left(-\frac{\pi}{6}\right)$
b) $\cot\left(-\frac{3\pi}{2}\right)$
c) $\cos\left(-\frac{\pi}{3}\right)$
d) $\tan\left(\frac{\pi}{4} - \frac{\pi}{2}\right)$
e) $\sec\left(\frac{2\pi}{3} + \frac{\pi}{2}\right)$
f) $\csc\left(-\frac{3\pi}{4} + \pi\right)$
g) $\cos(-315^\circ)$
h) $\sin\left(-\frac{5\pi}{3}\right)$
i) $\tan(-\pi)$

4. Evaluate the following expressions.

a) $(\cos 0^\circ)(\sin 450^\circ)(\tan 135^\circ)$

b) $\left(\sin\frac{\pi}{3}\right)\left(\cos\frac{\pi}{6}\right)\left(\tan\frac{5\pi}{4}\right)$

c) $\dfrac{(\sin 45^\circ)(\sec 120^\circ)(\sqrt{2})}{(\cos 300^\circ)(\tan 120^\circ)(\cot 240^\circ)}$

5. Express each of the following as functions of an acute angle.

a) $\sin(-312^\circ)$ b) $\cos 830^\circ$ c) $\tan 172^\circ$

d) $\sec\left(-\frac{7\pi}{6}\right)$ e) $\cot\frac{11\pi}{3}$ f) $\sin\frac{8\pi}{3}$

g) $\cos 115^\circ$ h) $\tan 318^\circ$ i) $\csc 243^\circ$

6. Sketch the following angles and find the values of their trigonometric functions.

a) $\frac{\pi}{2}$ b) $-\pi$ c) $-\frac{\pi}{2}$ d) 2π e) $\frac{5\pi}{2}$ f) -5π

7. Sketch the following angles and find the values of their trigonometric functions.

a) $\left\{\frac{\pi}{3} \pm 2n\pi\right\}$ b) $\left\{\frac{\pi}{6} \pm \frac{\pi}{2}\right\}$

8. How many quadrantal angles are there in each of the following intervals?

a) $\frac{2\pi}{3}$ to $\frac{7\pi}{4}$ b) $-\pi < \theta < 4\pi$

c) -4 radians to 4 radians d) $|\theta| \leq \frac{\pi}{2}$

e) -215° to 725° f) $-\frac{2\pi}{3} < \theta \leq \frac{3\pi}{2}$

9. Simplify the following expressions.

a) $\sin\frac{\pi}{4}\tan\frac{\pi}{4}\cos\frac{\pi}{4}$ b) $\cos\frac{\pi}{3}\sin\frac{\pi}{6}$

c) $\cos\left(-\frac{\pi}{3}\right)\sin\left(-\frac{\pi}{6}\right) - \sin\frac{\pi}{3}\cos\frac{\pi}{6}$

d) $\dfrac{\sin 30^\circ \tan 60^\circ}{\cot 30^\circ}$ e) $\dfrac{\cos\frac{\pi}{6}}{\tan\frac{\pi}{4}} + \dfrac{1 - \sin\frac{\pi}{6}}{\sec\frac{\pi}{6}}$

10. Find all distinct values of the function if n is a nonnegative integer.

a) $f(n) = \sin n\pi$ b) $f(n) = \cos\frac{n\pi}{2}$

c) $f(n) = 2\sin\left(\pm\frac{n\pi}{2}\right)$ d) $f(n) = 3\tan\frac{n\pi}{3}$

e) $f(n) = 2\sin\frac{n\pi}{4}$ f) $f(n) = \cos\frac{n\pi}{3}$

11. Find the values.

arcsin $\frac{1}{2}$ b) $\arctan\sqrt{3}$ c) $\arccos\frac{\sqrt{3}}{2}$

arcsin $\left(\frac{\sqrt{2}}{2}\right)$ e) $\arccos\left(-\frac{1}{2}\right)$ f) $\arccos\frac{1}{2}$

g) $\text{arccot}\ 1$ h) $\text{arcsec}\ 2$ i) $\text{arccsc}\sqrt{2}$

12. Find the values.

a) Arcsin $\frac{1}{2}$ b) Arccos $\frac{\sqrt{2}}{2}$ c) Arctan $(-\sqrt{3})$

d) Arccot 1 e) Arcsec 2 f) Arccsc 1

13. Find the following values.

a) sin arcsin $\frac{3}{5}$ b) cos arctan $\frac{3}{4}$

c) tan arcsin $\frac{7}{25}$ d) sec arcsin y

e) cos arctan 5 f) cot arccos n

14. Evaluate.

a) Arcsin $\left(\sin \frac{\pi}{3}\right)$ b) Arctan $\left(\tan \frac{3\pi}{4}\right)$

c) Arccos $[\cos(-30°)]$ d) Arcsec (sec 45°)

15. Express the following as a trigonometric function of an angle measured in degrees (let π be approximated by 3.14).

a) sin 2 b) cos 1.57 c) tan 3.14 d) cot 1

chapter 3

Properties of the Trigonometric Functions

22. INTRODUCTION

Let us consider two functions f and g defined over some domain for the argument x. The equation $f(x) = g(x)$ may be true for all values of x in the domain of definition, it may be true for only certain values of x, or it may not be true for any values of x. We distinguish among these possibilities by giving names to the conditions.

DEFINITION. *The equation* f(x) = g(x) *is called an* **identity** *if and only if it is true for every* x *in the domain of definition for the two functions.*

ILLUSTRATION. The equation

$$\frac{2}{x-1} + \frac{x}{x-3} = \frac{x^2 + x - 6}{(x-1)(x-3)}$$

is an identity since it is true for every value of x for which both sides are defined. Note that neither side is defined for $x = 1$ and $x = 3$.

DEFINITION. *The equation* f(x) = g(x) *is called a* **conditional equation** *if there exists at least one value of* x *for which the statement is true and at least one value of* x *for which* f(x) ≠ g(x) *with* x *in the domain of definition for both* f *and* g.

ILLUSTRATION. The equation

$$3x + 2 = x + 6$$

is a conditional equation since it is true for $x = 2$, but is false for $x = 1$ where both sides are defined for these values of x.

The values of x which make a conditional equation true are called **solutions** and they form the **solution set.** If there are no values of x in the domain of definition which make the statement of an equation true, then there are no solutions or, in other words, the solution set is the null set. An equation which has no solution is called a **null equation** and usually represents an impossible situation.

ILLUSTRATION. The equation $\sqrt{x+1} = -5$ has no solutions since there are no values of x for which a positive radical will equal a negative number.

DEFINITION. *If* f(x) *is any expression, then* f(x) = f(x) *is called a* **tautology**.

The equation $f(x) = g(x)$ will be an identity if either $f(x)$ can be reduced to $g(x)$ or $g(x)$ can be reduced to $f(x)$ by the use of permissible transformations; i.e., if $f(x) = g(x)$ can be transformed into a tautology.

ILLUSTRATION. Show that $(a+b)(a+b) = a^2 + 2ab + b^2$ is an identity.

SOLUTION. Perform the multiplication on the left side.

$$(a+b)(a+b) = a^2 + ab + ab + b^2 = a^2 + 2ab + b^2$$

Thus we have the tautology

$$a^2 + 2ab + b^2 = a^2 + 2ab + b^2$$

and the equation is an identity.

To prove that an equation is not an identity, it is only necessary to exhibit a value of the argument in the domain of definition for which the equation is not true.

ILLUSTRATION. Prove that $\sin^2 t = 1 + \cos^2 t$ is not an identity.

SOLUTION. Each side is defined for $t = 30^\circ$, but for this value of t the equation is not true since

$$\frac{1}{4} \neq 1 + \frac{3}{4}$$

In this chapter we shall concern ourselves with trigonometric expressions which are identities, and derive a number of formulas that will permit us to transform one trigonometric expression into another. Conditional trigonometric equations will be discussed in Chapter 5.

23. FUNDAMENTAL IDENTITIES

There are many relations among the trigonometric functions that are true for all values of the argument for which the functions are defined. The most important of these are called the fundamental identities. There are eight of them.

The reciprocal relations

$$\csc t = \frac{1}{\sin t} \qquad (1)$$

$$\sec t = \frac{1}{\cos t} \qquad (2)$$

$$\cot t = \frac{1}{\tan t} \qquad (3)$$

The quotient relations

$$\tan t = \frac{\sin t}{\cos t} \qquad (4)$$

$$\cot t = \frac{\cos t}{\sin t} \qquad (5)$$

The Pythagorean relations

$$\sin^2 t + \cos^2 t = 1 \qquad (6)$$

$$\tan^2 t + 1 = \sec^2 t \qquad (7)$$

$$1 + \cot^2 t = \csc^2 t \qquad (8)$$

These identities are proved directly from the definitions.

PROOF [*of* (1)]. By the definitions we have

$$\csc t = \frac{r}{y} \qquad \sin t = \frac{y}{r}$$

Therefore
$$\csc t = \frac{r}{y} = \frac{1}{y/r} = \frac{1}{\sin t}$$

PROOF [*of* (4)]. By the definitions we have

$$\sin t = \frac{y}{r} \qquad \cos t = \frac{x}{r} \qquad \tan t = \frac{y}{x}$$

Therefore
$$\frac{\sin t}{\cos t} = \frac{y/r}{x/r} = \frac{y}{r} \cdot \frac{r}{x} = \tan t$$

PROOF [*of* (7)]. By the Pythagorean theorem we have

$$y^2 + x^2 = r^2$$

Divide each term by x^2 to obtain

$$\frac{y^2}{x^2} + 1 = \frac{r^2}{x^2}$$

Then, from the definitions,

$$\tan^2 t + 1 = \sec^2 t$$

The reader should prove all the identities and then commit them to memory.

The fundamental identities may take different forms. The following are examples of variations.

$$\sin \alpha = \frac{1}{\csc \alpha} \qquad 1 - \cos^2 \alpha = \sin^2 \alpha \qquad \sin \alpha \cot \alpha = \cos \alpha$$

The first five identities permit us to transform all of the trigonometric functions into the sine and cosine functions. The first of the Pythagorean relations permits us to change from the sine function to the cosine function or vice versa. Thus *any trigonometric expression can be transformed into an expression involving only one trigonometric function.* It is also possible to express each trigonometric function in terms of each of the other five functions.

ILLUSTRATION. Express $\sin t$ in terms of each of the other functions.

SOLUTION. By (6), $\sin t = \pm\sqrt{1 - \cos^2 t}$

By (4), (2), and (7), $\sin t = \cos t \tan t = \dfrac{\tan t}{\sec t} = \dfrac{\tan t}{\pm\sqrt{1 + \tan^2 t}}$

By (1) and (8), $\sin t = \dfrac{1}{\csc t} = \dfrac{1}{\pm\sqrt{1 + \cot^2 t}}$

By (4), (2), and (7), $\sin t = \cos t \tan t = \dfrac{\pm\sqrt{\sec^2 t - 1}}{\sec t}$

By (1), $\sin t = \dfrac{1}{\csc t}$

In the above illustration the "$\pm$" is determined by the quadrant in which $T(1,t)$ falls.

24. FUNCTIONS OF THE COMPLEMENTARY ANGLES

We recall that two angles, α and β, are said to be complementary angles if their sum is $90°$; i.e., $\alpha + \beta = 90°$. The names of the trigonometric functions were chosen in such a way that they are related in pairs as function and cofunction. Thus cosine is the cofunction of sine; cotangent is the cofunction of tangent; cosecant is the cofunction of secant; and vice versa, sine is the cofunction of the cosine; etc.

THEOREM. *The trigonometric function of an angle is equal to the cofunction of its complementary angle.*

trigonometric function of α = cofunction of $(90° - \alpha)$

PROOF.* Consider the right triangle ABC with acute angles α and β, which are complementary angles by the property of the right triangle. Label the parts as shown in Figure 23; we then have

$$\text{side opposite } \alpha = a = \text{side adjacent to } \beta$$
$$\text{side adjacent to } \alpha = b = \text{side opposite } \beta$$

The definitions given in Section 13C can be used to obtain the following.

FIGURE 23

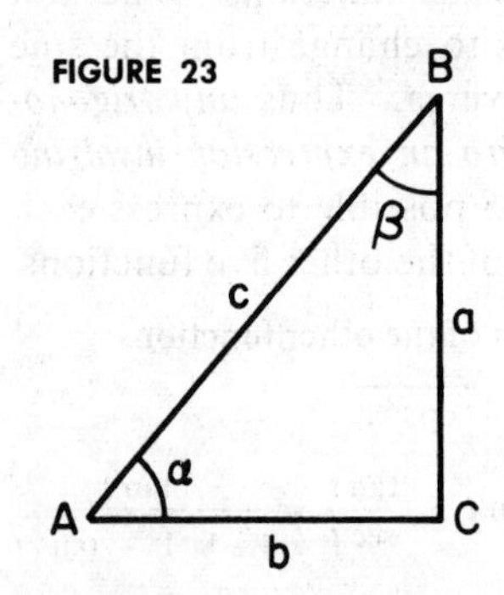

$$\sin\alpha = \frac{a}{c} = \cos\beta$$
$$\cos\alpha = \frac{b}{c} = \sin\beta$$
$$\tan\alpha = \frac{a}{b} = \cot\beta$$
$$\cot\alpha = \frac{b}{a} = \tan\beta$$
$$\sec\alpha = \frac{c}{b} = \csc\beta$$
$$\csc\alpha = \frac{c}{a} = \sec\beta$$

This theorem is used to transform the trigonometric functions of an angle to a function of its complementary angle.

ILLUSTRATIONS

1. Transform $\sin 30^\circ$ to a cosine function.

SOLUTION. $\sin 30^\circ = \cos(90^\circ - 30^\circ) = \cos 60^\circ$.

2. For what values of θ and ϕ do we have $\tan\theta = \cot\phi$?

SOLUTION. Since tangent and cotangent are complementary functions, they will be equal for all values of θ and ϕ for which $\theta + \phi = 90^\circ$; for example, $\theta = 27^\circ$ and $\phi = 63^\circ$, $\theta = 110^\circ$ and $\phi = -20^\circ$.

The relations among the trigonometric functions of **supplementary angles** (angles whose sum equals 180°) are derived from the theorem of Section 17.

25. PROOFS OF GENERAL IDENTITIES

If we are given two expressions $f(t)$ and $g(t)$, we can prove the identity $f(t) = g(t)$ providing $f(t)$ can be transformed into $g(t)$ or vice versa. We shall use this procedure to prove identities; i.e.,

to prove an identity, alter only one member until it assumes the same form as the member which has not been altered.

*The general proof is given in Section 27.

We shall make the transformations by using the fundamental identities (or any identity that has already been proved) and permissible algebraic operations. The properties of complementary angles may also be used, if convenient.

ILLUSTRATIONS

1. Prove the identity $\dfrac{1 - \tan^2 t}{1 + \tan^2 t} = 1 - 2\sin^2 t$.

SOLUTION. We shall transform the left member.

Since $\tan t = \dfrac{\sin t}{\cos t}$, $\qquad \dfrac{1 - \dfrac{\sin^2 t}{\cos^2 t}}{1 + \dfrac{\sin^2 t}{\cos^2 t}} =$

Simplifying, $\qquad \dfrac{\cos^2 t - \sin^2 t}{\cos^2 t} \cdot \dfrac{\cos^2 t}{\cos^2 t + \sin^2 t} =$

Since $\cos^2 t + \sin^2 t = 1$, $\qquad \cos^2 t - \sin^2 t =$

Since $\cos^2 t = 1 - \sin^2 t$, $\qquad 1 - 2\sin^2 t = 1 - 2\sin^2 t$

2. Prove the identity $\cos(90° - \theta)\sec\theta = \tan\theta$.

SOLUTION. Transform the left side.

By $\cos(90° - \theta) = \sin\theta$, $\qquad \sin\theta\sec\theta =$

By (2), $\qquad \dfrac{\sin\theta}{\cos\theta} =$

By (4), $\qquad \tan\theta = \tan\theta$

We have already stressed the fact that identities hold only for the values in the domain of definition for the trigonometric functions. We should also check the complete trigonometric expression; for example, the expression

$$\frac{1}{1 - \sin\theta}$$

is not defined for $\theta = 90°$ since then $1 - \sin\theta = 0$. A more troublesome expression is one which contains a radical, such as $\sqrt{f(t)} = g(t)$. The left member of this equation can never be negative since $\sqrt{a^2} = |a|$. The negative values are obtained from $-\sqrt{f(t)}$. Consequently, we must eliminate all the values of t for which $g(t)$ is negative as well as those for which the functions are not defined. If we can find a value of t in the domain of definition for which $\sqrt{f(t)} \neq g(t)$, the equation is not an identity but a conditional equation.

ILLUSTRATION. Is the following equation an identity?

$$\sqrt{\frac{1 - \sin t}{1 + \sin t}} = \sec t - \tan t$$

SOLUTION. The right side can be transformed, thus

$$\sec t - \tan t = \frac{1}{\cos t} - \frac{\sin t}{\cos t} = \frac{1 - \sin t}{\cos t}$$

Since $\sin t \leq 1$, the numerator is never negative. However, the denominator is negative for $\frac{\pi}{2} < t < \frac{3\pi}{2}$. Therefore the right side can be negative, and the equation is not an identity. It is true only for values of t such that $-\frac{\pi}{2} < t < \frac{\pi}{2}$.

Suggestions for proving identities

I. Have the fundamental identities well in mind. This should include variations such as $\sec t \cos t = 1$, $1 - \sin^2 t = \cos^2 t$.

II. Any function can be expressed in terms of any other function; in particular, all functions can be expressed in terms of sines and cosines.

III. If one member involves only one function it may be best to express everything on the other side in terms of this function.

IV. Look for the possibility of factoring one of the expressions.

V. Avoid the use of radicals whenever possible.

VI. A preliminary simplification of both sides may suggest a method of attack.

ILLUSTRATIONS. Prove the following identities.

1. $\sec^2 t - \csc^2 t = \tan^2 t - \cot^2 t$

SOLUTION. Change the left side.

By (7) and (8), $\tan^2 t + 1 - (1 + \cot^2 t) =$

Simplify, $\tan^2 t - \cot^2 t = \tan^2 t - \cot^2 t$

2. $\dfrac{\cos x}{1 - \sin x} = \dfrac{1 + \sin x}{\cos x}$

SOLUTION. A clearing of fractions would result in the fundamental identity (6). However, this is not a satisfactory proof. Let us work on the left side.

Multiply by $\frac{1 + \sin x}{1 + \sin x}$, $\dfrac{(1 + \sin x)(\cos x)}{(1 + \sin x)(1 - \sin x)} =$

Simplify, $\dfrac{(1 + \sin x)\cos x}{1 - \sin^2 x} =$

By (6), $$\frac{(1 + \sin x)\cos x}{\cos^2 x} =$$

Simplify, $$\frac{1 + \sin x}{\cos x} = \frac{1 + \sin x}{\cos x}$$

3. $\sin^4 t - \cos^4 t = 2\sin^2 t - 1$

SOLUTION. Factor the left side.

$$(\sin^2 t - \cos^2 t)(\sin^2 t + \cos^2 t) =$$

By (6), $$[\sin^2 t - (1 - \sin^2 t)][1] =$$

Simplify, $$2\sin^2 t - 1 = 2\sin^2 t - 1$$

4. $\dfrac{\tan^2 t + 1}{2\cos(-t) - \sin(90^\circ - t)} = \sec^3 t$

SOLUTION. Change the left side.

$$\frac{\tan^2 t + 1}{2\cos(-t) - \sin(90^\circ - t)} = \frac{\sec^2 t}{2\cos t - \cos t} = \frac{\sec^2 t}{\cos t} = \sec^3 t$$

5. $\dfrac{\cot t - \cos t}{\cos^3 t} = \dfrac{\csc t}{1 + \sin t}$

SOLUTION. Change the left side.

$$\frac{\cot t - \cos t}{\cos^3 t} = \frac{\dfrac{\cos t}{\sin t} - \cos t}{\cos^3 t} = \frac{\cos t - \sin t\cos t}{\sin t\cos^3 t}$$

$$= \frac{\cos t(1 - \sin t)}{\sin t\cos^3 t} = \frac{1 - \sin t}{\sin t(1 - \sin^2 t)}$$

$$= \frac{1}{\sin t} \cdot \frac{1}{1 + \sin t} = \frac{\csc t}{1 + \sin t}$$

26. EXERCISE 3

1. Prove the eight fundamental identities.
2. Complete the following table expressing each function in terms of the other functions.

	$\sin t$	$\cos t$	$\tan t$	$\cot t$	$\sec t$	$\csc t$
$\sin t$	$\sin t$	$\pm\sqrt{1 - \cos^2 t}$				$\dfrac{1}{\csc t}$
$\cos t$		$\cos t$				
$\tan t$			$\tan t$			
$\cot t$				$\cot t$		
$\sec t$					$\sec t$	
$\csc t$						$\csc t$

Prove the following identities.

3. $\dfrac{\sin x}{\csc x} + \dfrac{\cos x}{\sec x} = 1$
4. $\tan x + \cot x = \sec x \csc x$
5. $\dfrac{\sec x}{\cot x + \tan x} = \sin x$
6. $\cos^4 x - \sin^4 x = 1 - 2\sin^2 x$
7. $\sin^2 x \sec^2 x + \sin^2 x \csc^2 x = \sec^2 x$
8. $\sec^2 x (1 - \sin^2 x) = 1$
9. $\tan x \sin x + \cos x = \sec x$
10. $\sin x + \cos x = \dfrac{\sec x + \csc x}{\tan x + \cot x}$
11. $\dfrac{1 - \cos^6 t}{\sin^2 t} = 1 + \cos^2 t + \cos^4 t$
12. $\dfrac{\cos^2 x}{\sec x} + \dfrac{\tan^2 x \cos x}{1 + \tan^2 x} = \cos x$
13. $\cot^4 \theta + \cot^2 \theta = \csc^4 \theta - \csc^2 \theta$
14. $\dfrac{\sin t}{1 + \cos t} + \dfrac{1 + \cos t}{\sin t} = 2\csc t$
15. $\dfrac{\sin^3 t + \cos^3 t}{\sin t + \cos t} = 1 - \sin t \cos t$
16. $1 - \dfrac{\sin^2 x}{1 + \cos x} = \cos x$
17. $\dfrac{1 - 2\cos^2 t}{\sin t \cos t} = \tan t - \cot t$
18. $(\sec x + \tan x)(1 - \sin x) = \cos x$
19. $\dfrac{\sin x \cos x}{\cos^2 x - \sin^2 x} = \dfrac{\tan x}{1 - \tan^2 x}$
20. $\dfrac{\sec x - \csc x}{\sec x + \csc x} = \dfrac{\tan x - 1}{\tan x + 1}$
21. $(a \sin x + b \cos x)^2 + (b \sin x - a \cos x)^2 = a^2 + b^2$
22. $(\sin \alpha \cos \beta)^2 + (\sin \alpha \sin \beta)^2 + \cos^2 \alpha = 1$
23. $\dfrac{2 \sin x \cos x}{1 + \cos^2 x - \sin^2 x} = \tan x$
24. $2(1 + \sin x)(1 + \cos x) = (1 + \sin x + \cos x)^2$
25. $\tan^2 x - \sin^2 x = \sin^2 x \tan^2 x$

27. COSINE OF $(\alpha \pm \beta)$

In the next few sections we shall derive twenty-four more formulas which exhibit properties of the trigonometric functions. When these are combined with the fundamental identities, we will have thirty-two formulas and shall number them in sequence.

THEOREM. *Let α and β be any two real numbers, then*

$$\cos(\alpha + \beta) = \cos\alpha\cos\beta - \sin\alpha\sin\beta \tag{9}$$

$$\cos(\alpha - \beta) = \cos\alpha\cos\beta + \sin\alpha\sin\beta \tag{10}$$

PROOF. We shall develop the proof for formula (9) and then use the already established properties to prove the remaining formulas. Place the three angles α, $-\beta$, and $\alpha + \beta$ in standard

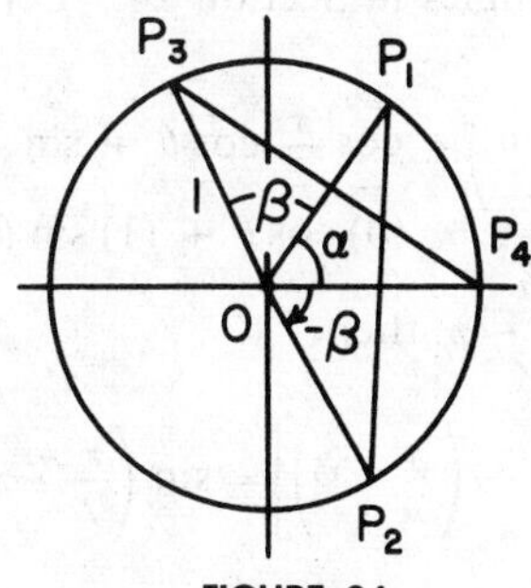

FIGURE 24

positions, and let their terminal sides intersect the unit circle at the respective points P_1, P_2, and P_3. See Figure 24. The coordinates of these points are:

$$\begin{aligned} x_1 &= \cos\alpha & y_1 &= \sin\alpha \\ x_2 &= \cos(-\beta) & y_2 &= \sin(-\beta) \\ x_3 &= \cos(\alpha + \beta) & y_3 &= \sin(\alpha + \beta) \end{aligned}$$

Let P_4 be the point $P_4(1,0)$. Then

$$\angle P_3OP_4 = |\alpha + \beta| \text{ and}$$
$$\angle P_1OP_2 = |\alpha| + |-\beta| = |\alpha + \beta|$$

Since these two angles are equal, $\triangle P_1OP_2 \cong \triangle P_3OP_4$ (s.a.s.) and $\overline{P_3P_4} = \overline{P_1P_2}$. We use the distance formula to obtain

$$(x_3 - x_4)^2 + (y_3 - y_4)^2 = (x_1 - x_2)^2 + (y_1 - y_2)^2$$

or

$$[\cos(\alpha + \beta) - 1]^2 + [\sin(\alpha + \beta)]^2$$
$$= (\cos\alpha - \cos\beta)^2 + (\sin\alpha + \sin\beta)^2$$

which simplifies to

$$2 - 2\cos(\alpha + \beta) = 2 - 2\cos\alpha\cos\beta + 2\sin\alpha\sin\beta$$

or $$\cos(\alpha + \beta) = \cos\alpha\cos\beta - \sin\alpha\sin\beta$$

The last formula on page 45 is true for all values of α and β, positive or negative. By writing $\alpha - \beta$ as $\alpha + (-\beta)$ we obtain

$$\begin{aligned}\cos(\alpha - \beta) &= \cos[\alpha + (-\beta)] = \cos\alpha\cos(-\beta) - \sin\alpha\sin(-\beta)\\ &= \cos\alpha\cos\beta + \sin\alpha\sin\beta\end{aligned}$$

We can now generalize the relations of complementary angles discussed for acute angles in Section 24. Let $\alpha = \frac{\pi}{2}$ and $\beta = \theta$, then

$$\begin{aligned}\cos\left(\frac{\pi}{2} - \theta\right) &= \cos\frac{\pi}{2}\cos\theta + \sin\frac{\pi}{2}\sin\theta\\ &= (0)\cos\theta + (1)\sin\theta = \sin\theta\end{aligned}$$

If we now let $\theta = \frac{\pi}{2} - \phi$, then

$$\cos\left[\frac{\pi}{2} - \left(\frac{\pi}{2} - \phi\right)\right] = \sin\left(\frac{\pi}{2} - \phi\right)$$

or

$$\cos\phi = \sin\left(\frac{\pi}{2} - \phi\right)$$

Thus the relations between complementary angles and their sines and cosines are true for any angles θ and ϕ.

In the above proofs we used the theorem concerning the functions of negative arguments (see Section 16) and the fundamental identities. Since we did not use the properties of Section 17, the theorem on page 45 can be applied to prove the properties of that section for a general angle α.

ILLUSTRATION. Prove $\cos(\pi - \alpha) = -\cos\alpha$.

SOLUTION. $$\begin{aligned}\cos(\pi - \alpha) &= \cos\pi\cos\alpha + \sin\pi\sin\alpha\\ &= (-1)\cos\alpha + (0)\sin\alpha = -\cos\alpha\end{aligned}$$

Formulas (9) and (10) can be used to find exact values for the cosines of arguments that can be expressed as the sum or difference of arguments for which the exact cosine and sine values are known.

ILLUSTRATIONS

1. Find the exact value of $\cos 105°$.

SOLUTION. $$\begin{aligned}\cos 105° &= \cos(60° + 45°)\\ &= \cos 60°\cos 45° - \sin 60°\sin 45°\\ &= \frac{1}{2}\cdot\frac{\sqrt{2}}{2} - \frac{\sqrt{3}}{2}\cdot\frac{\sqrt{2}}{2} = \frac{\sqrt{2} - \sqrt{6}}{4}\end{aligned}$$

2. Find the exact value of $\cos 15^\circ$.

SOLUTION.

$$\begin{aligned}\cos 15^\circ &= \cos(45^\circ - 30^\circ)\\ &= \cos 45^\circ \cos 30^\circ + \sin 45^\circ \sin 30^\circ\\ &= \frac{\sqrt{2}}{2}\cdot\frac{\sqrt{3}}{2} + \frac{\sqrt{2}}{2}\cdot\frac{1}{2} = \frac{\sqrt{6}+\sqrt{2}}{4}\end{aligned}$$

28. SINE OF $(\alpha \pm \beta)$

THEOREM. *Let α and β be any two real numbers, then*

$$\sin(\alpha+\beta) = \sin\alpha\cos\beta + \cos\alpha\sin\beta \qquad (11)$$

$$\sin(\alpha-\beta) = \sin\alpha\cos\beta - \cos\alpha\sin\beta \qquad (12)$$

PROOF. By making use of the properties of the preceding section, we have

$$\begin{aligned}\sin(\alpha+\beta) &= \cos\left[\frac{\pi}{2} - (\alpha+\beta)\right] = \cos\left[\left(\frac{\pi}{2}-\alpha\right) - \beta\right]\\ &= \cos\left(\frac{\pi}{2}-\alpha\right)\cos\beta + \sin\left(\frac{\pi}{2}-\alpha\right)\sin\beta\\ &= \sin\alpha\cos\beta + \cos\alpha\sin\beta\end{aligned}$$

Furthermore,

$$\begin{aligned}\sin(\alpha-\beta) &= \sin[\alpha + (-\beta)]\\ &= \sin\alpha\cos(-\beta) + \cos\alpha\sin(-\beta)\\ &= \sin\alpha\cos\beta - \cos\alpha\sin\beta\end{aligned}$$

29. TANGENT AND COTANGENT OF $(\alpha \pm \beta)$

THEOREM. *Let α and β be any two real numbers, then*

$$\tan(\alpha+\beta) = \frac{\tan\alpha + \tan\beta}{1 - \tan\alpha\tan\beta} \qquad (13)$$

$$\tan(\alpha-\beta) = \frac{\tan\alpha - \tan\beta}{1 + \tan\alpha\tan\beta} \qquad (14)$$

$$\cot(\alpha+\beta) = \frac{\cot\alpha\cot\beta - 1}{\cot\alpha + \cot\beta} \qquad (15)$$

$$\cot(\alpha-\beta) = \frac{\cot\alpha\cot\beta + 1}{\cot\beta - \cot\alpha} \qquad (16)$$

PROOF. We shall exhibit the proof of (13) and indicate the procedure for the other three.

$$\tan(\alpha+\beta) = \frac{\sin(\alpha+\beta)}{\cos(\alpha+\beta)}$$

By (11) and (9),
$$= \frac{\sin\alpha\cos\beta + \cos\alpha\sin\beta}{\cos\alpha\cos\beta - \sin\alpha\sin\beta}$$

Divide numerator and denominator by $\cos\alpha\cos\beta$,
$$= \frac{\dfrac{\sin\alpha\cos\beta}{\cos\alpha\cos\beta} + \dfrac{\cos\alpha\sin\beta}{\cos\alpha\cos\beta}}{\dfrac{\cos\alpha\cos\beta}{\cos\alpha\cos\beta} - \dfrac{\sin\alpha\sin\beta}{\cos\alpha\cos\beta}}$$

Simplify,
$$= \frac{\dfrac{\sin\alpha}{\cos\alpha} + \dfrac{\sin\beta}{\cos\beta}}{1 - \dfrac{\sin\alpha}{\cos\alpha}\dfrac{\sin\beta}{\cos\beta}}$$

By (4),
$$= \frac{\tan\alpha + \tan\beta}{1 - \tan\alpha\tan\beta}$$

The proof of (15) is accomplished in a similar manner by using $\sin\alpha\sin\beta$ as a divisor of the numerator and denominator. Formulas (14) and (16) are obtained from (13) and (15), respectively, by writing $(\alpha - \beta) = [\alpha + (-\beta)]$ and using the properties of the functions of negative arguments. Some applications of these formulas can be illustrated.

ILLUSTRATIONS

1. Find the exact values of $\sin 75°$ and $\tan 75°$.

SOLUTION

$$\sin 75° = \sin(45° + 30°) = \sin 45°\cos 30° + \cos 45°\sin 30°$$
$$= \frac{\sqrt{2}}{2}\cdot\frac{\sqrt{3}}{2} + \frac{\sqrt{2}}{2}\cdot\frac{1}{2} = \frac{\sqrt{6}+\sqrt{2}}{4}$$

$$\tan 75° = \tan(45° + 30°) = \frac{\tan 45° + \tan 30°}{1 - \tan 45°\tan 30°}$$

$$= \frac{1 + \dfrac{\sqrt{3}}{3}}{1 - \dfrac{\sqrt{3}}{3}} = \frac{3+\sqrt{3}}{3-\sqrt{3}} = \frac{(3+\sqrt{3})(3+\sqrt{3})}{(3-\sqrt{3})(3+\sqrt{3})}$$

$$= \frac{9 + 6\sqrt{3} + 3}{9 - 3} = \frac{12 + 6\sqrt{3}}{6} = 2 + \sqrt{3}$$

2. Prove $\tan(\pi - \alpha) = -\tan\alpha$.

SOLUTION. $\tan(\pi - \alpha) = \dfrac{\tan\pi - \tan\alpha}{1 + \tan\pi\tan\alpha} = \dfrac{0 - \tan\alpha}{1 - (0)\tan\alpha} = -\tan\alpha$

3. Prove $\cot\left(\frac{\pi}{2} - \theta\right) = \tan \theta$.

SOLUTION. $\cot\left(\frac{\pi}{2} - \theta\right) = \dfrac{\cot \frac{\pi}{2} \cot \theta + 1}{\cot \theta - \cot \frac{\pi}{2}} = \dfrac{1}{\cot \theta} = \tan \theta$

NOTE. The above formulas cannot be used to prove the properties of $\cot(\pi \pm \alpha)$ or $\tan\left(\frac{\pi}{2} - \alpha\right)$ for this would involve $\cot \pi$ and $\tan \frac{\pi}{2}$, which are undefined terms. This does not nullify the properties but simply indicates that other methods of proof must be developed. They can be proved by applying the fundamental identities first.

ILLUSTRATIONS

1. Find $\sin(\alpha + \beta)$ if $\sin \alpha = \frac{3}{5}$, $\cos \beta = \frac{12}{13}$, and α and β are both in in the first quadrant.

SOLUTION. Since $\sin \alpha = \frac{3}{5}$, we can take $y = 3$ and $r = 5$ from which we obtain $x = \pm \sqrt{25 - 9} = \pm \sqrt{16} = \pm 4$; and since α is in quadrant I, we have $x = 4$. From $\cos \beta = \frac{12}{13}$ and with β in quadrant I, we obtain $x = 12$, $r = 13$, and $y = 5$. We apply formula (11) to obtain

$$\begin{aligned} \sin(\alpha + \beta) &= \sin \alpha \cos \beta + \cos \alpha \sin \beta \\ &= \left(\frac{3}{5}\right)\left(\frac{12}{13}\right) + \left(\frac{4}{5}\right)\left(\frac{5}{13}\right) = \frac{36 + 20}{65} = \frac{56}{65} \end{aligned}$$

2. Find $\tan(\text{Arctan}\,\frac{5}{4} + \text{Arctan}\,\frac{3}{4})$.

SOLUTION. Let $\alpha = \text{Arctan}\,\frac{5}{4}$, then $\tan \alpha = \frac{5}{4}$. Let $\beta = \text{Arctan}\,\frac{3}{4}$, then $\tan \beta = \frac{3}{4}$. By substituting these values into formula (13), we obtain

$$\begin{aligned} \tan(\text{Arctan}\,\tfrac{5}{4} + \text{Arctan}\,\tfrac{3}{4}) &= \tan(\alpha + \beta) \\ &= \frac{\tan \alpha + \tan \beta}{1 - \tan \alpha \tan \beta} \\ &= \frac{\frac{5}{4} + \frac{3}{4}}{1 - \left(\frac{5}{4}\right)\left(\frac{3}{4}\right)} \\ &= \frac{8}{4} \cdot \frac{16}{16 - 15} = 32 \end{aligned}$$

3. Find $\sin\left[\text{Arccos}\,\frac{3}{5} - \text{Arcsin}\left(-\frac{5}{13}\right)\right]$.

SOLUTION. Let $\alpha = \text{Arccos}\ \frac{3}{5}$ and $\beta = \text{Arcsin}\left(-\frac{5}{13}\right)$. Since we are considering principal values, α is in quadrant I and β is in quadrant IV. (See Figure 25). Thus we have

$\cos\alpha = \frac{3}{5}$ and $\sin\alpha = \frac{4}{5}$
$\sin\beta = -\frac{5}{13}$ and $\cos\beta = \frac{12}{13}$

By formula (12) we have

$$\sin(\alpha - \beta) = \sin\alpha\cos\beta - \cos\alpha\sin\beta = \left(\frac{4}{5}\right)\left(\frac{12}{13}\right) - \left(\frac{3}{5}\right)\left(-\frac{5}{13}\right) = \frac{63}{65}$$

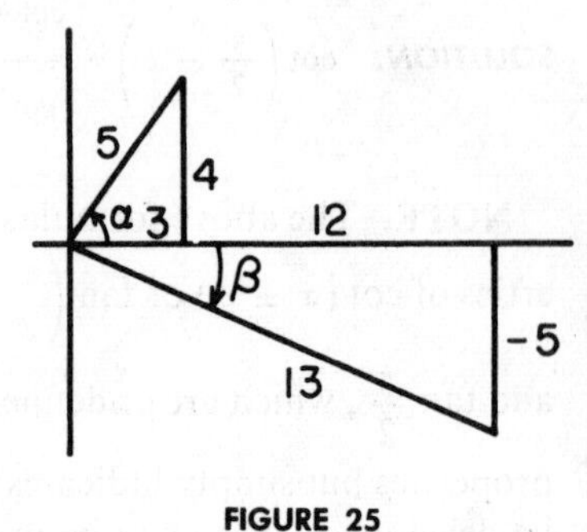

FIGURE 25

4. Find $\cos 140^\circ \cos 20^\circ + \sin 140^\circ \sin 20^\circ$.

SOLUTION. By formula (10),

$$\cos 140^\circ \cos 20^\circ + \sin 140^\circ \sin 20^\circ = \cos(140^\circ - 20^\circ) = \cos 120^\circ$$

Then by the reference angle theorem,

$$\cos 120^\circ = -\cos 60^\circ = -\tfrac{1}{2}$$

30. EXERCISE 4

1. Complete the following table with exact values.

	sin	cos	tan	cot	sec	csc
15°						
75°						
105°						

2. Verify each of the following.

a) $\cos 37^\circ \cos 23^\circ - \sin 37^\circ \sin 23^\circ = \frac{1}{2}$

b) $\sin 112^\circ \cos 68^\circ + \sin 68^\circ \cos 112^\circ = 0$

c) $\sin\frac{7\pi}{4}\cos\frac{5\pi}{4} - \cos\frac{7\pi}{4}\sin\frac{5\pi}{4} = 1$

d) $\cos\frac{7\pi}{12}\cos\frac{\pi}{4} + \sin\frac{7\pi}{12}\sin\frac{\pi}{4} = \frac{1}{2}$

e) $\sin 5x = \sin 2x \cos 3x + \cos 2x \sin 3x$

f) $\dfrac{\tan 22^\circ + \tan 38^\circ}{1 - \tan 22^\circ \tan 38^\circ} = \sqrt{3}$

3. Find $\sin(\alpha + \beta)$, $\tan(\alpha + \beta)$, and $\cot(\alpha - \beta)$ for each of the following.

a) $\sin\alpha = \frac{3}{5}$, $\sin\beta = \frac{12}{13}$, with α and β in quadrant II

b) $\tan\alpha = \frac{8}{15}$, $\sin\beta = -\frac{5}{13}$, with α and β in the same quadrant
c) $\cos\alpha = \frac{12}{37}$, $\sin\beta = \frac{9}{41}$, with α in IV and β in II

4. Find the following values.
a) $\sin\left(\text{Arcsin}\,\frac{8}{17} + \text{Arccos}\,\frac{7}{25}\right)$ b) $\cos\left(\text{Arctan}\,\frac{3}{4} - \text{Arccos}\,\frac{5}{13}\right)$
c) $\tan\left(\text{Arcsin}\,\frac{12}{37} - \text{Arccos}\,\frac{12}{13}\right)$ d) $\cot\left[\text{Arcsin}\,\frac{24}{25} + \text{Arcsin}\left(-\frac{3}{5}\right)\right]$
5. Prove that $\sin(\theta + 2n\pi) = \sin\theta$.
6. Express $\sin(\alpha + \beta + \gamma)$ in terms of sines and cosines of α, β, and γ. [Hint: $\alpha + \beta + \gamma = (\alpha + \beta) + \gamma$.]
7. Draw a unit circle and locate the three points P_1 $(\cos\gamma, \sin\gamma)$, P_2 $(\cos\beta, \sin\beta)$, and P_3 $(\cos\alpha, \sin\alpha)$ such that $\alpha = \beta + \gamma$. Set the distance $\overline{P_1A}$, where A is the point $A(1,0)$, equal to $\overline{P_2P_3}$ and derive the formula for $\cos(\alpha - \beta)$.

31. FORMULAS FOR DOUBLE ARGUMENTS

THEOREM. *Let α be any real number, then*

$$\sin 2\alpha = 2\sin\alpha\cos\alpha \tag{17}$$

$$\cos 2\alpha = \cos^2\alpha - \sin^2\alpha = 2\cos^2\alpha - 1 = 1 - 2\sin^2\alpha \tag{18}$$

$$\tan 2\alpha = \frac{2\tan\alpha}{1 - \tan^2\alpha} \tag{19}$$

$$\cot 2\alpha = \frac{\cot^2\alpha - 1}{2\cot\alpha} \tag{20}$$

PROOF. The proofs of these formulas are all obtained by letting $\beta = \alpha$ in formulas (11), (9), (13), and (15); for example,

$$\sin 2\alpha = \sin(\alpha + \alpha) = \sin\alpha\cos\alpha + \cos\alpha\sin\alpha$$
$$= 2\sin\alpha\cos\alpha$$

For $\cos 2\alpha$, we list three equivalent formulas which are obtained from formula (9) and then changed by the fundamental identity (6).

These formulas are usually used to change an expression from a function of a multiple argument to one of a single argument; however, it is sometimes convenient to use these formulas to go from the right side to the left side of an expression.

ILLUSTRATION. Show that $\dfrac{\cos^2 x - \sin^2 x}{\sin 2x} = \cot 2x$.

SOLUTION. By (18) and (5),

$$\frac{\cos^2 x - \sin^2 x}{\sin 2x} = \frac{\cos 2x}{\sin 2x} = \cot 2x$$

32. FORMULAS FOR HALF-ARGUMENTS

THEOREM. *Let θ be any real number, then*

$$\sin\frac{\theta}{2} = \pm\sqrt{\frac{1-\cos\theta}{2}} \tag{21}$$

$$\cos\frac{\theta}{2} = \pm\sqrt{\frac{1+\cos\theta}{2}} \tag{22}$$

$$\tan\frac{\theta}{2} = \pm\sqrt{\frac{1-\cos\theta}{1+\cos\theta}} = \frac{1-\cos\theta}{\sin\theta} = \frac{\sin\theta}{1+\cos\theta} \tag{23}$$

$$\cot\frac{\theta}{2} = \pm\sqrt{\frac{1+\cos\theta}{1-\cos\theta}} = \frac{\sin\theta}{1-\cos\theta} = \frac{1+\cos\theta}{\sin\theta} \tag{24}$$

PROOF. From formula (18) we have

$$\cos 2\alpha = 1 - 2\sin^2\alpha$$

Solve this equation for $\sin\alpha$ to obtain

$$\sin\alpha = \pm\sqrt{\frac{1-\cos 2\alpha}{2}}$$

If we now let $\theta = 2\alpha$, then $\alpha = \dfrac{\theta}{2}$ and we obtain (21). Formula (22) can be obtained from

$$\cos 2\alpha = 2\cos^2\alpha - 1$$

Formulas (23) and (24) can be obtained by using the fundamental identities and simplifying the resulting expressions. To obtain the variations, first rationalize the denominator and then, the numerator. Consider formula (23).

$$\tan\frac{\theta}{2} = \frac{\sin\frac{\theta}{2}}{\cos\frac{\theta}{2}} = \frac{\pm\sqrt{2}\sqrt{1-\cos\theta}}{\pm\sqrt{2}\sqrt{1+\cos\theta}} = \pm\sqrt{\frac{1-\cos\theta}{1+\cos\theta}}$$

$$= \pm\sqrt{\left(\frac{1-\cos\theta}{1+\cos\theta}\right)\left(\frac{1-\cos\theta}{1-\cos\theta}\right)} = \frac{1-\cos\theta}{\sin\theta}$$

There is no ambiguity in the use of the "$\pm$" since it will be determined by the quadrant in which $\dfrac{\theta}{2}$ falls. Furthermore, in the case of the tangent and cotangent, we have $1 - \cos\theta$ and $1 + \cos\theta$ which will always be nonnegative, and $\sin\theta$ will always have the same sign as the tangent or contangent of the half-argument.

We pause to emphasize again that all of the formulas are valid only in the domain of definition of the trigonometric functions. It should be very clear that if $\theta = \pi$, then the denominator of formula (23) is zero and we have an inadmissible operation; but, in this case, $\frac{1}{2}\theta = \frac{1}{2}\pi$ and we have already indicated that for this argument the tangent function is undefined.

33. ADDITIONAL FORMULAS FOR SINES AND COSINES

The remaining eight formulas concern the product and algebraic sums of the sine and cosine functions.

THEOREM. *Let α and β be any real numbers, then*

$$2 \sin \alpha \cos \beta = \sin(\alpha + \beta) + \sin(\alpha - \beta) \tag{25}$$

$$2 \cos \alpha \cos \beta = \cos(\alpha + \beta) + \cos(\alpha - \beta) \tag{26}$$

$$2 \sin \alpha \sin \beta = \cos(\alpha - \beta) - \cos(\alpha + \beta) \tag{27}$$

$$2 \sin \beta \cos \alpha = \sin(\alpha + \beta) - \sin(\alpha - \beta) \tag{28}$$

NOTE. Formulas (25) and (28) are equivalent since one can be obtained from the other by merely interchanging α and β. The placement of the coefficient is a matter of choice; we can place it as it is placed above, or eliminate it from the left side by placing $\frac{1}{2}$ on the right side.

PROOF. These formulas are obtained by adding and subtracting formulas (9), (10), (11), and (12); for example,

$$\begin{aligned} \sin(\alpha + \beta) &= \sin\alpha\cos\beta + \cos\alpha\sin\beta \\ \sin(\alpha - \beta) &= \sin\alpha\cos\beta - \cos\alpha\sin\beta \\ \hline \sin(\alpha + \beta) + \sin(\alpha - \beta) &= 2\sin\alpha\cos\beta \end{aligned}$$

which is formula (25).

THEOREM. *Let θ and ϕ be any real numbers, then*

$$\sin\theta + \sin\phi = 2\sin\frac{\theta + \phi}{2}\cos\frac{\theta - \phi}{2} \tag{29}$$

$$\sin\theta - \sin\phi = 2\cos\frac{\theta + \phi}{2}\sin\frac{\theta - \phi}{2} \tag{30}$$

$$\cos\theta + \cos\phi = 2\cos\frac{\theta + \phi}{2}\cos\frac{\theta - \phi}{2} \tag{31}$$

$$\cos\theta - \cos\phi = -2\sin\frac{\theta + \phi}{2}\sin\frac{\theta - \phi}{2} \tag{32}$$

PROOF. The proofs of these formulas are obtained from formulas (25), (26), (27), and (28) by letting

$$\theta = \alpha + \beta \qquad \phi = \alpha - \beta$$

It follows that $\theta + \phi = 2\alpha$ and $\theta - \phi = 2\beta$, or

$$\alpha = \frac{\theta + \phi}{2} \qquad \beta = \frac{\theta - \phi}{2}$$

When these values are substituted into formulas (25–28), we get the formulas of the theorem.

We are not noting addition formulas for the secant and cosecant functions. These functions can be expressed in terms of sines and cosines by the fundamental identities, and thus we can use the formulas for sines and cosines in transforming trigonometric expressions.

ILLUSTRATION. Prove $\csc 2x = \frac{1}{2}\sec x \csc x$.

SOLUTION.
$$\csc 2x = \frac{1}{\sin 2x} = \frac{1}{2\sin x\cos x} = \frac{1}{2}\cdot\frac{1}{\sin x}\cdot\frac{1}{\cos x} = \tfrac{1}{2}\sec x\csc x$$

34. IDENTITIES INVOLVING ADDITION FORMULAS

We shall now apply these formulas to transform trigonometric expressions and reduce $f(x)$ to $g(x)$ or vice versa in identities $f(x) = g(x)$. We offer the following suggestions.

I. *Express secant and cosecant in terms of sine and cosine.*

II. *Reduce multiple arguments by changing to functions of single arguments.*

III. *Avoid introducing radicals whenever possible.*

ILLUSTRATIONS

1. Prove $\tan x = \dfrac{\sin 2x}{1 + \cos 2x}$.

SOLUTION. Change the right member by using (17) and (18).

$$\tan x = \frac{2\sin x\cos x}{1 + 2\cos^2 x - 1} = \frac{2\sin x\cos x}{2\cos^2 x} = \frac{\sin x}{\cos x} = \tan x$$

2. Prove $\cos 3x = 4\cos^3 x - 3\cos x$.

SOLUTION.
$$\begin{aligned}\cos 3x &= \cos(2x + x) = \cos 2x\cos x - \sin 2x\sin x\\ &= (\cos^2 x - \sin^2 x)\cos x - (2\sin x\cos x)\sin x\\ &= \cos^3 x - \sin^2 x\cos x - 2\sin^2 x\cos x\\ &= \cos^3 x - 3\cos x(1 - \cos^2 x)\\ &= 4\cos^3 x - 3\cos x\end{aligned}$$

3. Prove $\tan\left(\frac{x}{2} + 45^\circ\right) - \tan x = \sec x$.

SOLUTION

$$\tan\left(\frac{x}{2} + 45^\circ\right) - \tan x$$

$$= \frac{\tan\frac{x}{2} + \tan 45^\circ}{1 - \tan\frac{x}{2}\tan 45^\circ} - \tan x$$

$$= \frac{\tan\frac{x}{2} + 1}{1 - \tan\frac{x}{2}} - \tan x = \frac{\frac{\sin x}{1 + \cos x} + 1}{1 - \frac{\sin x}{1 + \cos x}} - \frac{\sin x}{\cos x}$$

$$= \frac{\sin x + 1 + \cos x}{1 + \cos x - \sin x} - \frac{\sin x}{\cos x}$$

$$= \frac{\sin x\cos x + \cos x + \cos^2 x - \sin x - \sin x\cos x + \sin^2 x}{\cos x\,(1 + \cos x - \sin x)}$$

$$= \frac{\cos x - \sin x + 1}{\cos x\,(1 + \cos x - \sin x)}$$

$$= \frac{1}{\cos x} = \sec x$$

4. Prove $\dfrac{\cos 7x - \cos 5x}{\sin 5x - \sin 7x} = \tan 6x$.

SOLUTION

$$\frac{\cos 7x - \cos 5x}{\sin 5x - \sin 7x} = \frac{-2\sin\frac{7x + 5x}{2}\sin\frac{7x - 5x}{2}}{2\cos\frac{5x + 7x}{2}\sin\frac{5x - 7x}{2}}$$

$$= -\frac{\sin 6x\sin x}{\cos 6x\sin(-x)} = \frac{\sin 6x}{\cos 6x}$$

$$= \tan 6x$$

5. Evaluate $\tan[2\ \text{Arcsin}\ \frac{4}{5}]$.

SOLUTION. Let $\alpha = \text{Arcsin}\ \frac{4}{5}$, then $\sin\alpha = \frac{4}{5}$ and $0 \le \alpha \le 90^\circ$ by the principal value condition. Therefore $y = 4$, $r = 5$, and $x = 3$. Then by definition $\tan\alpha = \frac{4}{3}$. Proceeding with the evaluation,

$$\tan[2\ \text{Arcsin}\ \tfrac{4}{5}] = \tan 2\alpha = \frac{2\tan\alpha}{1 - \tan^2\alpha} = \frac{2(\frac{4}{3})}{1 - (\frac{4}{3})^2}$$

$$= \left(\frac{8}{3}\right)\left(\frac{9}{-7}\right) = -\frac{24}{7}$$

6. Find the exact value of $\tan 22\frac{1}{2}^\circ$.

SOLUTION

$$\tan 22\frac{1}{2}^\circ = \tan\frac{45^\circ}{2} = \frac{1 - \cos 45^\circ}{\sin 45^\circ} = \frac{1 - \dfrac{1}{\sqrt{2}}}{\dfrac{1}{\sqrt{2}}} = \frac{\sqrt{2} - 1}{\sqrt{2}} \cdot \frac{\sqrt{2}}{1}$$

$$= \sqrt{2} - 1$$

7. Prove that $2 \cos 2\theta \cos \theta - \cos 3\theta = \cos \theta$.

SOLUTION. By formula (26),

$$2 \cos 2\theta \cos \theta - \cos 3\theta = \cos 3\theta + \cos \theta - \cos 3\theta = \cos \theta$$

8. Evaluate $2 \cos\left(\text{Arcsin}\,\frac{1}{2}\right) \sin\left(\text{Arccos}\,\frac{\sqrt{3}}{2}\right)$.

SOLUTION. Let $\alpha = \text{Arcsin}\,\frac{1}{2}$, then $\sin \alpha = \frac{1}{2}$ and $\alpha = 30^\circ$.

Let $\beta = \text{Arccos}\,\frac{\sqrt{3}}{2}$, then $\cos \beta = \frac{\sqrt{3}}{2}$ and $\beta = 30^\circ$.

Then

$$2 \cos\left(\text{Arcsin}\,\frac{1}{2}\right) \sin\left(\text{Arccos}\,\frac{\sqrt{3}}{2}\right) = 2 \cos 30^\circ \sin 30^\circ$$

$$= \sin 2\,(30^\circ) = \sin 60^\circ = \frac{\sqrt{3}}{2}$$

9. Verify the identity

$$\frac{2 + \cos x - 2 \sin^2 x}{\sin 2x + \sin x} = \cot x$$

SOLUTION

$$\frac{2 + \cos x - 2 \sin^2 x}{\sin 2x + \sin x} = \frac{2\,(1 - \sin^2 x) + \cos x}{2 \sin x \cos x + \sin x}$$

$$= \frac{2 \cos^2 x + \cos x}{\sin x\,(2 \cos x + 1)}$$

$$= \frac{\cos x\,(2 \cos x + 1)}{\sin x\,(2 \cos x + 1)} = \cot x$$

10. Verify that $\cos 20^\circ + \cos 100^\circ + \cos 140^\circ = 0$.

SOLUTION. We shall use formula (31).

$$\cos 20^\circ + \cos 100^\circ + \cos 140^\circ = 2 \cos 60^\circ \cos(-40^\circ) + \cos 140^\circ$$

$$= 2(\tfrac{1}{2}) \cos 40^\circ + \cos 140^\circ$$

$$= 2 \cos 90^\circ \cos(-50^\circ)$$

$$= (2)(0) \cos 50^\circ = 0$$

35. EXERCISE 5

Verify the following identities.

1. $\dfrac{2 \tan x}{\tan 2x} = 1 - \tan^2 x$
2. $\sin 2x \sec x + \cos 2x \csc x = \csc x$
3. $\dfrac{\cos x - \cos 2x}{\sin x + \sin 2x} = \tan \dfrac{x}{2}$
4. $\sin 3x = 3 \sin x - 4 \sin^3 x$
5. $\tan\left(\dfrac{\pi}{4} - \dfrac{x}{2}\right) + \tan x = \sec x$
6. $\tan x \sin 2x = 2 \sin^2 x$
7. $\dfrac{\tan 150^\circ - \tan 15^\circ}{1 + \tan 150^\circ \tan 15^\circ} = -1$
8. $\dfrac{\cos 2x - \cos 2y}{\cos x - \cos y} = 2(\cos x + \cos y)$
9. $\dfrac{3 \cos x + \cos 3x}{3 \sin x - \sin 3x} = \cot^3 x$
10. $\dfrac{2(\cos x - \cos^2 x)}{\cos 2x} = \tan 2x \tan \dfrac{x}{2}$
11. $\sin 3x + \sin x = 2 \sin 2x \cos x$
12. $\cos^6 x - \sin^6 x = \cos 2x (1 - \frac{1}{4} \sin^2 2x)$
13. $\dfrac{\sin 5x + \sin 3x}{\cos 5x + \cos 3x} = \tan 4x$
14. $\cos^2(45^\circ - x) - \sin^2(45^\circ - x) = \sin 2x$
15. $\dfrac{\sin 7x - \sin 9x}{\cos 9x - \cos 7x} = \cot 8x$
16. $\sin 2x \tan 2x = \dfrac{4 \tan^2 x}{1 - \tan^4 x}$
17. $\dfrac{\sin 3x}{\sin x} - \dfrac{\cos 3x}{\cos x} = 2$
18. $\sin 4x = 4 \sin x \cos^3 x - 4 \cos x \sin^3 x$
19. $\cos^4 \dfrac{\pi}{8} + \cos^4 \dfrac{3\pi}{8} + \cos^4 \dfrac{5\pi}{8} + \cos^4 \dfrac{7\pi}{8} = \dfrac{3}{2}$
20. $\cos^6 x + \sin^6 x = \cos 2x + \frac{1}{2} \sin^2 2x$
21. $\sin 36^\circ + \sin 54^\circ = \sqrt{2} \cos 9^\circ$
22. $\dfrac{\sin x + \sin y}{\cos x + \cos y} = \tan \dfrac{x + y}{2}$
23. Evaluate the following.
    a) $\cos[2 \operatorname{Arcsin} \frac{12}{13}]$
    b) $2 \sin[\operatorname{Arccos} \frac{1}{2}] \cos[\operatorname{Arcsin} \frac{1}{2}]$
    c) $\sin[\frac{1}{2} \operatorname{Arccos} \frac{24}{25}]$
    d) $\tan[2 \operatorname{Arccos} \frac{12}{37}]$
    e) $\tan[\frac{1}{2} \operatorname{Arcsin} \frac{3}{5}]$
    f) $\cot[\frac{1}{2} \operatorname{Arctan} \frac{5}{12}]$
24. Find the exact values.
    a) $\sin 75^\circ \sin 15^\circ$
    b) $\cos 135^\circ + \cos 45^\circ$
    c) $\sin 45^\circ \cos 15^\circ$
    d) $\sec 22^\circ 30'$

e) $\sin 75^\circ + \sin 30^\circ$ f) $\cot 52^\circ 30'$

g) Find the exact value of $\cos 15^\circ$ by $15^\circ = 45^\circ - 30^\circ$ and by $15^\circ = \frac{1}{2}(30^\circ)$. Show that the two answers are identical.

25. a) Show that $\sin \frac{x}{2} + \cos \frac{x}{2} = \pm \sqrt{1 + \sin x}$.

[Hint: Square left side and then take square root.]

b) Show that $\sin \frac{x}{2} - \cos \frac{x}{2} = \pm \sqrt{1 - \sin x}$.

c) Use (a) and (b) to find an expression for $\sin \frac{x}{2}$ in terms of $\sin x$.

☆36. GENERALIZATION OF ADDITION FORMULAS

We have already used the addition formulas to write $\sin (A + B + C)$ in terms of the functions of A, B, and C, and we have proved identities for $\sin 3x$ and $\cos 3x$. The addition formulas can be generalized; many years ago the generalizations were an integral part of a course in trigonometry. The applications however are limited, and the generalization is only of interest as a mathematical exercise. We shall present a brief discussion for the benefit of the more inquisitive reader.

Let us first çonsider a generalization of $\tan (\alpha + \beta)$ by developing a formula for $\tan (A_1 + A_2 + A_3 + \cdots + A_n)$ in terms of symbols defined by

$$s_1 = \tan A_1 + \tan A_2 + \cdots + \tan A_n$$
$$= \text{sum of the tangents of the separate angles}$$
$$s_2 = \tan A_1 \tan A_2 + \tan A_1 \tan A_3 + \cdots + \tan A_{n-1} \tan A_n$$
$$= \text{sum of the products of the tangents taken two at a time}$$
$$s_3 = \text{sum of the products of the tangents taken three at a time}$$
$$\vdots$$
$$s_n = \tan A_1 \tan A_2 \tan A_3 \ldots \tan A_n$$
$$= \text{product of the } n \text{ tangents}$$

THEOREM. *Let* $A_1, A_2, \ldots, A_n$ *be any set of* n *real numbers, then*

$$\tan (A_1 + A_2 + \cdots + A_n) = \frac{s_1 - s_3 + s_5 - s_7 + \cdots + (-1)^{k+1} s_{2k-1}}{1 - s_2 + s_4 - s_6 + \cdots + (-1)^m s_{2m}}$$

where $s_i(i = 1, \ldots, n)$ *are defined as above, and*

if n *is even, then* $2m = n$ *and* $2k - 1 = n - 1$
if n *is odd, then* $2k - 1 = n$ *and* $2m = n - 1$

This theorem can be proved by mathematical induction but we

shall not exhibit the proof here. Let us consider instead some examples.

When $n = 2$, we have $\tan(A_1 + A_2)$, $s_1 = \tan A_1 + \tan A_2$, and $s_2 = \tan A_1 \tan A_2$. If we substitute these values into the theorem we obtain formula (13).

When $n = 3$, we have $\tan(A_1 + A_2 + A_3)$ and

$$\begin{aligned} s_1 &= \tan A_1 + \tan A_2 + \tan A_3 \\ s_2 &= \tan A_1 \tan A_2 + \tan A_1 \tan A_3 + \tan A_2 \tan A_3 \\ s_3 &= \tan A_1 \tan A_2 \tan A_3 \end{aligned}$$

The theorem then states

$$\tan(A_1 + A_2 + A_3) = \frac{\tan A_1 + \tan A_2 + \tan A_3 - \tan A_1 \tan A_2 \tan A_3}{1 - \tan A_1 \tan A_2 - \tan A_1 \tan A_3 - \tan A_2 \tan A_3}$$

If all the A_i are equal, we obtain a formula for the tangent of multiple arguments. The quantities s_i now take a special form which may be expressed by using the binomial coefficients.*

$$\binom{n}{r} = \frac{n(n-1)(n-2)\cdots(n-r+1)}{r!}$$

For example

$$\binom{4}{2} = \frac{(4)(3)}{(1)(2)} = 6 \quad \binom{5}{3} = \frac{(5)(4)(3)}{(1)(2)(3)} = 10 \quad \binom{6}{4} = \frac{(6)(5)(4)(3)}{(1)(2)(3)(4)} = 15$$

When the A_i are equal, we have

$$\begin{aligned} s_1 &= n \tan A \\ s_2 &= \binom{n}{2} \tan^2 A = \frac{n(n-1)}{2} \tan^2 A \\ s_3 &= \binom{n}{3} \tan^3 A \\ s_k &= \binom{n}{k} \tan^k A \\ s_n &= \binom{n}{n} \tan^n A = \tan^n A \end{aligned}$$

*For a discussion of the binomial formula see Kaj L. Nielsen, *College Mathematics* (New York: Barnes and Noble, 1958), pp 66–69.

ILLUSTRATION. Use the theorem on page 58 to write a formula for $\tan 5x$.

SOLUTION. Since $5x = x + x + x + x + x$, we have $n = 5$, then

$$\binom{n}{2} = \frac{(5)(4)}{2} = 10 \qquad \binom{n}{3} = \frac{(5)(4)(3)}{(1)(2)(3)} = 10$$

$$\binom{n}{4} = 5 \qquad \binom{n}{5} = 1$$

and

$$s_1 = 5 \tan x \qquad s_2 = 10 \tan^2 x \qquad s_3 = 10 \tan^3 x$$

$$s_4 = 5 \tan^4 x \qquad s_5 = \tan^5 x$$

Consequently,

$$\tan 5x = \frac{5 \tan x - 10 \tan^3 x + \tan^5 x}{1 - 10 \tan^2 x + 5 \tan^4 x}$$

Let us now consider $\sin (A_1 + A_2 + \cdots + A_n)$. The answer to problem 6, Exercise 4, is

$$\begin{aligned}\sin (\alpha + \beta + \gamma) &= \sin \alpha \cos \beta \cos \gamma + \cos \alpha \sin \beta \cos \gamma \\ &\quad + \cos \alpha \cos \beta \sin \gamma - \sin \alpha \sin \beta \sin \gamma\end{aligned}$$

This answer can also be obtained as follows.

$$\begin{aligned}\sin (\alpha + \beta + \gamma) &= \cos \alpha \cos \beta \cos \gamma \,(s_1 - s_3) \\ &= \cos \alpha \cos \beta \cos \gamma \,(\tan \alpha + \tan \beta + \tan \gamma \\ &\quad - \tan \alpha \tan \beta \tan \gamma) \\ &= \text{answer given above}\end{aligned}$$

We present the general formulas for the sine and cosine functions where s_i are defined as above, and the conditions for the terminal s_i are the same as in the theorem for the tangent function.

$$\begin{aligned}&\sin (A_1 + A_2 + \cdots + A_n) \\ &\quad = \cos A_1 \cos A_2 \cdots \cos A_n \,[s_1 - s_3 + s_5 - \cdots + (-1)^{k+1} s_{2k-1}]\end{aligned}$$

and

$$\begin{aligned}&\cos (A_1 + A_2 + \cdots + A_n) \\ &\quad = \cos A_1 \cos A_2 \cdots \cos A_n \,[1 - s_2 + s_4 - \cdots + (-1)^m s_{2m}]\end{aligned}$$

ILLUSTRATION. Find an expression for $\sin 7x$ and $\cos 7x$.

SOLUTION. Since the argument is $7x$ we have $n = 7$ and

$$s_1 = 7\tan x \qquad s_2 = 21\tan^2 x \qquad s_3 = 35\tan^3 x$$
$$s_4 = 35\tan^4 x \qquad s_5 = 21\tan^5 x \qquad s_6 = 7\tan^6 x$$
$$s_7 = \tan^7 x$$

Then

$$\begin{aligned}\sin 7x &= \cos^7 x\,[s_1 - s_3 + s_5 - s_7]\\ &= \cos^7 x\,(7\tan x - 35\tan^3 x + 21\tan^5 x - \tan^7 x)\\ &= 7\cos^6 x\sin x - 35\cos^4 x\sin^3 x + 21\cos^2 x\sin^5 x - \sin^7 x\\ \cos 7x &= \cos^7 x\,(1 - s_2 + s_4 - s_6)\\ &= \cos^7 x\,(1 - 21\tan^2 x + 35\tan^4 x - 7\tan^6 x)\\ &= \cos^7 x - 21\cos^5 x\sin^2 x + 35\cos^3 x\sin^4 x - 7\cos x\sin^6 x\end{aligned}$$

☆37. EXACT VALUES

The values of the trigonometric functions are usually irrational numbers which are approximated by decimal fractions. However, we have seen (Section 18) that for certain values of the argument the trigonometric functions can be expressed in exact numbers, whenever necessary, by the use of radicals. In this section we shall consider more values of the argument.

Consider the number $x = \dfrac{\pi}{10}$ so that $2x = \dfrac{\pi}{5}$ and $3x = \dfrac{3\pi}{10}$.

Then

$$2x = \frac{\pi}{2} - 3x = \frac{\pi}{2} - \frac{3\pi}{10} = \frac{5\pi - 3\pi}{10} = \frac{\pi}{5} = 2x$$

so that
$$\sin 2x = \sin\left(\frac{\pi}{2} - 3x\right) = \cos 3x$$

This equation can be transformed, by formula (17) and Illustration 2, Section 34, to

$$2\sin x\cos x = 4\cos^3 x - 3\cos x$$

and simplified to

$$\cos x\,(2\sin x + 3 - 4\cos^2 x) = 0$$

This is a conditional equation which is true for certain values of x. In particular, it should be true for $x = \dfrac{\pi}{10}$ since the equation was derived from this fact. The first factor yields $\cos x = 0$, which is true for $x = 90^\circ$, but we are not interested in this solution. Consider the second factor and transform it in the following manner.

$$\begin{aligned} 2\sin x + 3 - 4\cos^2 x &= 2\sin x + 3 - 4(1 - \sin^2 x) \\ &= 2\sin x - 1 + 4\sin^2 x \end{aligned}$$

Now we have $\quad 4\sin^2 x + 2\sin x - 1 = 0$

This is a quadratic equation in $\sin x$ that can be solved by applying the quadratic formula to obtain

$$\sin x = \frac{-2 \pm \sqrt{4 + 16}}{8} = \frac{\pm\sqrt{5} - 1}{4}$$

Since the argument $x = \frac{\pi}{10}$ is less than $\frac{\pi}{2}$, the sine must be positive, and we have

$$\sin\frac{\pi}{10} = \frac{\sqrt{5} - 1}{4} = \sin 18^\circ$$

Since $\cos 2x = 1 - 2\sin^2 x$, we have

$$\cos\frac{\pi}{5} = 1 - 2\left(\frac{\sqrt{5} - 1}{4}\right)^2 = \frac{\sqrt{5} + 1}{4} = \cos 36^\circ$$

We can now use the fundamental identities to find the values of the other trigonometric functions of 18° and 36°; for example,

$$\sin\frac{\pi}{5} = \sqrt{1 - \cos^2\left(\frac{\pi}{5}\right)} = \sqrt{1 - \frac{6 + 2\sqrt{5}}{16}} = \frac{1}{4}\sqrt{10 - 2\sqrt{5}}$$

The formulas of this chapter permit us to obtain the exact values of the trigonometric functions of a large class of arguments. Let us see if we can define this class in a systematic manner. The following are the fundamental members of the class.

π	$\frac{\pi}{2}$	$\frac{\pi}{3}$	$\frac{\pi}{4}$	$\frac{\pi}{5}$	$\frac{\pi}{6}$
180°	90°	60°	45°	36°	30°

Let n be a positive integer ($n = 1, 2, 3, \ldots$), then we can define other members of the class as follows. A and B are members of the class.

$$\frac{\pi}{2n} \qquad (A \pm B) \qquad nA$$

ILLUSTRATION. Find the exact value of $\sin 7\frac{1}{2}^\circ$.

SOLUTION. From problem 24(g), Exercise 5, we have

$$\cos 15^\circ = \tfrac{1}{2}\sqrt{2 + \sqrt{3}}$$

Then we have

$$\sin 7^\circ 30' = \sin \frac{15^\circ}{2} = \sqrt{\frac{1 - \cos 15^\circ}{2}} = \frac{1}{2}\sqrt{2 - \sqrt{2 + \sqrt{3}}}$$

Although we can find the exact values of the trigonometric functions of a large class of arguments, they do not have much practical value since only a few of them are of a simple nature. It should also be noted that the above class does not contain all possible arguments for which there exist exact values of the trigonometric functions.

☆38. EXERCISE 6

1. Find an expression for $\sin 5x$, $\cos 5x$, and $\tan 5x$ in terms of functions of the argument x.

2. Find $\sin 75^\circ$ in terms of sines and cosines of 15°. Use the results of problem 1, Exercise 4, for functions of 15°, to find the value of $\sin 75^\circ$. Compare this answer with that obtained in problem 1, Exercise 4.

3. Use the exact values of $\tan 15^\circ$ to find

a) $\tan 45^\circ$ b) $\tan 75^\circ$

4. Complete the following table with exact values.

	$7\frac{1}{2}^\circ$	9°	18°	$22\frac{1}{2}^\circ$	27°	36°
sin						
cos						

5. Prove the following.

a) $\sin^2 72^\circ - \sin^2 60^\circ = \dfrac{\sqrt{5} - 1}{8}$

b) $\sin \dfrac{\pi}{5} \sin \dfrac{2\pi}{5} \sin \dfrac{3\pi}{5} \sin \dfrac{4\pi}{5} = \dfrac{5}{16}$

c) $\cos^2 48^\circ - \sin^2 12^\circ = \dfrac{\sqrt{5} + 1}{8}$ [Hint: $48^\circ = 30^\circ + 18^\circ$.]

☆6. Define c_i in terms of $\cot A_j$ as we did s_i (in terms of $\tan A_j$), then show that

$$\cot(A_1 + A_2 + A_3 + A_4) = \frac{c_4 - c_2 + 1}{c_3 - c_1}$$

chapter 4

Graphs of the Trigonometric Functions

39. INTRODUCTION

We have defined a function as a set of ordered pairs. The set is usually generated by some law of formation which is often expressed in terms of an equation. The set of points in the cartesian coordinate plane can also be associated with a set of ordered pairs (x,y), which we have called the coordinates of a point. Combining these two ideas we see that the set of points which satisfy the law of formation of a function form a particular subset of the cartesian plane; we call this set a **locus** or a **curve** and emphasize that it is the set of points which satisfy the equation.

DEFINITION. *The* **graph** *of a locus is the geometric representation of the set of points of the locus.*

The graph of a function is also called the geometric image. It may be defined by a one-to-one correspondence between the set of ordered pairs and the set of points whose coordinates are given by ordered pairs and satisfy the equation of the function.

To obtain the graph of a function, we plot the points whose coordinates satisfy the equation of the function. However, since there are usually an infinite number of points, the procedure is reduced to finding a reasonable number of points and joining these by a smooth curve. The determination of what is a reasonable number depends upon the function and our ability to obtain characteristic points, such as the largest or smallest value. Some knowledge of the domain and range of the function helps to obtain the graph, and the plotting of a graph aids in understanding the domain and range of a function. In this chapter we shall consider the graphs of the trigonometric functions.

40. PERIODICITY

In Section 17 we showed that any trigonometric function of $(\theta + 2n\pi)$ equals the same function of θ. Any function, not equal to a constant, which repeats its values at a regular interval is said to be **periodic,** and the interval is called the **period;** thus if

$$f(x) = f(x + k)$$

for every value of x in the domain of definition, then $f(x)$ is said to be periodic with period k.

It is easily seen that all the trigonometric functions are periodic with period 2π; however, the two functions $\tan x$ and $\cot x$ repeat themselves at intervals of π so that they have a period of π as well as 2π. The smallest possible period of a periodic function is often referred to as the **primitive period**. The property of periodicity permits us to obtain all the information about the values of the trigonometric functions by considering the argument in the interval from 0 to 2π.

41. VARIATIONS OF THE TRIGONOMETRIC FUNCTIONS

Let us consider the function y given by

$$y = \sin x$$

which is defined for every value of x in the domain of real numbers. We can calculate a set of ordered pairs by assigning values to x and calculating the corresponding value of y.

x	0°	30°	60°	90°	120°	150°	180°	210°	240°	270°	300°	330°	360°
y	0	.5	.87	1	.87	.5	0	−.5	−.87	−1	−.87	−.50	0

A close study of this table reveals that as x varies from 0° to 90°, y varies from 0 to 1; then as x lies in the interval $90° \leq x \leq 180°$, we see that y lies in the interval $1 \geq y \geq 0$; furthermore, for $180° \leq x \leq 270°$, we have $0 \geq y \geq -1$; and for $270° \leq x \leq 360°$, we have $-1 \leq y \leq 0$. Thus as x increases, we notice that $\sin x$ increases to 1, then decreases to -1, and then increases again. The precise variation is summarized in the following table.

values of x	0° to 90°	90° to 180°	180° to 270°	270° to 360°
$\sin x$	inc. 0 to 1	dec. 1 to 0	dec. 0 to −1	inc. −1 to 0

Let us now consider the function

$$f(t) = \tan t = \frac{y}{x}$$

When $t = 0$ we have $y = 0$, and the value of the function is $f(0) = 0$. When $t = 45^\circ$ we have $y = x$ and $f(45^\circ) = 1$. As t approaches 90°, the value of x gets smaller and the value of $f(t)$ gets larger. As soon as t passes 90°, x becomes a small negative number and the tangent function becomes a very large negative number. Now as t approaches and becomes 180°, the tangent increases to zero. Even though there is no value for $\tan t$ at $t = 90^\circ$ (i.e., $\tan 90^\circ$ does not exist), we can express its behavior near 90° symbolically by

$$\tan\left(\frac{\pi}{2}\right)^{-} = +\infty$$

and

$$\tan\left(\frac{\pi}{2}\right)^{+} = -\infty$$

where ∞ is the symbol called *infinity*. The above expression does not say that infinity exists; it is merely a convenient way to express the fact that the absolute value of a function will become larger than any preassigned number. With this symbolic concept, we can summarize the variations of the six trigonometric functions as follows.

θ	0° to 90°	90° to 180°	180° to 270°	270° to 360°
$\sin\theta$	0 to 1	1 to 0	0 to -1	-1 to 0
$\cos\theta$	1 to 0	0 to -1	-1 to 0	0 to 1
$\tan\theta$	0 to $+\infty$	$-\infty$ to 0	0 to $+\infty$	$-\infty$ to 0
$\cot\theta$	$+\infty$ to 0	0 to $-\infty$	$+\infty$ to 0	0 to $-\infty$
$\sec\theta$	1 to $+\infty$	$-\infty$ to -1	-1 to $-\infty$	$+\infty$ to 1
$\csc\theta$	$+\infty$ to 1	1 to $+\infty$	$-\infty$ to -1	-1 to $-\infty$

42. GRAPHS OF THE TRIGONOMETRIC FUNCTIONS

We shall now consider the graph of the function

$$y = \text{trigonometric function of } x$$

in the (x,y)-plane. That is, we shall obtain a set of ordered pairs (x,y), consider them as the coordinates of points, and plot the points on a rectangular coordinate system. The graph will be obtained by connecting a finite number of points by a smooth curve. In order to obtain a true picture we shall take the same units on each of the axes; this will require that the argument x be a real number; we shall change any angular measure to radian measure when calculating the pairs (x,y). We calculate a table of values at intervals of $\pi/6$ for $0 \leq x \leq 2\pi$, and also insert the values at $n\pi/4 (n = 1,3,5,7)$; see page 68. The argument is listed in exact form in the left column and in decimal form in the extreme right column. All values have been approximated, correct to two decimal places.

Since the numerical values of the trigonometric functions may be determined from the values of the argument in the interval $0 \leq x \leq \pi/2$, we exhibit a table of values with $\Delta x = 0.1$.

x	$\sin x$	$\cos x$	$\tan x$	$\cot x$
.1	.10	1.00	.10	9.97
.2	.20	.98	.20	4.93
.3	.30	.96	.31	3.23
.4	.39	.92	.42	2.37
.5	.48	.88	.55	1.83
.6	.56	.83	.68	1.46
.7	.64	.76	.84	1.19
.8	.72	.70	1.03	.97
.9	.78	.62	1.26	.79
1.0	.84	.54	1.56	.64
1.1	.89	.45	1.97	.51
1.2	.93	.36	2.57	.39
1.3	.96	.27	3.60	.28
1.4	.99	.17	5.80	.17
1.5	1.00	.07	14.1	.07
1.6	1.00	−.03	−34.2	−.03

The graphs of the trigonometric functions are shown in Figures 26–31.

x	$\sin x$	$\cos x$	$\tan x$	$\cot x$	$\sec x$	$\csc x$	x
0	0	1.00	0	$\pm\infty$	1.00	$\pm\infty$	0
$\frac{\pi}{6}$	0.50	0.87	0.58	1.73	1.15	2.00	0.52
$\frac{\pi}{4}$	0.71	0.71	1	1	1.41	1.41	0.79
$\frac{\pi}{3}$	0.87	0.50	1.73	0.58	2.00	1.15	1.05
$\frac{\pi}{2}$	1.00	0	$\pm\infty$	0	$\pm\infty$	1.00	1.57
$\frac{2\pi}{3}$	0.87	−0.50	−1.73	−0.58	−2.00	1.15	2.09
$\frac{3\pi}{4}$	0.71	−0.71	−1	−1	−1.41	1.41	2.36
$\frac{5\pi}{6}$	0.50	−0.87	−0.58	−1.73	−1.15	2.00	2.62
π	0	−1.00	0	$\pm\infty$	−1.00	$\pm\infty$	3.14
$\frac{7\pi}{6}$	−0.50	−0.87	0.58	1.73	−1.15	−2.00	3.67
$\frac{5\pi}{4}$	−0.71	−0.71	1	1	−1.41	−1.41	3.93
$\frac{4\pi}{3}$	−0.87	−0.50	1.73	0.58	−2.00	−1.15	4.19
$\frac{3\pi}{2}$	−1.00	0	$\pm\infty$	0	$\pm\infty$	−1.00	4.71
$\frac{5\pi}{3}$	−0.87	0.50	−1.73	−0.58	2.00	−1.15	5.24
$\frac{7\pi}{4}$	−0.71	0.71	−1	−1	1.41	−1.41	5.50
$\frac{11\pi}{6}$	−0.50	0.87	−0.58	−1.73	1.15	−2.00	5.76
2π	0	1.00	0	$\pm\infty$	1.00	$\pm\infty$	6.28

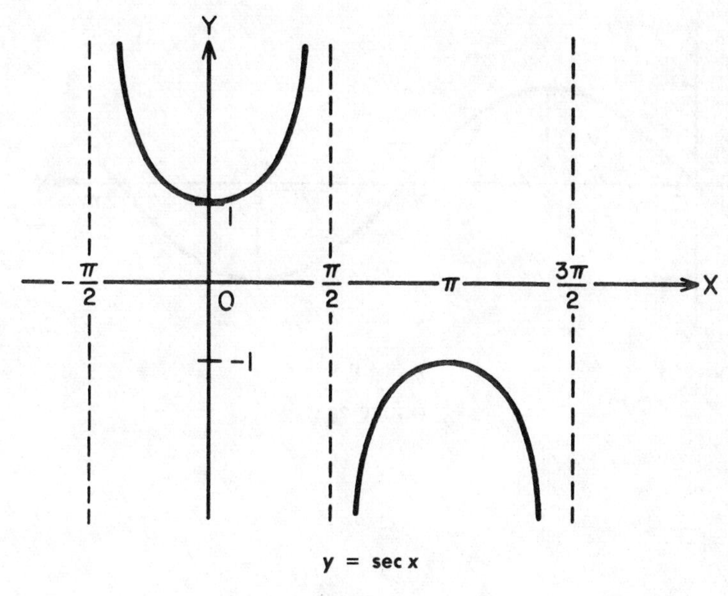

$y = \sec x$

FIGURE 26

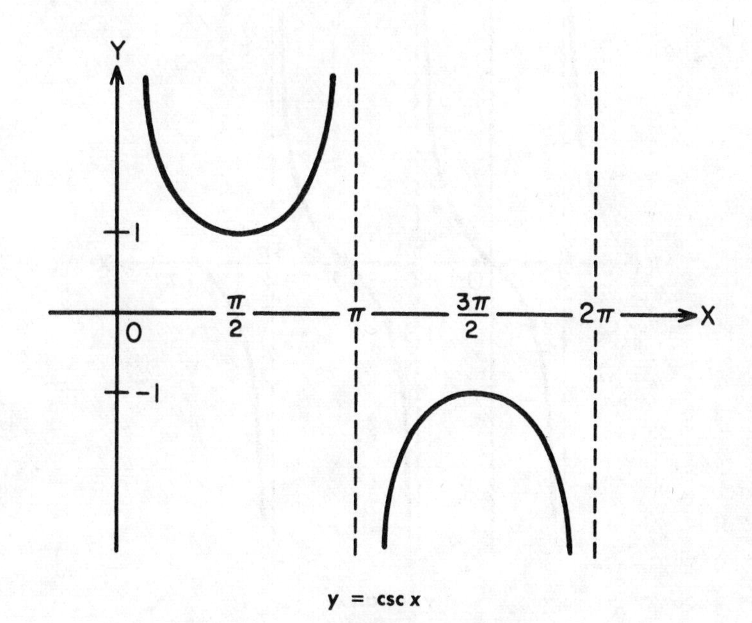

$y = \csc x$

FIGURE 27

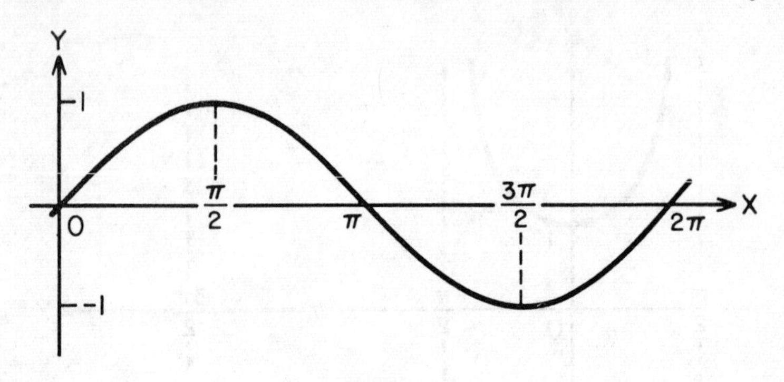

$y = \sin x$

FIGURE 28

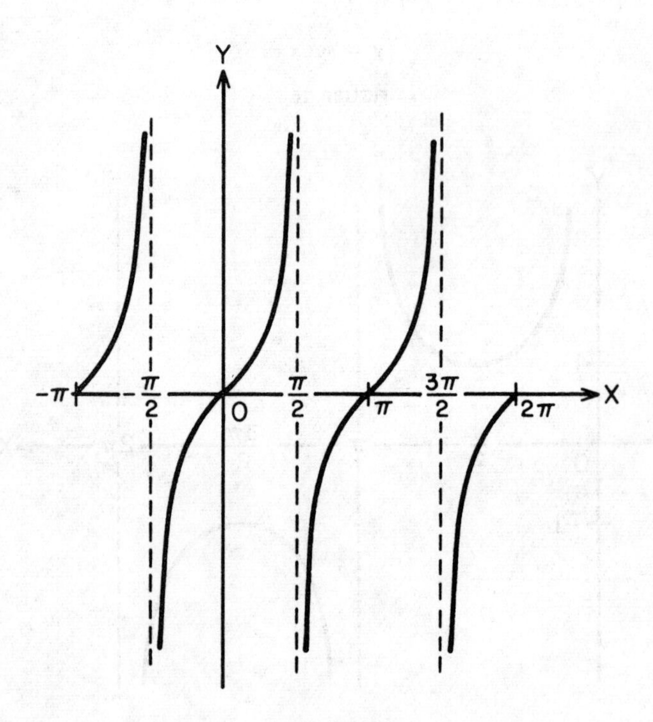

$y = \tan x$

FIGURE 30

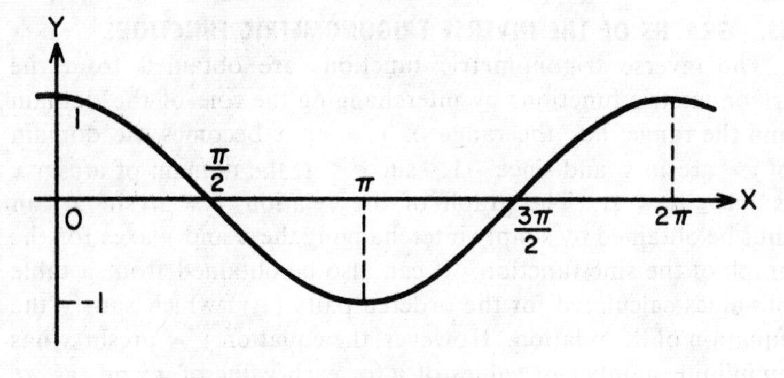

$y = \cos x$

FIGURE 29

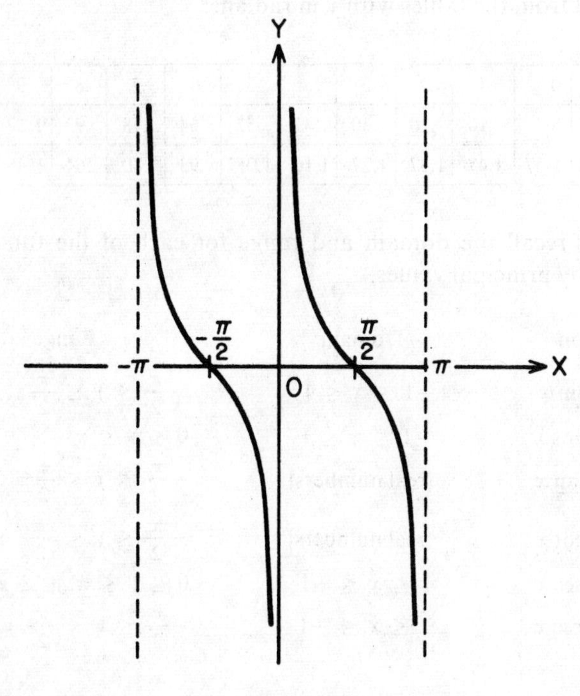

$y = \cot x$

FIGURE 31

43. GRAPHS OF THE INVERSE TRIGONOMETRIC FUNCTIONS

The inverse trigonometric functions are obtained from the trigonometric functions by interchanging the role of the domain and the range; i.e., the range of $y = \sin x$ becomes the domain of $y = \arcsin x$, and since $-1 \leq \sin x \leq 1$, the domain of arcsin x is $-1 \leq x \leq 1$. The graph of the relation $y = \arcsin x$ can thus be obtained by simply interchanging the x and y axes for the graph of the sine function. It can also be obtained from a table of values calculated for the ordered pairs (x,y) which satisfy the equation of the relation. However, the equation $y = \arcsin x$ has an infinite number of values of y for each value of x and, as we have seen, represents a relation and not a single valued function. In order to obtain a function, we define the principal value of the inverse relation (see Section 20). For this function, we can obtain a unique table of values for (x,y). We now present a small table of values for the inverse sine and inverse cosine; these values are obtained from the tables with y in radians.

x	0	.1	.2	.3	.4	.5	.6	.7	.8	.9	1
Arcsin x	0	.10	.20	.30	.41	.52	.64	.78	.93	1.12	1.57
Arccos x	1.57	1.47	1.37	1.27	1.16	1.05	.93	.80	.64	.45	0

Let us recall the domain and range for each of the functions defined by principal values.

Function	Domain	Range
$y = \text{Arcsin } x$	$-1 \leq x \leq 1$	$-\frac{\pi}{2} \leq y \leq \frac{\pi}{2}$
$y = \text{Arccos } x$	$-1 \leq x \leq 1$	$0 \leq y \leq \pi$
$y = \text{Arctan } x$	{real numbers}	$-\frac{\pi}{2} \leq y \leq \frac{\pi}{2}$
$y = \text{Arccot } x$	{real numbers}	$-\frac{\pi}{2} \leq y \leq \frac{\pi}{2}, y \neq 0$
$y = \text{Arcsec } x$	$1 \leq x \leq -1$	$0 \leq y \leq \pi, y \neq \pi/2$
$y = \text{Arccsc } x$	$1 \leq x \leq -1$	$-\frac{\pi}{2} \leq y \leq \frac{\pi}{2}, y \neq 0$

The graphs of the inverse relations for sine, cosine, and tangent are shown in Figures 32–34. The principal values are indicated by the heavy portion of the curve.

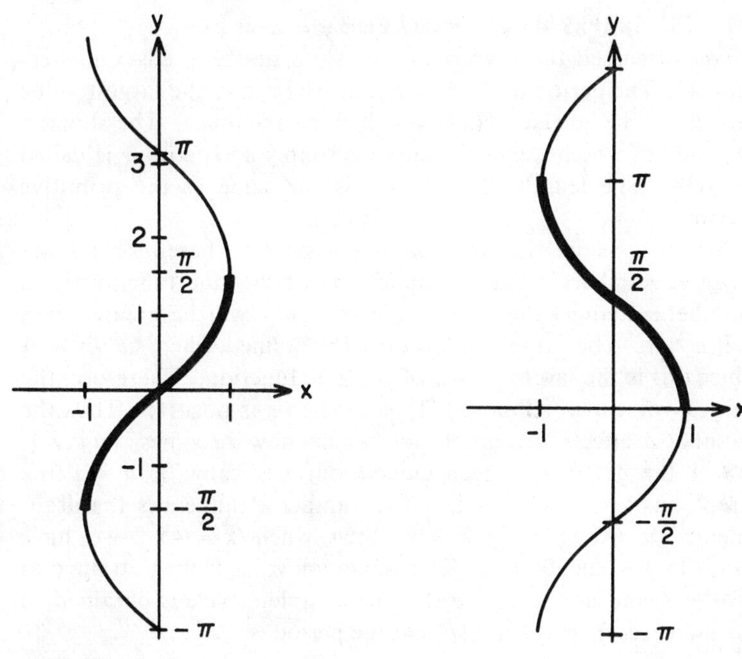

FIGURE 32

FIGURE 33

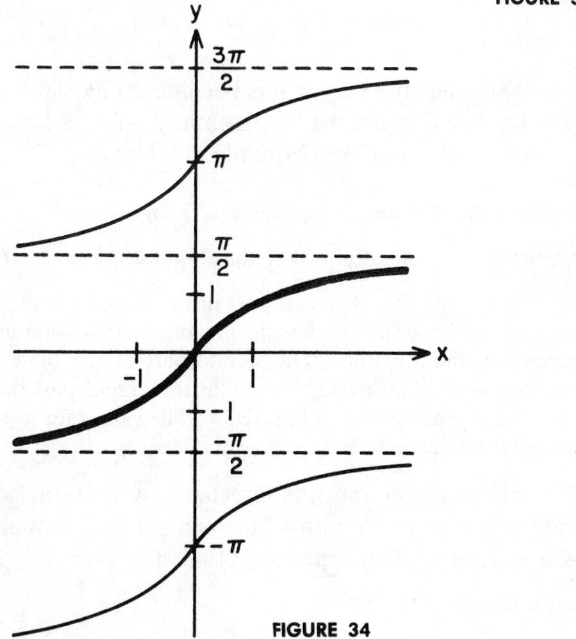

FIGURE 34

44. THE GRAPHS OF $y = a \sin kx$ and $y = a \cos kx$

We discussed the graphs of $y = \sin x$ and $y = \cos x$ in Section 42. The period of these functions is 2π, and the largest value of y is 1. The largest $|f(x)|$ is called the **amplitude**. The shortest segment of which the entire curve is simply a repetition is called a **cycle.** The length of the cycle is the same as the primitive period.

Let us consider the function $y = a \sin kx$ where a and k are positive numbers. The multiplication of the sine function by a number a changes the values of y so that y will be greater than 1 if $a > 1$. The largest value of y is obtained when $\sin kx = 1$ since this is the largest value of the sine function. Therefore, the largest value of y will be $(a)(1) = a$ when a is positive. Thus the number a affects the amplitude which now becomes $|a|$. If $a > 1$ the graph has been pulled out vertically; if $a < 1$ the graph has been "squashed." The number k influences the argument; for example, if $k = 2$, then when $x = 45°$, we have $\sin 2(45°) = \sin 90° = 1$. The maximum value is then attained at $x = \pi/4$ and not at $x = \pi/2$. The complete cycle is obtained as kx increases from 0 to 2π so that the period is

$$kx = 2\pi \qquad \text{or} \qquad x = \frac{2\pi}{|k|}$$

This means that the period has become smaller if $k > 1$ and the graph has been "squeezed" horizontally. If $k < 1$ then the graph has been "pulled out" horizontally.

ILLUSTRATION. Graph the function $y = 3 \sin 2x$.

SOLUTION. The amplitude $= 3$, the period $= \dfrac{2\pi}{2} = \pi$. Calculate the table (top of page 75). The first part of the table gives the characteristic points where the graph reaches the maximum and minimum points and the zeros of the function. The second part of the table is a detailed evaluation of the first part. To obtain the graph, plot the set of pairs (x,y). The graph is shown in Figure 35. The values for $\pi \le x \le 2\pi$ are repeated from the cycle, $0 \le x \le \pi$.

The above discussion was directed to $y = a \sin kx$ but is also valid for $y = a \cos kx$ with adjustments for the actual values; i.e., $\cos 0 = 1$, etc. Thus the amplitude for each is $|a|$ and the period is $\dfrac{2\pi}{|k|}$.

x	$2x$	$\sin 2x$	$y = 3 \sin 2x$
0	0	0	0
$\pi/4$	$\pi/2$	1	3
$\pi/2$	π	0	0
$3\pi/4$	$3\pi/2$	-1	-3
π	2π	0	0
.1	.2	.20	.6
.2	.4	.39	1.17
.3	.6	.56	1.68
.4	.8	.72	2.16
.5	1.0	.84	2.52
.6	1.2	.93	2.79
.7	1.4	.99	2.97
.8	1.6	1.0	3.0

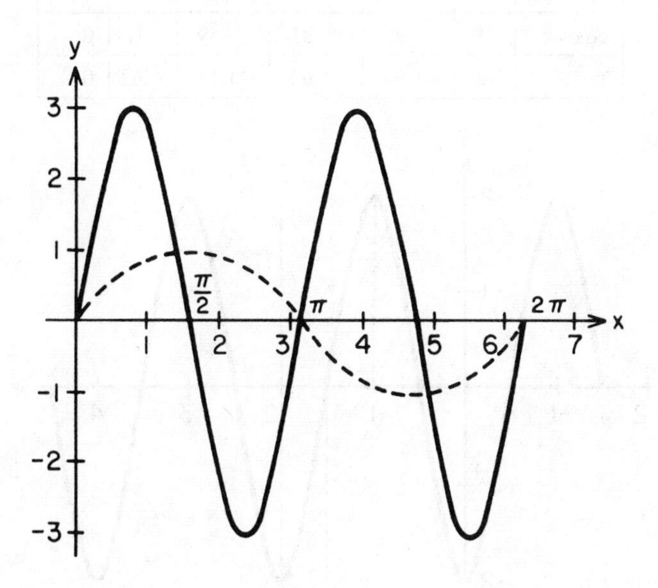

FIGURE 35

Frequently, it is desired to have only a sketch of the curve rather than an accurate graph. A sketch should show the characteristic values, in particular, the largest and smallest values, the intersections with the x-axis, and at least one cycle. To obtain a sketch: calculate the period and, starting at the origin, lay this length off on the x-axis; divide the period into four quarter-periods (the characteristic values will occur at these values of x); locate the values $y = a$ and $y = -a$, and draw horizontal lines through these values; draw the curve in the first quarter-period and use this branch to complete the first cycle; finally, sketch as many cycles as desired.

ILLUSTRATION. Sketch the graph of $y = -2 \cos \pi x$.

SOLUTION. The period is $\frac{2\pi}{\pi} = 2$; the amplitude is $|-2| = 2$. The first period is the interval [0,2] which makes each quarter-period equal to $\frac{1}{2}$. Draw the horizontal lines $y = 2$ and $y = -2$. The first branch is in the interval $[0,\frac{1}{2}]$; some values can be obtained from the following table. The graph is shown in Figure 36.

x	0	.1	.2	.3	.4	.5
πx	0	.31	.63	.94	1.26	1.57
$\cos \pi x$	1	.95	.81	.59	.31	0
y	-2	-1.90	-1.62	-1.18	$-.62$	0

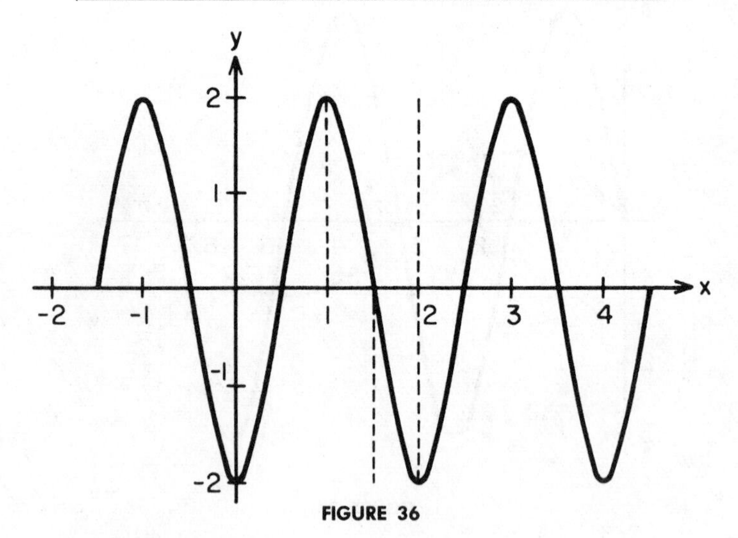

FIGURE 36

45. THE GRAPHS OF $y = a \sin (kx + b)$

The graphs of the two functions $y = \cos x$ and $y = \sin x$ are identical except that one is shifted to the left (or right) a distance of $\pi/2$ with respect to the other. This can also be seen from the fact that $\sin (x + \pi/2) = \cos x$. Another method of illustrating this shift is to consider the function $y = \sin (x - 1)$. An abbreviated table of values shows that the intersections with the x-axis occur at $\{1 \pm n\pi\}$ and that one cycle of the sine function is displayed in the interval $1 \leq x \leq 2\pi + 1$. In other words, the graph has been shifted one unit to the right; see Figure 37.

x	1	$\pi/2 + 1$	$\pi + 1$	$3\pi/2 + 1$	$2\pi + 1$
$x - 1$	0	$\pi/2$	π	$3\pi/2$	2π
y	0	1	0	-1	0

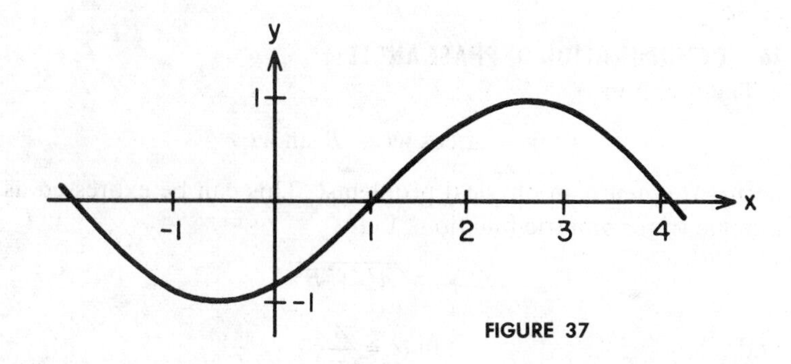

FIGURE 37

The general function $y = a \sin (kx + b) = a \sin k (x + b/k)$ has a **phase shift** (also called *phase displacement*) of $|b/k|$; the constant b is called the **phase constant** (or *phase difference*). The amplitude is $|a|$ and the period is $2\pi/k$.

The graph of $y = a \cos (kx + b)$ can be analyzed the same way.

ILLUSTRATION. Find the amplitude, period, and phase displacement, and sketch the graph of $y = 4 \sin (2x - \pi/2)$.

SOLUTION. The amplitude is 4; the period $= \frac{2\pi}{2} = \pi$; the phase displacement is $\pi/2 \div 2 = \pi/4$. The graph crosses the axis at $(\pi/4, 0)$ and completes one cycle at $\frac{\pi}{4} + \pi = \frac{5\pi}{4}$. The graph is shown in Figure 38, on page 78.

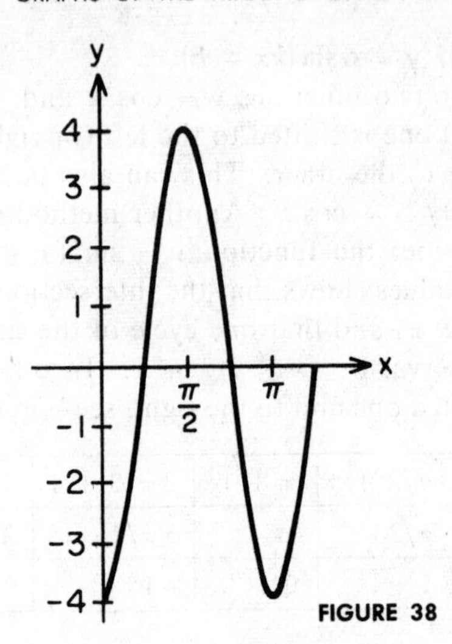

FIGURE 38

46. DETERMINATION OF PHASE ANGLE

The function

$$y = A \cos wt + B \sin wt$$

occurs frequently in physical problems. This can be expressed as a single trigonometric function. Let

$$C = \sqrt{A^2 + B^2}$$

and
$$\tan \alpha = \frac{A}{B}$$

Then we have
$$\sin \alpha = \frac{A}{C}$$

and
$$\cos \alpha = \frac{B}{C}$$

The above function can then be written in the form

$$\begin{aligned} y &= C\left[\frac{A}{C} \cos wt + \frac{B}{C} \sin wt\right] \\ &= C(\sin \alpha \cos wt + \cos \alpha \sin wt) \\ &= C \sin (wt + \alpha) \end{aligned}$$

We note that the sum of the two trigonometric functions is a function with amplitude C and a phase angle α; the period of the new function is $\frac{2\pi}{w}$, which is the same as the period of each of the original trigonometric functions.

If we had chosen $\tan\theta = \frac{B}{A}$, then $\sin\theta = \frac{B}{C}$ and $\cos\theta = \frac{A}{C}$.

Then

$$\begin{aligned} y &= C(\cos\theta\cos wt + \sin\theta\sin wt) \\ &= C\cos(wt - \theta) \end{aligned}$$

This can also be obtained from

$$\sin(wt + \alpha) = \cos\left(\frac{\pi}{2} - wt - \alpha\right)$$

$$= \cos\left(wt - \left[\frac{\pi}{2} - \alpha\right]\right) = \cos(wt - \theta)$$

and we see that

$$\alpha + \theta = \frac{\pi}{2}$$

ILLUSTRATIONS

1. Express the function $y = \cos x + \sqrt{3}\sin x$ as a single trigonometric function.

SOLUTION. $A = 1$ and $B = \sqrt{3}$ so that $C = \sqrt{1 + 3} = 2$. We choose α such that

$$\sin\alpha = \frac{1}{2} \text{ and } \cos\alpha = \frac{\sqrt{3}}{2} \quad \text{or} \quad \alpha = \frac{\pi}{6}$$

Then

$$y = C\sin(x + \alpha) = 2\sin\left(x + \frac{\pi}{6}\right)$$

2. Express the function $y = 4\cos 2x + 3\sin 2x$ in terms of a phase angle.

SOLUTION. $A = 4, B = 3$, and $C = \sqrt{16 + 9} = 5$. Then

$$y = 5\sin(2x + \theta)$$

where $\theta_1 = \arctan\frac{4}{3} \doteq 0.93$ radians.

47. COMPOSITION OF ORDINATES

Whenever a function is expressed as an algebraic sum of simpler functions, a sketch of the locus may be obtained by using a method known as the *composition of ordinates*. The procedure is as follows.

1. Sketch the graph of each function on the same coordinate system.

2. Use a handy device (ruler, dividers, or the straight edge of a piece of paper) to add the line segments of corresponding (same value of x) ordinates taking into account their respective signs.

ILLUSTRATIONS

1. Sketch the locus of $y = \frac{1}{2}x + \cos x$.

SOLUTION. Draw the line $y_1 = \frac{1}{2}x$ and the curve $y_2 = \cos x$. Choose some values of x, and on vertical lines through these values of x measure y_1 and y_2; then add these segments algebraically. For example, on Figure 39 at $x = A$, we have $y_1 = \overline{AB}$ and $y_2 = \overline{AC}$ from which we obtain $\overline{AD} = \overline{AB} + \overline{AC}$.

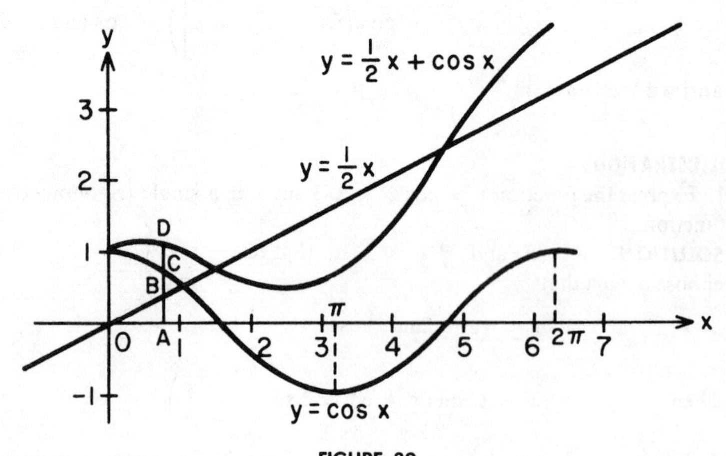

FIGURE 39

2. Sketch the graph of $y = 3 \sin x + \cos 2x$.

SOLUTION. Sketch the curves $y_1 = 3 \sin x$ and $y_2 = \cos 2x$. Choose values of x, and on vertical lines through these values of x measure y_1 and y_2; then add these segments algebraically. The sketch is shown in Figure 40.

48. GRAPHS IN POLAR COORDINATES

The concept of polar coordinates was discussed in Section 9. In this section we shall consider relations which are stated in terms of polar coordinates.

ILLUSTRATION. Draw the graph of $r = 3 \cos \theta$.

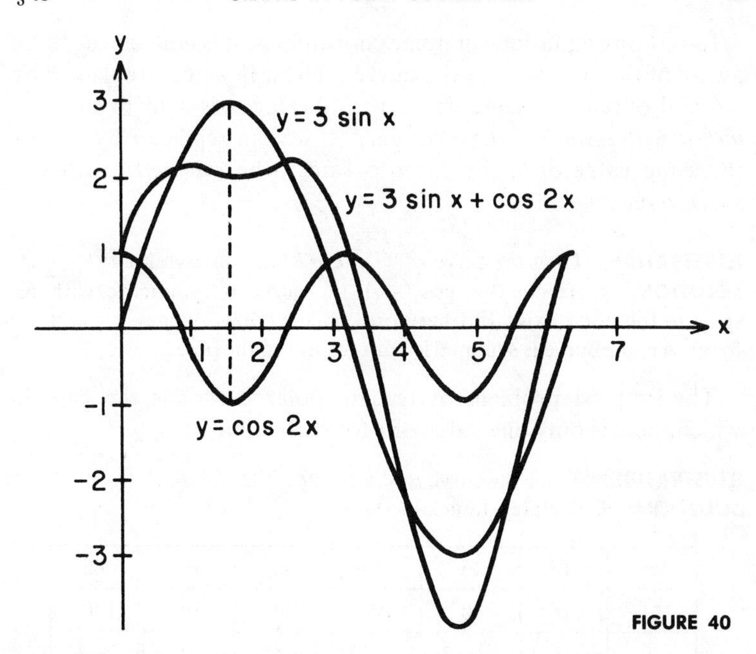

FIGURE 40

SOLUTION. Assign values to θ and calculate the corresponding values of r. The points are plotted and joined by a smooth curve obtaining the graph of Figure 41.

θ	$\cos\theta$	$r = 3\cos\theta$
0	1.00	3
30°	.87	2.61
60°	.50	1.50
90°	0	0
120°	−.50	−1.50
150°	−.87	−2.61
180°	−1.00	−3

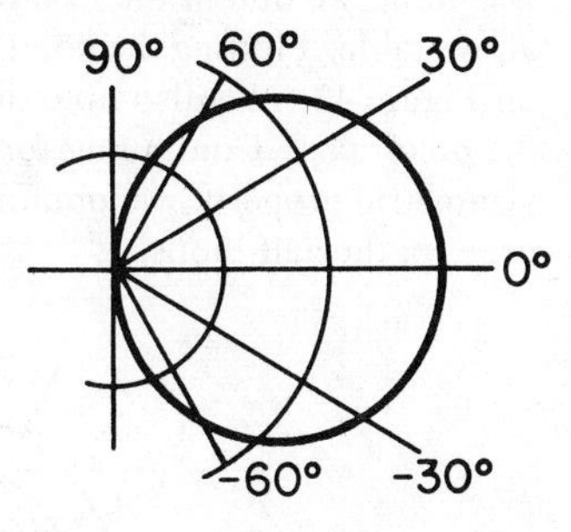

FIGURE 41

If an equation contains only *one* of the variables (r,θ), it means that the other variable may have any and all values.

ILLUSTRATION. What is the graph of $r = 5$?

SOLUTION. This equation says that for *all values of* θ, $r = 5$; i.e., (5,0°), (5,30°), etc., are all points on the graph, which is then a circle of radius 5 with its center at the origin.

In plotting equations in polar coordinates, it is convenient to be aware of the *symmetry* of the curve. Thus, if we can replace θ by $-\theta$ and obtain the same value of r, the locus is said to be *symmetric with respect to the polar axis.* If we can replace r by $-r$ for the same value of θ, the curve is said to be *symmetric with respect to the pole.*

ILLUSTRATION. Does the curve $r^2 = a^2 \cos \theta$ have any symmetry?
SOLUTION. Since $\cos \theta = \cos(-\theta)$, the curve is symmetric with respect to the polar axis. Furthermore, since we have values $+r$ and $-r$ for each θ, the curve is symmetric with respect to the pole.

The simplest problem of tracing polar curves is the case in which there is only one value of r for each value of θ.

ILLUSTRATION. Trace the curve $r = 5 \cos 2\theta$.
SOLUTION. Calculate a table of values.

θ	$0°$	$15°$	$22\frac{1}{2}°$	$30°$	$45°$	$60°$
2θ	$0°$	$30°$	$45°$	$60°$	$90°$	$120°$
r	5	$\frac{5}{2}\sqrt{3}$	$\frac{5}{2}\sqrt{2}$	$\frac{5}{2}$	0	$-\frac{5}{2}$

It is easily seen that as θ varies from $45°$ to $90°$, 2θ varies from $90°$ to $180°$, and we obtain the negative values of r in reverse order from the above table. Plotting the points we obtain the half-loops marked 1 and 2 on Figure 42. We also note that the curve is symmetric with respect to the polar axis. Continuing for all values of θ up to $360°$ and using the symmetric property, we obtain the entire curve of Figure 42. Note the order of the half-loops.

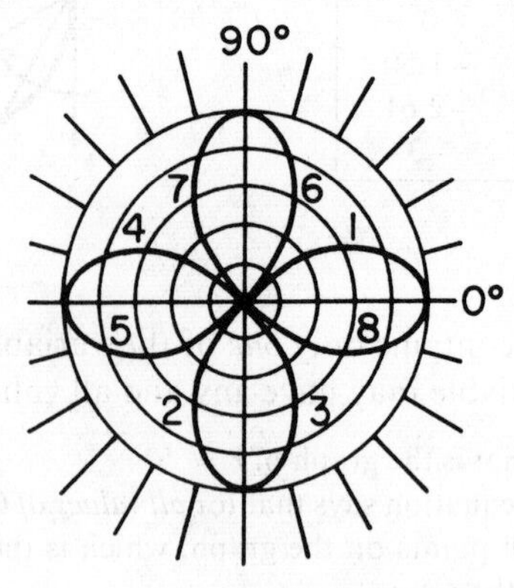

FIGURE 42

49. TRANSFORMATION OF COORDINATES

It is often desired to transform the equation of a curve from its given form in rectangular coordinates to a form in polar coordinates or vice versa. Let the pole be at the origin and let the polar axis coincide with the positive x-axis (see Figure 43). Then we have

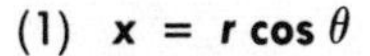

(1) $x = r\cos\theta$

(2) $y = r\sin\theta$

(3) $r^2 = x^2 + y^2$

(4) $\tan\theta = \dfrac{y}{x}$

(5) $\sin\theta = \dfrac{y}{\pm\sqrt{x^2 + y^2}}$

(6) $\cos\theta = \dfrac{x}{\pm\sqrt{x^2 + y^2}}$

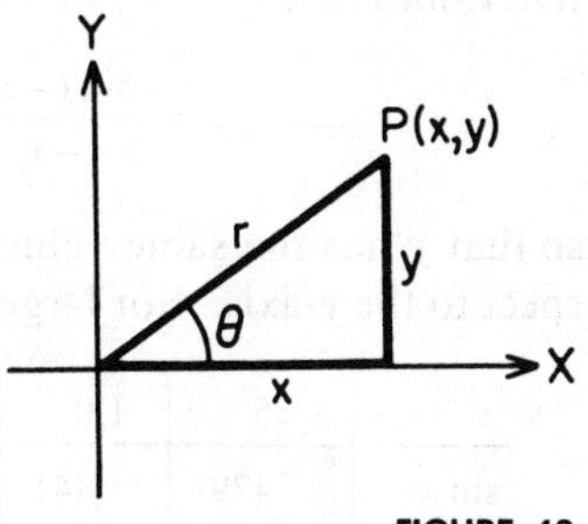

FIGURE 43

ILLUSTRATIONS

1. Change the equation

$$x^2 + y^2 - 2x - 6y = 6$$

to polar coordinates.

SOLUTION. From equations (1), (2), and (3), we have

$$\begin{aligned} x^2 + y^2 &= r^2 \\ -2x &= -2r\cos\theta \\ -6y &= -6r\sin\theta \end{aligned}$$

Thus the polar equation is

$$r^2 - 2r\cos\theta - 6r\sin\theta = 6$$

or

$$r^2 - 2r(\cos\theta + 3\sin\theta) - 6 = 0$$

2. Change the equation

$$r + 2\cos\theta = 0$$

to rectangular coordinates.

SOLUTION. From equations (3) and (6), we have

$$\sqrt{x^2 + y^2} + 2\frac{x}{\sqrt{x^2 + y^2}} = 0$$

which upon simplifying becomes

$$x^2 + y^2 + 2x = 0 \qquad \text{(a circle)}$$

☆50. THE FUNCTION $y = \frac{\sin x}{x}$

This particular function is defined everywhere except at $x = 0$. In advanced mathematics we are very interested in this function as x approaches 0. Let us consider the graph of this function by calculating a table of ordered pairs. First we notice that for negative values of $x = -x_i$, we have

$$y = \frac{\sin(-x_i)}{-x_i} = \frac{-\sin x_i}{-x_i} = \frac{\sin x_i}{x_i}$$

so that y has the same values for $\pm\, x$; i.e., is symmetric with respect to the y-axis. For large intervals of x, we have

x	.5	1.0	1.5	2.0	π	$3\pi/2$
$\sin x$	.479	.841	.997	.909	0	−1
y	.96	.84	.66	.45	0	−.21

For small values of x, we have

x	.01	.05	.1	.2	.3
$\sin x$	.01000	.04998	.09983	.1987	.2955
y	1*	.999	.998	.993	.98

At $x = .01$, we also have $\sin x = .01000$ to the degree of accuracy of a five-place table; however,* if we had tables to a greater number of decimal places, we would see that $\sin .01 < .01$. The graph of these values is shown in Figure 44.

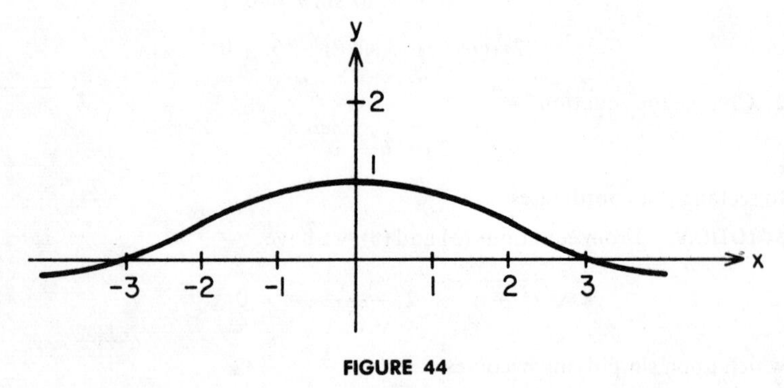

FIGURE 44

*See footnote on page 85.

If we compare the function with $y = a \sin x$ where $a = 1/x$ is a variable, we may think of the function as having a variable amplitude; and since $1/x$ gets smaller as x gets larger, the amplitude is diminishing. In physical analysis this is referred to as a **damping factor**.

The function is undefined at $x = 0$, but we notice that as we let x approach zero from either the positive or negative side, the graph gets closer to the value $y = 1$. The graph would be a complete graph if we added one more value to our definition of the function; this is accomplished by

$$y = \begin{cases} \dfrac{\sin x}{x} & \text{if } x \neq 0 \\ 1 & \text{if } x = 0 \end{cases}$$

The analysis of this function as x approaches zero can also be made by considering comparative sizes of the sine function, the arc on the unit circle, and the tangent function. Consider the unit circle and an angle θ measured in radians whose vertex is at the center of the unit circle, and $0 < \theta < \pi/2$. Let the terminal side be extended to the intersection with the tangent line at A. See Figure 45. We have already seen that $\sin\theta = y$, $\theta = t$, $\tan\theta = z$. Since PBA is a right triangle with PA as hypotenuse, we have $y < PA$, and since the chord is smaller than the arc, we have $y < t$. The area of the section $OPA <$ area of $\triangle OAR$ or

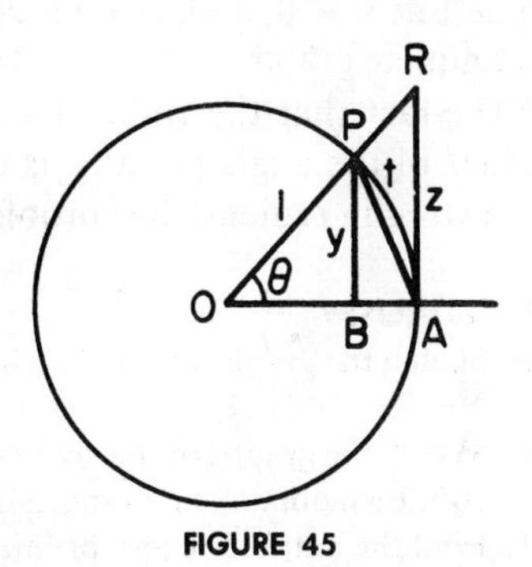

FIGURE 45

$$\tfrac{1}{2}r^2\theta = \tfrac{1}{2}\theta < \tfrac{1}{2}\overline{OA}\,z = \tfrac{1}{2}z$$

so that $\theta < z$. Thus we have

$$y < t < z$$

or*

$$\sin\theta < \theta < \tan\theta$$

*For one second of arc, we have
sin 1″ = 0.00000 48481 36811 076
1″ = 0.00000 48481 36811 095 radians
tan 1″ = 0.00000 48481 36811 133

If we divide this inequality by the positive number $\sin\theta$ (positive since $0 < \theta < \pi/2$), we have

$$1 < \frac{\theta}{\sin\theta} < \frac{1}{\cos\theta}$$

or the reciprocal relation, which states

$$\cos\theta < \frac{\sin\theta}{\theta} < 1$$

Now as θ approaches zero the cosine approaches 1, in fact at $\theta = 0$, $\cos\theta = 1$. Thus the quantity $\frac{\sin\theta}{\theta}$ is "squeezed" between 1 and 1 as θ approaches zero, and consequently this fraction itself must approach 1. In higher mathematics we express this in terms of a limit by saying that as θ approaches 0, the limit of $\frac{\sin\theta}{\theta}$ becomes 1. It is logical that we define the function $y = \frac{\sin x}{x}$ to be 1 at $x = 0$, and, as we have seen in Figure 44, we then have a complete graph.

The fact that the ratio of $\sin\theta$ to θ is close to one for small values of the angle permits us to approximate the sine of θ by θ measured in radians. See problem 12, Exercise 7.

51. EXERCISE 7

1. Sketch the graphs of the six trigonometric functions for $-2\pi \leq x \leq 2\pi$.
2. Sketch the graphs of the six inverse trigonometric relations. Choose your own domain and range. Indicate the principal values.
3. Find the amplitude and primitive periods, and sketch the graphs of the following functions.
   a) $y = \sin 2x$ b) $y = 2\cos 2x$
   c) $y = 5\sin 3x$ d) $y = 10\sin 60t$
4. Find the phase angle and sketch the graphs of the following functions.
   a) $y = 2\sin(x + \pi/4)$ b) $y = 3\cos(2x - 1)$
   c) $y = \sqrt{3}\cos x + \sin x$ d) $y = \cos x + \sin x$
   e) $y = 5\cos x + 12\sin x$ f) $y = 2\cos t + 2\sin t$
5. Sketch the graphs of the following functions by composition of ordinates.
   a) $y = \cos x + \sin x$ b) $y = \sin x - \cos x$
   c) $y = x + \sin x$ d) $y = 2x - \cos x$

6. Sketch the following graphs (see Section 12 for definitions).

a) $y = \text{vers}\, x$ b) $y = \text{covers}\, x$

c) $y = \text{hav}\, x$ d) $y = \text{ex sec}\, x$

7. Sketch the graph of $y = x \sin x$ for $-2\pi \leq x \leq 2\pi$. (On the same graph draw the lines $y = x$ and $y = -x$.)

8. Sketch the following functions y for $-2\pi \leq x \leq 2\pi$.

a) $y = \sin^2 x$ b) $y = \sin |x|$

c) $y = \frac{1}{2}(1 - \cos 2x)$ d) $y = |1 - \sin x|$

9. Sketch the loci of the following polar equations.

a) $r = 3 \cos 3\theta$ b) $r = 3 \sin 3\theta$

c) $r = 2 \cos 2\theta$ d) $r = \sin \frac{1}{2}\theta$

10. Transform the following equations into polar equations.

a) $x^2 + y^2 + 2x - 4y = 0$ b) $y^2 = x^3$

11. Transform the following equations into rectangular equations.

a) $r(2 \cos \theta - 3 \sin \theta) + 5 = 0$ b) $r = \dfrac{3}{2 - 2 \cos \theta}$

c) $r^2 = 9 \tan \theta$ d) $r = 5$

12. Make a table of values for θ in radians and $\sin \theta$, accurate to 2 decimal places, beginning with $\theta = 0$, and at intervals of every 0.1. Stop when they no longer agree to 2 decimal places. Convert the largest angle of agreement to degrees.

chapter 5

Trigonometric Equations

52. SOLUTIONS OF EQUATIONS

We discussed the three types of equations in Section 22 and considered the identities in Chapter 3. In this chapter we shall consider the conditional equations and in particular conditional trigonometric equations. The **solution** of a conditional equation is the set of numbers which satisfies the equation; that is, the solution set, when substituted into the equation, will reduce the equation to an easily recognizable true statement. For example, consider the equation

$$3x - 5 = 7$$

The solution of this equation is $x = 4$, because when x is replaced by the number 4, the equation is reduced to $7 = 7$, which is a known true statement. In this example there is only one value of x that will satisfy the equation. Many equations, however, have more than one solution; for example, a quadratic equation has two solutions and an algebraic equation of degree n has n solutions. The solutions of an equation are called the **roots** of the equation, the **solution set**, and sometimes, the **zeros** of the function.

53. SIMPLE TRIGONOMETRIC EQUATIONS

Let us consider the simple trigonometric equation

$$\sin x = \tfrac{1}{2}$$

We could consult a table to find the values of x for which the sine of x is equal to 0.5, or we could recall from Section 18 that when $x = 30°$, $\sin 30° = \frac{1}{2}$. However, from the properties of the trigonometric functions we know that

$$\sin (180° - 30°) = \sin 150° = \tfrac{1}{2}$$

thus $x = 150^\circ$ will also satisfy the equation. Furthermore, $\sin(30^\circ + n\,360^\circ) = \sin(150^\circ + n\,360^\circ) = \frac{1}{2}$, where n is any integer. Consequently, there are an infinite number of values for x which will satisfy this equation; however, they form a specific set. This solution set may be written in a number of ways since the values may be stated in terms of angles in radian measure or degrees, or as numbers, which can be found from tables or graphs.* Thus the solution set can be written

$$x = \left\{\frac{\pi}{6} + 2n\pi, \frac{5\pi}{6} + 2n\pi\right\}$$

or

$$x = \left\{30^\circ + n\,360^\circ, 150^\circ + n\,360^\circ\right\}$$

where n is any integer.

All the roots of the equations of type $\sin x = k$ and $\cos x = k$, where $-1 \le k \le 1$, can be found by inspection, or from tables for any interval covering a complete period, adding integral multiples of 2π when necessary.

ILLUSTRATIONS

1. Solve the equation $\sin x = \dfrac{\sqrt{3}}{2}$.

SOLUTION. From Section 18 we know that

$$\sin x = \frac{\sqrt{3}}{2}$$

when $x = \dfrac{\pi}{3}$ and $x = \dfrac{2\pi}{3}$. Thus the solution set is

$$x = \left\{\frac{\pi}{3} + 2n\pi, \frac{2\pi}{3} + 2n\pi\right\}$$

2. Solve the equation $\cos x = -\frac{1}{2}$.

SOLUTION. By inspection (see Sections 18 and 20),

$$x = \frac{2\pi}{3} \quad \text{and} \quad x = \frac{4\pi}{3}$$

Thus the complete solution is

$$x = \frac{2\pi}{3} + 2n\pi \quad \text{and} \quad x = \frac{4\pi}{3} + 2n\pi$$

*The graphical solution is obtained by drawing the graph of $y = \sin x$ and $y = 0.5$ (for the above example), and reading off the points of intersection of the two curves.

Refer to the Appendix for examples involving the use of tables.

Equations of the type

$$\tan x = k \quad \text{and} \quad \cot x = k$$

are solved in the same manner, except that the period for these two functions is π. Thus only one fundamental solution in the interval $0 \leq x \leq \pi$ need be found; the complete solution is then obtained by adding integral multiples of π to this fundamental solution.

ILLUSTRATIONS

1. Solve the equation $\tan x = \sqrt{3}$.

SOLUTION. By inspection $x = \frac{\pi}{3}$. Therefore, the complete solution is

$$x = \frac{\pi}{3} + n\pi, \quad n = 0, \pm 1, \pm 2, \ldots$$

2. Solve the equation $\cot x = -1$.

SOLUTION. By inspection $x = 135^\circ$ and the complete solution is

$$x = 135^\circ + n\,180^\circ, \quad n = 0, \pm 1, \pm 2, \ldots$$

The solution of a trigonometric equation may also be expressed in terms of the inverse function. Thus the solution of $\sin x = k$ may be written as $x = \arcsin k$. This, however, is a symbolic method of denoting the solution set; it is usually preferable to express x in terms of numbers since, in many cases, they will be the measures of angles.

Let us now consider a simple trigonometric equation of the form

$$(\text{trigonometric function of } n\,x) = k$$

The solution set is obtained by getting the complete solution for $n\,x$ and then, dividing by n.

ILLUSTRATION. Solve the equation $\sin 3x = -\frac{1}{2}$.

SOLUTION. The fundamental solutions are

$$3x = 210^\circ \quad \text{and} \quad 3x = 330^\circ$$

Consequently,

$$3x = 210^\circ + n\,360^\circ \quad \text{and} \quad 3x = 330^\circ + n\,360^\circ$$

Now divide by 3 to obtain the following, where n is an integer.

$$x = 70^\circ + n\,120^\circ \quad \text{and} \quad x = 110^\circ + n\,120^\circ$$

In the last illustration, the second solution could be written $3x = -30° + n\,360°$, from which we would obtain $x = -10° + n\,120°$. This, however, is the same solution set as the one shown, since the same elements are obtained from different values of n; for example, $x = 110°$ for $n = 0$ in the set $\{110° + n\,120°\}$ and for $n = 1$ in the set $\{-10° + n\,120°\}$.

In some cases it is desired to find only those members of the solution set which fall in some interval. These can be obtained by choosing values of n such that the values of x fall in the specified interval. If an equation involves a multiple angle, be sure to find all the members which fall in the interval.

ILLUSTRATION. Find all the values of x in the interval $-\pi \le x \le \pi$ which satisfy the equation $\cos 3x = \frac{1}{2}$.

SOLUTION. By inspection we have

$$3x = \frac{\pi}{3} + 2n\pi \quad \text{and} \quad 3x = \frac{5\pi}{3} + 2n\pi$$

therefore

$$x = \frac{\pi + 6n\pi}{9} \quad \text{and} \quad x = \frac{5\pi + 6n\pi}{9}$$

Now when

$$n = 0, \quad x = \frac{\pi}{9} \quad \text{and} \quad x = \frac{5\pi}{9}$$

$$n = 1, \quad x = \frac{7\pi}{9} \quad \text{and} \quad x = \frac{11\pi}{9}$$

$$n = -1, \quad x = -\frac{5\pi}{9} \quad \text{and} \quad x = -\frac{\pi}{9}$$

$$n = -2, \quad x = -\frac{11\pi}{9} \quad \text{and} \quad x = -\frac{7\pi}{9}$$

All other values of n will yield values of x outside the interval; we note that two of the above values are also outside the interval. The desired solution set is

$$\left\{-\frac{7\pi}{9}, -\frac{5\pi}{9}, -\frac{\pi}{9}, \frac{\pi}{9}, \frac{5\pi}{9}, \frac{7\pi}{9}\right\}$$

Equations involving $\sec x$ and $\csc x$ are usually transformed to equations involving $\sin x$ and $\cos x$ by use of the trigonometric identities. Thus the solutions of the equation $\sec x = 2$ can be obtained from the equation $\cos x = \frac{1}{2}$.

Since the ranges of some trigonometric functions are limited, it is easy to write null equations involving trigonometric functions. The equation $\sin x = 2$ is a null equation since there are no values of x which would satisfy this equation.

54. EXERCISE 8

1. Find all the roots of each of the following equations.

a) $\sin x = \frac{1}{2}$ b) $\cos x = \frac{1}{2}$ c) $\tan x = 1$

d) $\sin x = -\frac{1}{2}$ e) $\cos x = -\frac{\sqrt{3}}{2}$ f) $\tan x = -\frac{\sqrt{3}}{3}$

g) $\sin x = -1$ h) $\cos x = 2$ i) $\tan x = 0$

j) $\sin x = 0$ k) $\cos x = -1$ l) $\tan x = \sqrt{3}$

m) $\sin 4x = \frac{\sqrt{3}}{2}$ n) $\cos 6x = 0$ o) $\tan 3x = 1$

2. Find the solutions of the following equations for the interval $0 \leq x < 2\pi$.

a) $\sin x = 1$ b) $\cos 2x = 0$ c) $\tan 3x = 0$

d) $\sin 5x = \frac{1}{2}$ e) $\cos x = -1$ f) $\tan 2x = -\sqrt{3}$

3. Find the solutions of the following equations for the interval $-180° \leq x \leq 180°$.

a) $\sin x = 0$ b) $\cos x = \frac{1}{2}$ c) $\cot(3x + 27°) = 1$

d) $\sin 3x = -1$ e) $\cos 5x = 0$ f) $\tan 4x = \sqrt{3}$

4. Using tables, find the approximate solutions.

a) $\sin 2x = .6$ b) $\cos x = \frac{1}{3}$ c) $\tan x = 2$

d) $\cot 2x = 1.5$ e) $\sin x = \sqrt{5}$ f) $\cos 3x = \frac{2}{3}$

55. THE EQUATION $\sin x = \sin y$

The equation $\sin x = \sin y$ raises the question, "What is the relationship between x and y?" It is clear that the equation is satisfied if $y = x$, for then we have $\sin x = \sin x$. However, the equation is also satisfied if $y = \pi - x$, $y = x + 2n\pi$, or $y = (\pi - x) + 2n\pi$, where n is an integer. The same question is proposed by the equations $\cos x = \cos y$, $\sin x = -\sin y$, and $\cos x = -\cos y$. By recalling the concept of the reference angle (see Section 17), we can summarize the analysis in the following table where $n = 0, \pm 1, \pm 2, \pm 3$, etc.

equation	implies	or
$\sin x = \sin y$	$y = x + 2n\pi$	$y = (\pi - x) + 2n\pi$
$\sin x = -\sin y$	$y = (x + \pi) + 2n\pi$	$y = -x + 2n\pi$
$\cos x = \cos y$	$y = x + 2n\pi$	$y = -x + 2n\pi$
$\cos x = -\cos y$	$y = (\pi - x) + 2n\pi$	$y = (\pi + x) + 2n\pi$

The same kind of analysis can be applied to the other four trigonometric functions. The property of the complementary angles can be used to transform equations of the form $\sin x = \pm \cos y$ to one of the types shown in the table on page 92; for example,

$$\sin x = \cos y = \sin\left(\frac{\pi}{2} - y\right)$$

ILLUSTRATION. Solve the equation $\sin 2x = \cos 7x$.

SOLUTION. Transform the equation to the form

$$\sin 2x = \sin\left(\frac{\pi}{2} - 7x\right)$$

Then

$$2x = \frac{\pi}{2} - 7x + 2n\pi \qquad\qquad 2x = \pi - \left(\frac{\pi}{2} - 7x\right) - 2n\pi$$

$$9x = \frac{\pi}{2} + 2n\pi \qquad\qquad -5x = \frac{\pi}{2} - 2n\pi$$

$$x = \frac{\pi + 4n\pi}{18} \qquad\qquad x = -\frac{\pi}{10} + \frac{2n\pi}{5}$$

$$= \frac{1 + 4n}{18}\pi \qquad\qquad = \frac{4n - 1}{10}\pi$$

Note that in the second part of the solution, we wrote the right-hand side in the form of $\theta - 2n\pi$. This is permissible since n assumes the values of the positive and negative integers as well as zero. Thus the same solution set would be generated whether we wrote $\theta + 2n\pi$ or $\theta - 2n\pi$; the identical members occurring for values of n with opposite signs.

56. EQUATIONS INVOLVING MORE THAN ONE FUNCTION

If the equation involves more than one trigonometric function, we shall employ algebraic techniques to obtain the solutions. Let us review the operations which lead to equivalent equations (that is, equations which have the same roots as the given equation); they are:

a) addition or subtraction of a term, to or from both members;
b) multiplying or dividing both members by a nonzero *constant*.

The multiplication of both members by a term involving the unknown will lead to an equation which may have more roots than the original equation; thus, it is necessary to check the roots and determine if extraneous solutions have been added. If the equation is divided by a term containing the unknown, the resulting equation may contain fewer roots than the original equation. This operation should be avoided.

The following suggestions are useful in reducing trigonometric equations to elementary types.

a) Express all functions involved in terms of a single function of some angle.
b) Collect all terms on one side and thereby equate the terms to zero.
c) Factor whenever possible.
d) When $XY = 0$, then $X = 0$, or $Y = 0$, and both should be investigated.

Let us now consider some examples.

ILLUSTRATIONS

1. Solve the following equation for roots in the interval $0 \leq x \leq 2\pi$.

$$2 \cos x \sin x + 2 \cos x - \sin x = 1$$

SOLUTION. First transpose the number 1 to the left-hand side and equate to zero, then factor by grouping terms to obtain

$$(2 \cos x - 1)(\sin x + 1) = 0$$

We now have two factors, each containing only one trigonometric function, and we set each factor equal to zero to obtain

$$\begin{array}{l|l} 2 \cos x - 1 = 0 & \sin x + 1 = 0 \\ \cos x = \frac{1}{2} & \sin x = -1 \\ x = \dfrac{\pi}{3}, \dfrac{5\pi}{3} & x = \dfrac{3\pi}{2} \end{array}$$

in the interval specified.

2. Find the complete solution for

$$\sec^2 x - \tan x - 1 = 0$$

SOLUTION. This equation may be transformed to one involving a single trigonometric function by using a fundamental identity (see Section 23). The result is

$$\tan^2 x + 1 - \tan x - 1 = 0$$

or

$$\tan^2 x - \tan x = 0$$

which may be factored to yield the product

$$\tan x (\tan x - 1) = 0$$

Set each factor equal to zero.

$\tan x = 0$	$\tan x - 1 = 0$
$x = 0 + n\pi$	$\tan x = 1$
$= n\pi$	$x = \dfrac{\pi}{4} + n\pi$

3. Find the values of $0^\circ \le x < 360^\circ$ which satisfy the equation

$$\sin x - \cos x = 1$$

SOLUTION. The equation can be transformed to one containing a single trigonometric function by using the identity

$$\cos x = \sqrt{1 - \sin^2 x}$$

to obtain
$$\sin x - \sqrt{1 - \sin^2 x} = 1$$

The algebraic procedure in solving an equation of this type is to isolate the radical and square both sides, obtaining

$$(\sin x - 1)^2 = 1 - \sin^2 x$$

or
$$\sin^2 x - 2\sin x + 1 = 1 - \sin^2 x$$

which reduces to
$$2\sin^2 x - 2\sin x = 0$$

This equation may be factored to yield

$\sin x = 0$	$\sin x = 1$
$x = 0, 180^\circ$	$x = 90^\circ$

However, in squaring both sides of an equation we are multiplying by the unknown, and the resulting equation may contain more roots than the original. It is therefore necessary to substitute the roots into the original equation.

$x = 0^\circ$:	$0 - 1 \neq 1,$	$\therefore$	$x = 0^\circ$ is extraneous
$x = 180^\circ$:	$0 - (-1) = 1,$	$\therefore$	$x = 180^\circ$ is a root
$x = 90^\circ$:	$1 - 0 = 1,$	$\therefore$	$x = 90^\circ$ is a root

57. EQUATIONS INVOLVING MULTIPLE ARGUMENTS

The solution of trigonometric equations that involve multiple arguments is accomplished by reducing the equation to one or more equations involving only one function of one argument and whose solutions include all solutions of the original equation. The procedure can be explained by considering some examples.

ILLUSTRATIONS

1. Solve the equation $\cos 3x = \frac{1}{2}$.

SOLUTION. Let $3x = y$, then we know that $y = \frac{\pi}{3} + 2n\pi$ and $y = \frac{5\pi}{3} + 2n\pi$ will satisfy the given equation. The solution set for x is then obtained by dividing by 3, thus

$$x = \frac{\pi}{9} + \frac{2n\pi}{3} \quad \text{and} \quad x = \frac{5\pi}{9} + \frac{2n\pi}{3}$$

2. Solve the equation $3 \tan x = \tan 2x$.

SOLUTION. This equation is transformed into one involving one argument by using the formula for $\tan 2x$ to obtain

$$3 \tan x = \frac{2 \tan x}{1 - \tan^2 x}$$

which is simplified to

$$3 \tan x (1 - \tan^2 x) - 2 \tan x = 0$$

or

$$\tan x (1 - 3 \tan^2 x) = 0$$

We now have the two equations

$\tan x = 0$	$\tan^2 x = \frac{1}{3}$
$x = 0 + n\pi$	$x = \frac{\pi}{6} + n\pi$ and $\frac{5\pi}{6} + n\pi$

However, in the simplification, we multiplied by the unknown and so the solutions must be checked. It turns out that all of the above values form a solution set.

3. Solve the equation $\cos 3x + \sin 2x - \sin 6x + \cos 5x = 0$.

SOLUTION. The equation is transformed by the use of formulas (30) and (31) of Section 33.

$$2 \cos 4x \cos x + 2 \cos 4x \sin(-2x) = 0$$

or

$$\cos 4x (\cos x - \sin 2x) = 0$$

We then have two equations and, by formula (17) of Section 31, we may write

$\cos 4x = 0$	$\cos x - 2 \sin x \cos x = 0$	
	$\cos x (1 - 2 \sin x) = 0$	
$\cos 4x = 0$	$\cos x = 0$	$\sin x = \frac{1}{2}$
$4x = 90^\circ + n\,360^\circ$	$x = 90^\circ + n\,360^\circ$	$x = 30^\circ + n\,360^\circ$
$4x = 270^\circ + n\,360^\circ$	$x = 270^\circ + n\,360^\circ$	$x = 150^\circ + n\,360^\circ$

The solution set for $0^\circ \leq x < 360^\circ$ is

$$x = 30^\circ, 90^\circ, 150^\circ, 270^\circ \; (22^\circ 30' + n\,45^\circ, n = 0, 1, 2, 3)$$

4. Solve the equation $\dfrac{\tan x + \tan 2x}{1 - \tan x \tan 2x} = 1.$

SOLUTION. Although the solution of most problems employs the addition formula to reduce the argument to a single angle, this problem is an illustration of the use of the addition formula to increase the size of the unknown, but still reduce the equation to one involving a single trigonometric function and argument. Using formula (13) of Section 29, we obtain

$$\tan 3x = 1$$

and
$$3x = \frac{\pi}{4} + n\pi$$

so that
$$x = \frac{\pi}{12} + \frac{n\pi}{3}$$

58. EQUATIONS INVOLVING INVERSE FUNCTIONS

We have already seen that the solution set of a trigonometric equation may be written in the form of the inverse relations. Thus the solution of the equation

$$\tan \theta = k$$

may be written in the form

$$\theta = \arctan k$$

We may also have equations involving the inverse functions; the procedure for obtaining the solutions can be explained by considering an example.

ILLUSTRATION. Solve the equation

$$\text{Arcsin } u + \text{Arcsin } 2u = \frac{\pi}{2}$$

SOLUTION. Let $\alpha = \text{Arcsin } u$ and $\beta = \text{Arcsin } 2u$. Then the equation reduces to

$$\alpha + \beta = \frac{\pi}{2}$$

and we may write

$$\cos(\alpha + \beta) = \cos\frac{\pi}{2} = 0$$

or
$$\cos\alpha\cos\beta - \sin\alpha\sin\beta = 0$$

We now transform this trigonometric equation into an algebraic equation by the properties of the inverse function (see Section 20). Using the diagrams of Figure 46, we obtain

$$\sqrt{1 - u^2}\sqrt{1 - 4u^2} - 2u^2 = 0$$
$$1 - 5u^2 + 4u^4 = 4u^4$$
$$u^2 = \frac{1}{5} \quad \text{or} \quad u = \pm\frac{\sqrt{5}}{5}$$

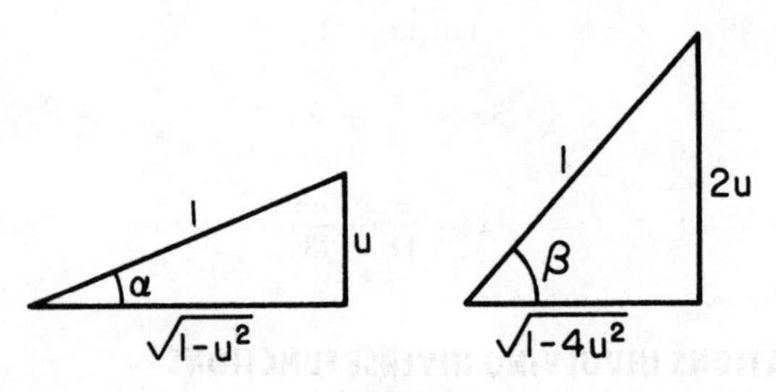

FIGURE 46

These solutions must be checked, since in the process of solving the algebraic equation we squared both sides. We see that the negative value of u would result in the algebraic sum of two negative values being equal to a positive $\frac{\pi}{2}$, which is not true; hence, the negative value of u is extraneous. To show that the positive value satisfies the equation, we see that

$$\sqrt{1 - u^2} = 2/\sqrt{5} \quad \text{and} \quad \sqrt{1 - 4u^2} = 1/\sqrt{5}$$

and hence $\cos(\alpha + \beta) = \cos\alpha\cos\beta - \sin\alpha\sin\beta$

$$= \left(\frac{2}{\sqrt{5}}\right)\left(\frac{1}{\sqrt{5}}\right) - \left(\frac{1}{\sqrt{5}}\right)\left(\frac{2}{\sqrt{5}}\right) = 0$$

and $\alpha + \beta = \pi/2$.

59. EXERCISE 9

1. Find the complete solution to the equations.

a) $\sin 2x = \sin 7x$ b) $\sin 3x = -\sin 5x$
c) $\cos x = \cos 4x$ d) $\cos 2x = -\cos 5x$
e) $\sin 3x = \cos 4x$ f) $\sin 2x = -\cos 3x$
g) $\tan 3x = \tan 4x$ h) $\cot 2x = -\cot 3x$

2. Find the complete solution of the equations.

a) $\sin x = 2\sin x\cos x$ b) $\sin x\sec^2 x = 2\sin x$
c) $\sec x\tan x = \tan x$ d) $\sin x + 1 = \cos x$
e) $2\sin^2 x + \cos x = 1$ f) $2\sin^2 x - \cos x = 1$

g) $\sin x + \cos x = 1$
h) $2 \sin x - 1 = 0$
i) $2 \sin x + \csc x = 3$
j) $\sec x - 1 = \tan x$
k) $\sin x \cos x = 0$
l) $\sin^2 x + \sin x = 2$
m) $3 \cos^2 x = \sin^2 x$
n) $2 \sin x - \csc x = 1$
o) $2 \sin^2 x - \sin x = 1$
p) $2 \tan x \sin x = \tan x$
q) $2 \tan^2 x + \sec^2 x = 2$
r) $2 \sin^2 x - 3 \cos x = 3$
s) $\sqrt{3} \cot x + 1 = \csc^2 x$
t) $3 \cos x = \sqrt{3} \sin x$
u) $\csc^2 x - \sqrt{2} = (\sqrt{2} - 1) \csc x$
v) $\csc x + \cot x = \sqrt{3}$
w) $\tan x + \cot x = 2 \sec x$
x) $\cos x - \sqrt{3} \sin x = 1$
y) $\cot x + \tan x \csc x - \tan x = 0$
z) $3 \cot x - \cos^2 x \csc x - 2 \csc x + \sin x = 0$

3. Find the complete solution of the equations.

a) $\cos 2x = \sin x$
b) $\sin \dfrac{x}{2} = \sin x$
c) $2 \sin x \cos x = \frac{1}{2}$
d) $\sin 2x = \sin x$
e) $\cos^2 x = \cos 2x$
f) $\cos^2 x - \sin^2 x = \frac{1}{2}$
g) $\cos 2x - \cos x = \sin \dfrac{x}{2}$
h) $3 \sin x - 4 \sin^3 x = \frac{1}{2}$
i) $\cos 2x + \cos x = -1$
j) $\tan 2x = -2 \sin x$

4. Solve the following equations.

a) $\operatorname{Arccos} 2x = \operatorname{Arcsin} x$

b) $\operatorname{Arccos} 2x - \operatorname{Arccos} x = \dfrac{\pi}{3}$

c) $\operatorname{Arcsin} x = \operatorname{Arcsin} \left(\dfrac{1}{x}\right)$

d) $\operatorname{Arccos} \left(\dfrac{\sqrt{3}}{2}\right) + \operatorname{Arcsin} 3x = \dfrac{\pi}{2}$

e) $\operatorname{Arcsin} x + \operatorname{Arcsin} 2x = \dfrac{\pi}{2}$

f) $\operatorname{Arcsin} x + \operatorname{Arccos} 2x = \dfrac{\pi}{4}$

☆**5.** Using the quadratic formula, solve the equation

$$a \cos x + b \sin x = c$$

for $\sin x$ in terms of a, b, c. Find the condition on a, b, c that will make the roots for $\sin x$ equal.

☆**6.** Solve problem 5, using the phase angle (see Section 46).

☆60. MIXED EQUATIONS

Equations which involve both algebraic and trigonometric terms occur in scientific analyses. The elementary procedure for solving such equations is to approximate the roots by graphical methods. If it is desired to have the roots accurate to a fairly high degree, iteration methods are usually employed; these, however, require the use of automatic calculators since the arithmetic is rather cumbersome.* We shall illustrate the graphical method by an example.

ILLUSTRATION. Solve the equation

$$3 \sin x - \tfrac{1}{2}x - 1 = 0$$

SOLUTION. Group the trigonometric terms on one side of the equality sign and the algebraic terms on the other to obtain

$$3 \sin x = \tfrac{1}{2}x + 1$$

Let $$y_1 = \tfrac{1}{2}x + 1 \qquad y_2 = 3 \sin x$$

and draw the graphs of these two functions. The abscissas of the points of intersection are the required solutions. The algebraic function is usually plotted first to give an indication of the number of cycles which may be necessary for the graphing of the trigonometric function. In this example, the algebraic function is a straight line passing through the two points (0,1) and (2,2). The trigonometric function is a sine curve with amplitude 3 and period 2 π. The solution is shown in Figure 47. The approximate solutions are $x = .43$ and $x = 2.35$.

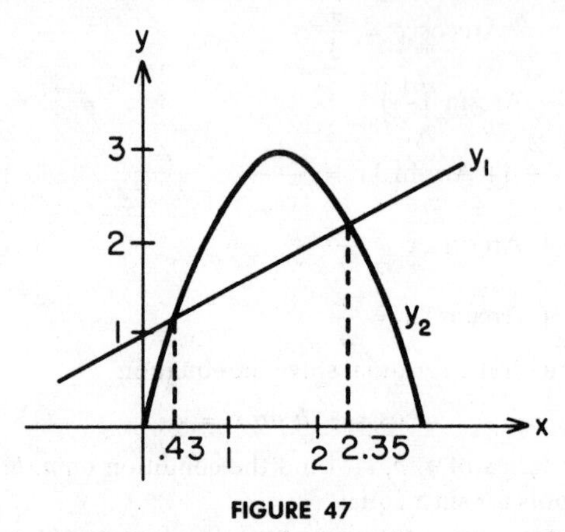

FIGURE 47

☆61. SYSTEMS OF TRIGONOMETRIC EQUATIONS

The solution of systems of trigonometric equations is usually accomplished by combining the equations and using the trigonometric identities. Let us consider some examples.

*See Kaj L. Nielsen, *Methods in Numerical Analysis* (2nd ed.; New York: Macmillan, 1964).

ILLUSTRATIONS

1. Solve the system (for $0 \leq x, y < 360°$).

$$2 \sin x - 2 \sin y = 1$$
$$2 \cos x + 2 \cos y = \sqrt{3}$$

SOLUTION. The equations may be transformed by the use of formulas (30) and (31), Section 33, to obtain

$$4 \cos \tfrac{1}{2}(x + y) \sin \tfrac{1}{2}(x - y) = 1$$
$$4 \cos \tfrac{1}{2}(x + y) \cos \tfrac{1}{2}(x - y) = \sqrt{3}$$

Dividing the first of these equations by the second yields

$$\tan \frac{1}{2}(x - y) = \frac{\sqrt{3}}{3}$$

Therefore $$\tfrac{1}{2}(x - y) = 30° \text{ or } 210°$$

and $$\sin \tfrac{1}{2}(x - y) = \tfrac{1}{2} \text{ or } -\tfrac{1}{2}$$

A substitution of these values into the first of the transformed equations yields

$$\cos \tfrac{1}{2}(x + y) = \tfrac{1}{2} \text{ or } -\tfrac{1}{2}$$

and $$\tfrac{1}{2}(x + y) = 60° \, (-60°) \text{ or } 120° \, (240°)$$

Let us write all possible combinations.

$\tfrac{1}{2}(x - y) = 30°$ $\tfrac{1}{2}(x + y) = 60°$	$\tfrac{1}{2}(x - y) = 30°$ $\tfrac{1}{2}(x + y) = -60°$	$\tfrac{1}{2}(x - y) = 210°$ $\tfrac{1}{2}(x + y) = 120°$	$\tfrac{1}{2}(x - y) = 210°$ $\tfrac{1}{2}(x + y) = 240°$
$x = 90°$ $y = 30°$	$x = -30°$ $y = -90°$	$x = 330°$ $y = -90°$	$x = 450°$ $y = 30°$

We see that these reduce to the two solutions $x = 90°$, $y = 30°$ and $x = 330°, y = 270°$.

2. Solve the system:

$$r \sin \theta = 3$$
$$r \cos \theta = 1$$

for $r > 0$ and $0 \leq \theta < 2\pi$.

SOLUTION. The value of r is obtained by adding the squares of the two equations, thus

$$r^2 (\sin^2 \theta + \cos^2 \theta) = r^2 = 9 + 1 = 10$$

so that $r = \sqrt{10}$.

Since $r > 0$, the original equations make $\sin \theta$ and $\cos \theta$ positive, and θ is acute. If we now divide the first equation by the second, we obtain $\tan \theta = 3$ and, from the tables, $\theta \doteq 1.25$

☆62. EXERCISE 10

1. Find the approximate solution of the following equations.
   a) $\cos 2x = x - 1$
   b) $\sin x = x - \frac{\pi}{2} + 1$
   c) $\tan x - x = 0$ for $-\frac{\pi}{2} < x < \frac{\pi}{2}$
   d) $\operatorname{Arcsin} x - \frac{\pi}{3}\left(\frac{x}{2} + 1\right) = 0$
   e) $\sin x = x^2$
2. Solve the following systems of equations.
   a) $\sin x + \sin y = 1$ and $\cos x + \cos y = 1$
   b) $\sin x + \sin y = 1$ and $\sin x - \sin y = 0$. Hint: treat as algebraic equations.
   c) $\sin x + \sin y = \frac{1}{2}$, and $\cos x - \cos y = \frac{1}{2}$
   d) $r \cos 2\theta = 1$ and $r \sin 2\theta = 1$
   e) $r \sin 2\theta = 1$ and $r \cos \theta = 1$
   f) $r \cos \theta = 2$ and $r \sin \theta = 3$

chapter 6

Exponential and Logarithmic Functions

63. INTRODUCTION

It is assumed that the reader is familiar with powers, roots, exponents, radicals, and their basic laws of operation.* In this chapter we shall extend these concepts to two special functions for the real number system. These two important transcendental functions are the exponential function and its inverse, the logarithmic function.

64. THE EXPONENTIAL FUNCTION

In order to limit our discussion to the real number system, we shall restrict the parameter a to positive values, i.e., $a > 0$, and define the exponential function.

DEFINITION. *The function* f *defined by the equation* $y = a^x$, $(a > 0)$, *is called the* **exponential function** *with base* a.

The domain of definition is the set of real numbers and, as we shall see, the range is $0 < y < \infty$. To clarify this definition let us consider the graphs for two values of a, say $a = 2$ and $a = 3$. First calculate the following table of values.

$y = 2^x$

x	-3	-2	-1	0	1	2	3
y	$\frac{1}{8}$	$\frac{1}{4}$	$\frac{1}{2}$	1	2	4	8

$y = 3^x$

x	-2	-1	0	1	2
y	$\frac{1}{9}$	$\frac{1}{3}$	1	3	9

By plotting these points and joining them by a smooth curve,† we

*See Kaj L. Nielsen, *College Mathematics* (New York: Barnes and Noble, 1958), pp. 7–11.

†This assumes that for irrational values of x the values of y will be such that the corresponding points (x,y) will lie on this smooth curve. The existence of values for y can be shown by the use of the least upper-bound principle of a sequence of numbers. The author prefers to delay such detailed discussions for more advanced courses.

obtain the graphs shown in Figure 48. A study of these graphs permits us to infer the following properties.

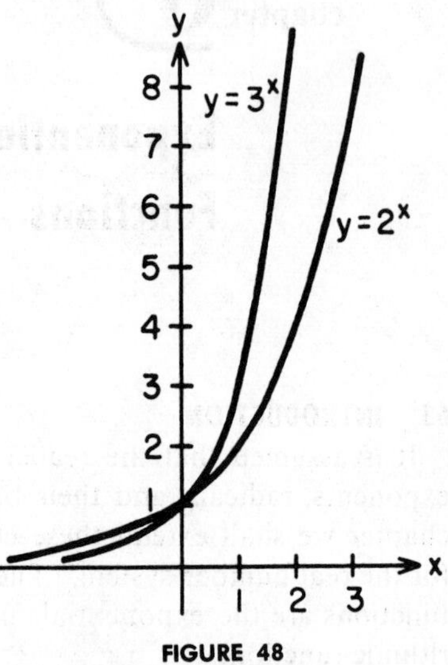

FIGURE 48

I. The function is positive ($y > 0$) for all values of x.

II. If $a > 1$, the function is an increasing function, that is, as x increases so does y; such functions are termed **monotone increasing**. Furthermore, as x decreases algebraically (takes on negative values), the function approaches but never becomes zero. Thus the range is $0 < y < \infty$.

III. When $x = 0$, the function has the value 1, and $y = 1$ for all values of a.

IV. Since $y > 0$ there are no zeros of the function.

Let us now concentrate on values for a, which are already restricted to $a > 0$. If $a > 1$, we have graphs of the function of the type shown in Figure 48. If $a = 1$,

$$y = a^x = 1^x = 1$$

for all values of x, and the graph is a straight line. If $a < 1$, the graph is slightly different; let us consider an example, say $a = \frac{1}{2}$. We obtain the graph of Figure 49.

x	y
-3	8
-2	4
-1	2
0	1
1	$\frac{1}{2}$
2	$\frac{1}{4}$
3	$\frac{1}{8}$

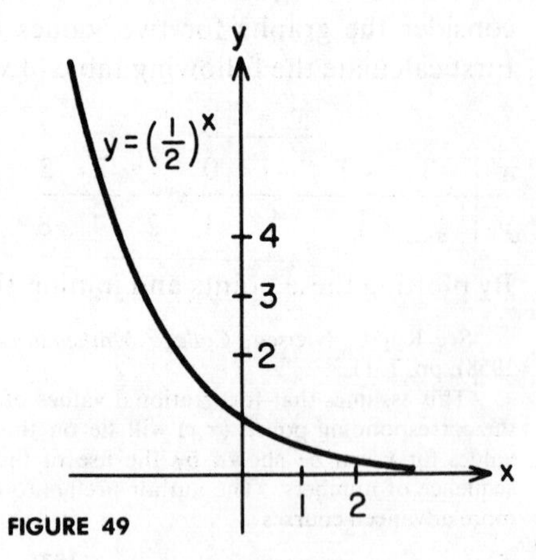

FIGURE 49

The properties we inferred earlier are still valid except that as x increases, the function decreases (termed **monotone decreasing**) and approaches zero for large values of x, but still does not touch the x-axis.

There is a very special value of a which makes the function

$$y = a^x$$

extremely important in physical applications. This value of a is an irrational number, symbolized by the letter e.

$$e \doteq 2.71828\ 18285$$

This value of a is so important that the function

$$y = e^x$$

is usually referred to as *the* exponential function, and the irrational number e is placed next to π as an important physical constant. The graph of $y = e^x$ lies "between" the two pictured in Figure 48. Values of this function have been calculated and many extensive tables have been published.*

65. EXERCISE II. Plot the graphs of the following functions. Use the abbreviated table in the back of this book whenever necessary.

1. $y = e^x$
2. $y = (\frac{3}{2})^x$
3. $y = \frac{1}{2}e^x$
4. $y = \frac{1}{3}e^{-3x}$
5. $y = -e^x$
6. $y = -10e^{x/2}$
7. $y = \frac{1}{2}(e^x + e^{-x})$
8. $y = \frac{1}{2}(e^x - e^{-x})$
9. $y = 2^{-x}$
10. $y = 10^x$

66. THE LOGARITHMIC FUNCTION

We have observed that in considering a function we also concern ourselves with the existence of the inverse of a function. The exponential function $y = a^x$ is a strictly monotone function for all positive values of a except $a = 1$. Consequently, if we restrict the values of a to $a > 0$ and $a \neq 1$, we have a unique value of y for each value of x, and the inverse function exists under these restrictions. However, in order to find an expression for this inverse function it is necessary to solve the equation $x = a^y$ for y, and, again, we must introduce new terminology as we did for the inverse of the trigonometric functions.

*For a medium table see Kaj L. Nielsen, *Logarithmic and Trigonometric Tables* (rev. ed.; New York: Barnes and Noble, 1961).

DEFINITION.* *The inverse of the function* $y = a^x$, $a > 0$, $a \neq 1$, *is given by the expression* $y = \log_a x$ *and is called the logarithm of* x *to the base* a.

The definition may be paraphrased in the sentence, "The logarithm of a number x to the base a is the exponent y of the power to which the base must be raised to equal the number x." The definition states

$$\text{if } x = a^y,\ y = \log a_x$$

Provided a is positive and not equal to one, the logarithm is unique; i.e., every positive number has one and only one logarithm, and every logarithm represents one and only one number. Let us consider some equivalent statements.

ILLUSTRATIONS

Exponential Form	Logarithmic Form
$2^3 = 8$	$\log_2 8 = 3$
$16^{1/2} = 4$	$\log_{16} 4 = \frac{1}{2}$
$3^{-2} = \frac{1}{9}$	$\log_3 \frac{1}{9} = -2$

The graph of $y = \log_a x$ can be obtained for a specified value of a by calculating a table of values for the pairs (x,y) or, since it is the inverse of $y = a^x$, it may be obtained by reflecting the graph of $y = a^x$ on the line $y = x$. The calculation of values for (x,y) may depend upon a logarithmic table, or they may be calculated by assigning values to y and finding x from the equation $x = a^y$. The graphs for $a = 3$ and $a = \frac{1}{2}$ are shown in Figure 50.

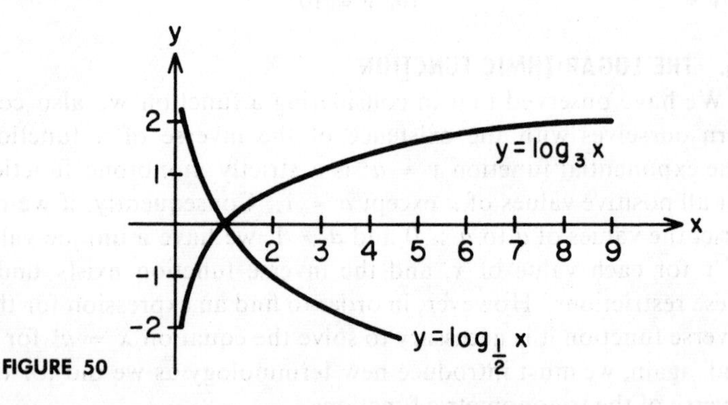

FIGURE 50

*The logarithmic function can also be defined in terms of the area bounded by a curve, the x-axis, and specific values of x. However, this is also the consequence of the process of integration encountered in the calculus and the author prefers that treatment to this concept.

A study of these graphs shows that the domain is the set of positive real numbers and the range is the set of all real numbers. This is true for all values of a for which we have defined the logarithmic function.

There are two special values of a which have wide application,* namely $a = 10$ and $a = e$. When $a = 10$, the logarithms are called **common logarithms** and are indicated by omitting the base and writing $\log x$. If $a = e$, the logarithms are called **natural logarithms**, and we write $\ln x$ for the symbolic notation. Extensive tables for the logarithms of numbers to these two bases have been published. Logarithmic tables and their uses are discussed in the Appendix.

67. PROPERTIES OF LOGARITHMS

Let us recall three of the laws of exponents with which the reader should be familiar from his study of algebra.

$$\text{I. } a^m a^n = a^{m+n} \qquad \text{II. } \frac{a^m}{a^n} = a^{m-n}$$

$$\text{III. } (a^m)^k = a^{km}$$

By making use of these laws of exponents, we can prove three important properties of logarithms.

PROPERTY I. *The logarithm of a product is equal to the sum of the logarithms of its factors.*

$$\log_a MN = \log_a M + \log_a N$$

PROOF

Let $\quad m = \log_a M \quad \text{and} \quad n = \log_a N$

By definition, $\quad M = a^m \quad \text{and} \quad N = a^n$

Then $\quad MN = a^m a^n = a^{m+n}$

By definition, $\quad \log_a MN = m + n = \log_a M + \log_a N$

This property is true for any number of factors; i.e.,

$$\log_a MNR = \log_a M + \log_a N + \log_a R$$

etc. for more factors.

PROPERTY II. *The logarithm of a quotient is equal to the logarithm of the numerator minus the logarithm of the denominator.*

$$\log_a \frac{M}{N} = \log_a M - \log_a N$$

*In recent years the logarithm to the base 2 has had considerable application, but few tables of logarithms to this base have been published.

Prove this by using formula (II) of the laws of exponents given on page 107.

PROPERTY III. *The logarithm of the* k*th power of a number equals* k *times the logarithm of the number.*

$$\log_a N^k = k \log_a N$$

PROOF

Let $n = \log_a N$, then $\qquad N = a^n$

By the laws of exponents, $\qquad N^k = (a^n)^k = a^{kn}$

By definition, $\qquad \log_a N^k = kn = k \log_a N$

ILLUSTRATION. Express the following as the algebraic sum of logarithms.

$$\log_a \frac{y^2\sqrt{x}}{zw^3}$$

SOLUTION. By Property II,

$$\log_a \frac{y^2\sqrt{x}}{zw^3} = \log_a (y^2\sqrt{x}) - \log_a (zw^3)$$

By Property I, $\qquad = \log_a y^2 + \log_a \sqrt{x} - [\log_a z + \log_a w^3]$

By Property III, $\qquad = 2\log_a y + \frac{1}{2}\log_a x - \log_a z - 3\log_a w$

68. CHANGE OF BASE

The definition of a logarithm clearly indicates its dependence upon a base. Since different bases may be used, we need a relationship between logarithms to different bases.

PROPERTY. *The logarithm of a number* N *to the base* b *is equal to the quotient of its logarithm to the base* a *divided by the logarithm of* b *to the base* a; i.e.,

$$\log_b N = \frac{\log_a N}{\log_a b}$$

PROOF. Let $x = \log_b N$, then by definition,

$$N = b^x$$

Take the logarithm of both members to the base a.

$$\log_a N = \log_a b^x = x \log_a b$$

Solve for x. $\qquad x = \frac{\log_a N}{\log_a b}$

Replace x by $\log_b N$. $\qquad \log_b N = \frac{\log_a N}{\log_a b}$

This property is most frequently used to change from common logarithms to natural logarithms and vice versa. Thus,

$$\log_{10} N = \frac{\log_e N}{\log_e 10}$$

and

$$\log_e N = \frac{\log_{10} N}{\log_{10} e}$$

The values of the constants to four decimals are

$$\log_{10} e = 0.4343 \quad \text{and} \quad \log_e 10 = 2.3026$$

which are *reciprocals* of each other. Thus we can write

$$\log_{10} N = 0.4343 \log_e N$$

and

$$\log_e N = 2.3026 \log_{10} N$$

ILLUSTRATIONS

1. Find $\log_9 6$.

SOLUTION

$$\log_9 6 = \frac{\log_{10} 6}{\log_{10} 9} = \frac{0.7782}{0.9542}$$
$$= 0.8156$$

2. Find $\log_e 14$.

SOLUTION

$$\log_e 14 = 2.3026 \log_{10} 14 = 2.3026(1.1461)$$
$$= 2.6390$$

69. EXPONENTIAL AND LOGARITHMIC EQUATIONS

The equations which we are about to discuss can be very difficult to solve; we shall consider only the simple ones.

> **DEFINITIONS.** *If the unknown is involved in an equation as an exponent, the equation is called an* **exponential equation.**
>
> *If the unknown is involved in an expression whose logarithm is indicated in an equation, the equation is called a* **logarithmic equation.**

EXAMPLES

$2^x = 28$ and $3^{x^2} = 5$ are exponential equations.
$\log(x^3 + 2x) = 18$ is a logarithmic equation.

The solution of an exponential equation is usually accomplished by taking the logarithms of both sides and applying the properties of logarithms.

ILLUSTRATIONS

1. Solve $5^{x+3} = 625$.

SOLUTION $$5^{x+3} = 625 = 5^4$$

Taking the logarithms of both sides,

$$\log(5^{x+3}) = \log 5^4$$

or $$(x + 3)\log 5 = 4 \log 5$$

and $$x + 3 = 4 \quad \text{or} \quad x = 1$$

2. Solve $(2^{x+2})(6^x) = 3^{x+1}$.

SOLUTION. Take logarithm of each member.

$$\log[(2^{x+2})(6^x)] = \log(3^{x+1})$$

By properties, $$(x + 2)\log 2 + x \log 6 = (x + 1)\log 3$$

Solving for x, $$x = \frac{\log 3 - 2 \log 2}{\log 2 + \log 6 - \log 3} = \frac{\log \frac{3}{4}}{\log 4}$$

$$= \frac{9.8751 - 10}{0.6021} = -.2074$$

Some logarithmic equations can be solved by applying the properties of logarithms and the definition.

ILLUSTRATION. Solve $\log x + \log(x + 21) = 2$.

SOLUTION

By Property I, $$\log[x(x + 21)] = 2$$

By definition, $$x^2 + 21x = 10^2 = 100$$

Solving for x, $$x^2 + 21x - 100 = 0$$

$$(x + 25)(x - 4) = 0$$

$$x = -25 \text{ and } x = 4$$

Since we have not defined logarithms for negative numbers, $x = -25$ is not a solution.

Check: $$\log 4 + \log 25 = 2$$

$$0.6021 + 1.3979 = 2$$

*70. LOGARITHMIC GRAPH PAPER

There are many applications of logarithms and their properties. Two logarithmic scales placed on rules may be combined to perform multiplication and division; this is the basis for the slide rule.*

If we scale the perpendicular axes of a coordinate system logarithmically, or one axis logarithmically and the other linearly, we form two special types of graph paper. The first, on which

*See Calvin C. Bishop, *Slide Rule: How to Use It* (New York: Barnes and Noble, 1955).

each axis is marked with a logarithmic scale, is called **log-log paper**; and the second, on which one axis has a logarithmic scale and the other a linear scale, is called **semilog paper**. See Figures 51 and 52 on pages 112 and 113.

Now consider the three functions:

linear function, $y = ax + b$

power function, $y = ax^n$

exponential function, $y = ae^{bx}$

The first function was studied in algebra. It was shown that when plotted on regular graph paper the graph was a straight line. Let us apply the properties of logarithms to the other two functions. ($Y = \log y$, $X = \log x$, $A = \log a$, and $B = b \log e$.)

$$\begin{aligned} y &= ax^n \\ \log y &= \log a + n \log x \\ Y &= A + nX \end{aligned} \qquad \Bigg| \qquad \begin{aligned} y &= ae^{bx} \\ \log y &= \log a + bx \log e \\ Y &= A + Bx \end{aligned}$$

It is easily seen that since a, b, and e are constants, their logarithms are known numbers, and thus A and B are constants. The equations, in terms of the new variables X and Y, are therefore linear equations, and their graphs will be straight lines in their respective coordinate systems. The new variables are related to the old variables by logarithms, and thus by the use of appropriately scaled axes we can plot these straight lines directly.

The graph of the equation $y = ax^n$ will be a straight line when plotted on log-log paper.

The graph of the equation $y = ae^{bx}$ will be a straight line when plotted on semilog paper.

ILLUSTRATIONS

1. Obtain a straight line graph of the equation $y = 3x^2$.

SOLUTION. Calculate the table of values.

x	1	2	3	5	10
y	3	12	27	75	300

Plot these values on log-log paper. The graph is shown in Figure 51.

2. Obtain a straight line graph of the equation $y = 2e^{x/2}$.

SOLUTION. By using the exponential table, we calculate the values.

x	0	1	2	3	5
y	2	3.3	5.4	9.0	24.4

Plot these values on semilog paper. The graph is shown in Figure 52.

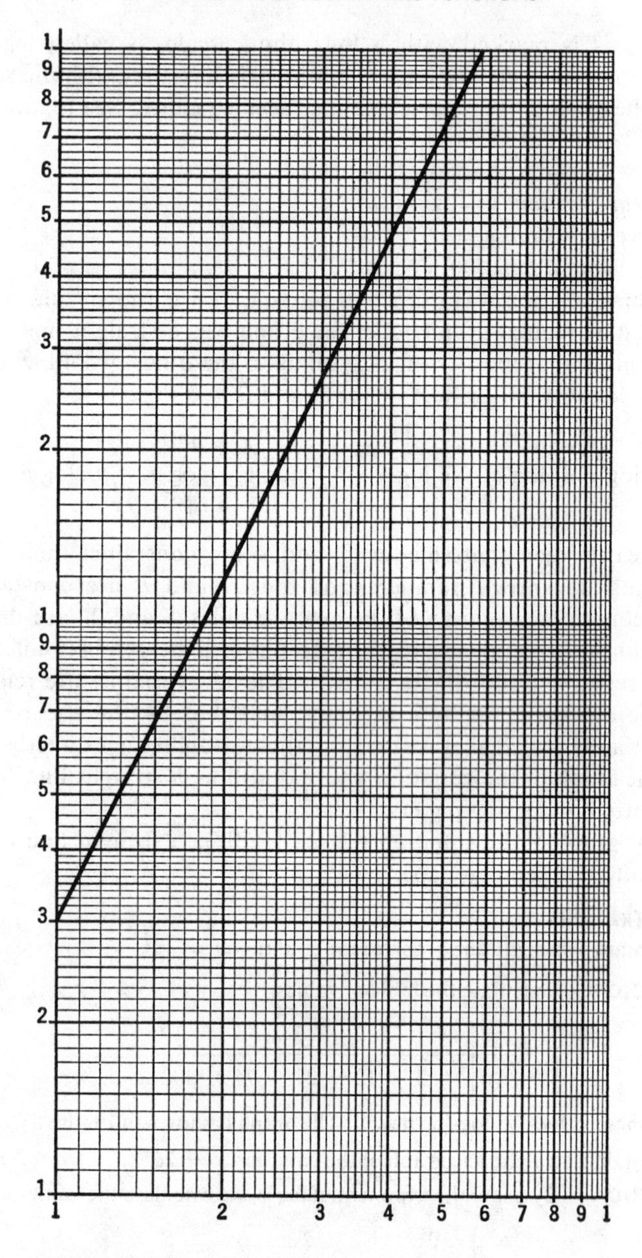

FIGURE 51 (above) and FIGURE 52 (on the opposite page) have been reproduced through the courtesy of Keuffel and Esser Company.

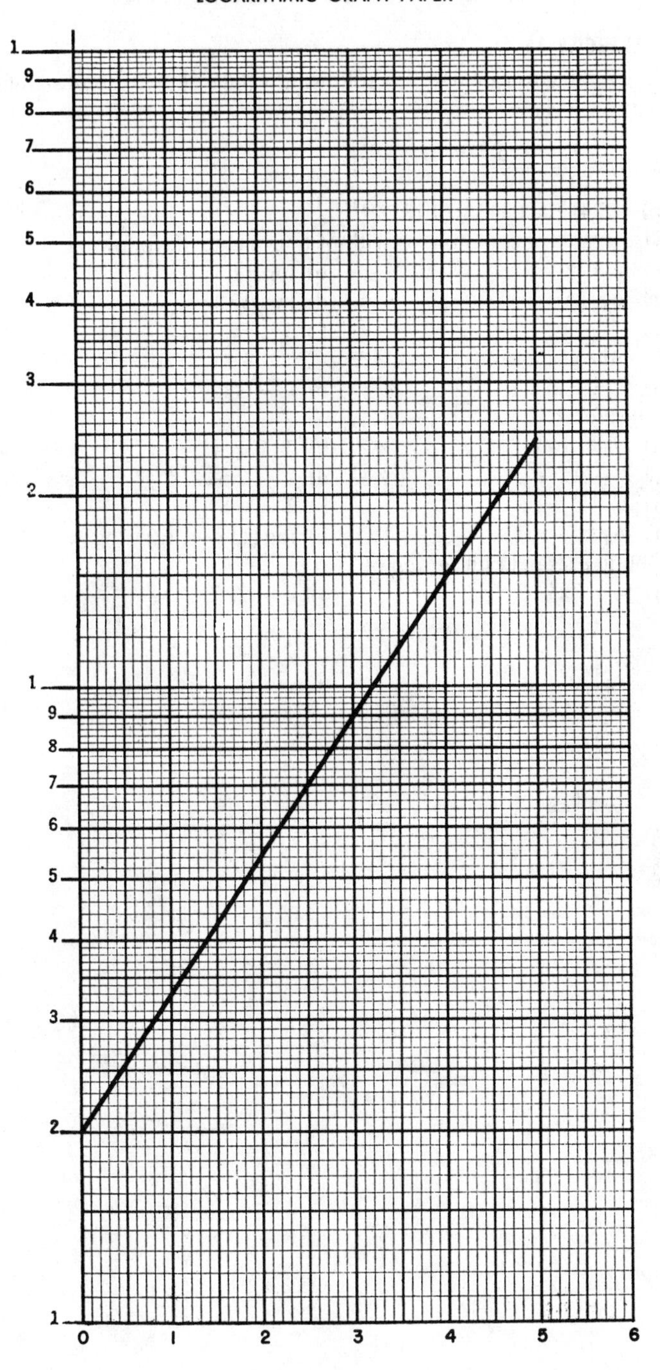
1
9
8
7
6
5
4
3
2
1
9
8
7
6
5
4
3
2
1
0
1
2
3
4
5
6

71. EXERCISE 12

1. Use a table to calculate the following values of N to four significant digits.

   a) $\sqrt{\dfrac{(32.2)\sqrt{5.789}}{(0.0345)(218)}}$

   b) $(18.4)^2(0.008432)(0.1812)^{1/2}$

2. Use a table of common logarithms to evaluate the following.

   a) $\ln 3.18$ b) $\log_3 12$

   c) $\log_8 25$ d) $\log_2 12.18$

   e) $5^{1.2}$ f) $3^{3.3}$

3. Plot the graphs of

   a) $y = 2 \ln x$ b) $y = 3 \log x$

4. Solve for x.

   a) $2^x = 10$ b) $2^{-x} = 4$

   c) $2 = 1.5\, e^{0.1x}$ d) $1 = 2.5\, e^{-1.2x}$

   e) $2^{x^2-3x} = \frac{1}{2}$ f) $3^{x^2} = 81$

   g) $\log(x - 5) + \log 3 = 1$ h) $\log(11x - 12) - \log(x - 1) = 1$

5. Obtain a straight line graph of

   a) $y = 2x^4$ b) $y = 2e^{2x}$

chapter 7

Geometric Applications

72. LINE VALUES

We shall begin this chapter by showing that the trigonometric functions represent the lengths of line segments when the argument is associated with the unit circle. Let us study the geometric configuration shown in Figure 53. The figure shows a portion of the unit circle (radius $= OB = 1$), and an angle θ with an initial side along OA and a terminal side that intersects the circle at B. The construction is also such that $GO \perp OA$, $BC \parallel DA \perp OA$, and $IG \parallel EF \perp OG$; from which it follows that $\angle GIO = \theta$, $EA = BC = FO$, $OC = FB = GH$, and $OG = OB = OA = 1$. By using these geometric properties and the definitions for the trigonometric functions of the angle θ, we have

$$\begin{aligned} \sin\theta &= BC \\ \cos\theta &= OC \\ \tan\theta &= DA \\ \cot\theta &= GI \\ \sec\theta &= OD \\ \csc\theta &= OI \\ \text{versin}\,\theta &= AC \\ \text{covers}\,\theta &= GF \\ \text{ex sec}\,\theta &= BD \end{aligned}$$

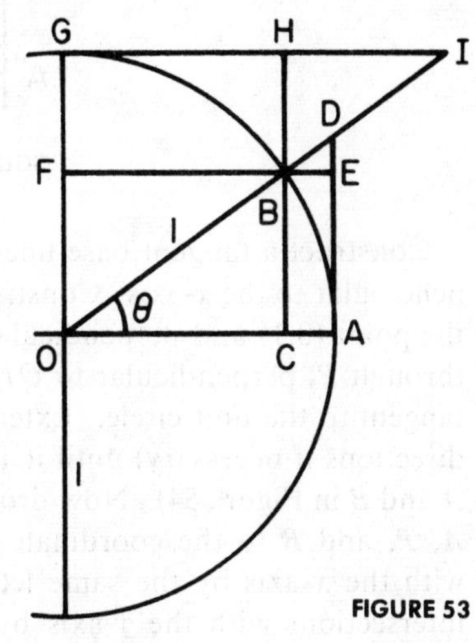

FIGURE 53

Let us illustrate a couple of these developments.

$$\sin\theta = \frac{BC}{OB} = \frac{BC}{1} = BC$$

$$\csc\theta = \csc\angle OIG = \frac{OI}{OG} = \frac{OI}{1} = OI$$

The reader should check the other statements.

Let us now consider a similar configuration by placing an angle in standard position on a rectangular coordinate system and letting the terminal side intersect the unit circle at the point $P(x,y)$. Figure 54 shows an angle whose terminal side falls in the second quadrant; however, the following statements are true for a general angle regardless of the quadrant in which the terminal side falls.*

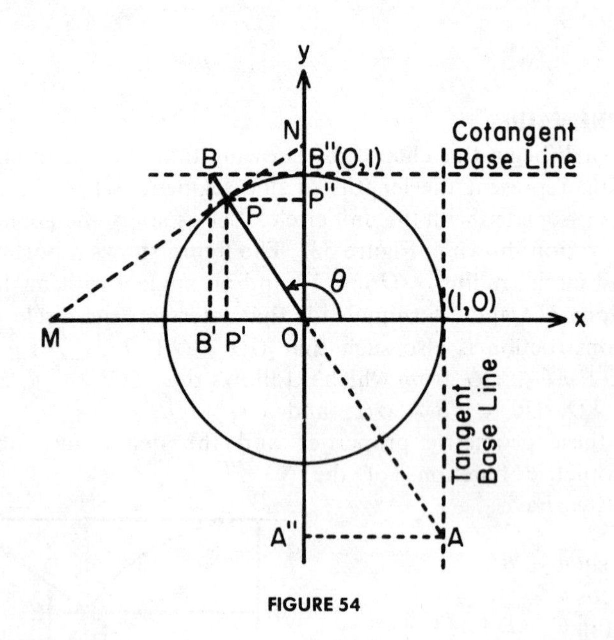

FIGURE 54

Construct a tangent base line through the point (1,0) and perpendicular to the x-axis. Construct a cotangent base line through the point (0,1) and perpendicular to the y-axis. Construct a line through P, perpendicular to OP. Notice that these lines are all tangent to the unit circle. Extend the terminal side of θ (in both directions if necessary) until it intersects these base lines (points A and B in Figure 54). Now drop perpendiculars from the points A, P, and B to the coordinate axes, denoting the intersections with the x-axis by the same letters and a single prime and the intersections with the y-axis by the same letters and a double

*For quadrantal angles, some of the points of intersection cannot be determined because of parallel lines in the construction, but these cases occur only when the corresponding trigonometric functions are undefined.

prime. We can then show that the coordinates of these points are given in terms of the trigonometric functions.

$$P''(0,\sin\theta) \quad A''(0,\tan\theta) \quad M(\sec\theta,0)$$
$$P'(\cos\theta,0) \quad B'(\cot\theta,0) \quad N(0,\csc\theta)$$

Let us indicate the development for B'. The coordinates of B' are $-|OB'|$ and 0; however,

$$\cot\theta = -\cot\angle B'OB = -\frac{|OB'|}{|BB'|} = -\frac{|OB'|}{1} = -|OB'|$$

The reader should check the remaining five statements.

73. SOLUTIONS OF TRIANGLES

Trigonometry had its foundation in the solution of triangles; this topic still forms an important application for the trigonometric functions. A triangle is a plane figure having six parts, namely three angles and three sides. If we are given any three parts of which at least one is a side, we can find the unknown parts. This is called the *solution* of the triangle. We recommend the following procedure.

I. *Draw the figure and label it.*

II. *To find any unknown part, use a formula which involves that part but no other unknown. In order to simplify the arithmetic select, whenever possible, a formula which leads to multiplication instead of division.*

III. *As a check, substitute the results in any one of the formulas which was not used in finding the unknown.*

IV. *Use a systematic and neat form in solving triangles.*

Figures 55 and 56 show the symbols commonly used to designate the parts of the right and oblique triangles.

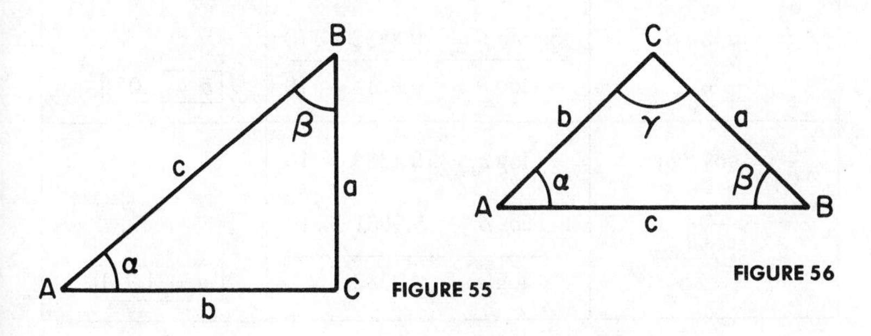

FIGURE 55

FIGURE 56

74. SOLUTIONS OF A RIGHT TRIANGLE

A right triangle is one in which one angle is 90°. Thus one part is already known, and it is only necessary to have given two sides, or an acute angle and a side, in order to solve the triangle. Recall the following formulas.

$$\sin\alpha = \frac{a}{c} = \cos\beta \quad (1) \qquad \cos\alpha = \frac{b}{c} = \sin\beta \quad (2)$$

$$\tan\alpha = \frac{a}{b} = \cot\beta \quad (3) \qquad \cot\alpha = \frac{b}{a} = \tan\beta \quad (4)$$

$$a^2 + b^2 = c^2 \quad (5) \qquad \alpha + \beta = 90^\circ \quad (6)$$

In any of these formulas we can solve for our variable in terms of the others. For example, from (1), we have $a = c \sin\alpha$. Formula (5) can be changed to the forms

$$a = \sqrt{(c-b)(c+b)} \quad (7) \qquad b = \sqrt{(c-a)(c+a)} \quad (8)$$

These formulas are mainly used in checking the solutions and are readily adapted to logarithmic computation. We cannot stress too strongly the use of a computing form. The student should write down the complete form before looking up any logarithms and doing the arithmetic.

ILLUSTRATIONS

1. Solve the right triangle ABC if $\beta = 37°23'$ and $a = 1.375$.

SOLUTION

Data: $a = 1.375$, $\beta = 37°23'$		
Formulas	Computation	
$\alpha = 90° - \beta$	$\alpha = 90° - 37°23' =$	$52°37'$
$\frac{b}{a} = \tan\beta$ or $b = a\tan\beta$	$\log a = 0.1383$ $\log\tan\beta = 9.8832 - 10$ $\log b = 10.0215 - 10$	$b = 1.051$
$\frac{a}{c} = \cos\beta$ or $c = \frac{a}{\cos\beta}$	$\log a = 10.1383 - 10$ $\log\cos\beta = 9.9001 - 10$ $\log c = 0.2382$	$c = 1.731$

Check: $a = \sqrt{(c - b)(c + b)}$		$\log(c - b) = 9.8325 - 10$
$c - b = 0.680$		$\log(c + b) = 0.4443$
$c + b = 2.782$		$\text{sum} = 0.2768$
$\log a = 0.1383$	←*Check*→	$\frac{1}{2}\text{sum} = 0.1384$

2. Solve the right triangle ABC if $b = 21.63$ and $c = 47.18$.

SOLUTION

Data: $b = 21.63$, $c = 47.18$		
Formulas	Computation	
$\cos\alpha = \dfrac{b}{c}$	$\log b = 1.3351$ $\log c = 1.6737$	
	$\log\cos\alpha = 9.6614 - 10$	$\boxed{\alpha = 62^\circ 42'}$
$\sin\alpha = \dfrac{a}{c}$ or $a = c\sin\alpha$	$\log c = 1.6737$ $\log\sin\alpha = 9.9487 - 10$	
	$\log a = 11.6224 - 10$	$\boxed{a = 41.92}$
$\beta = 90^\circ - \alpha$	$\beta = 90^\circ - 62^\circ 42' = \boxed{27^\circ 18'}$	

Check: $b = \sqrt{(c - a)(c + a)}$		$\log(c - a) = 0.7210$
$c - a = 5.26$		$\log(c + a) = 1.9499$
$c + a = 89.10$		$\text{sum} = 2.6709$
$\log b = 1.3351$	←——— ? ———→	$\frac{1}{2}\text{sum} = 1.3354$

The check indicates that the agreement is not too good in spite of the fact that there are no computing errors. The disagreement is the result of using four-place tables which yield an answer correct only to ± 1 in the last place. We must accept this kind of discrepancy. If more extensive tables had been used, we would find that a is closer to 41.93 than to 41.92.

There are many applications of the right triangle; only a few will be discussed here. In solving any problem from descriptive data, we suggest the following steps.

I. *Construct a figure.*

II. *Label the figure, introducing single letters to represent unknown angles or lengths.*

III. *Outline the solution and clearly indicate the formulas which are used.*

IV. *Arrange a good computational scheme and perform the arithmetic.*

Let us consider the configuration of Figure 57. Let O be the point of an observer who is sighting an object C. If the observer is **below** the object, the angle E made by the *line of sight* and the *horizontal* is called the **angle of elevation.** If the observer is **above** the object, the angle D made by the *line of sight* and the *horizontal* is called the **angle of depression.**

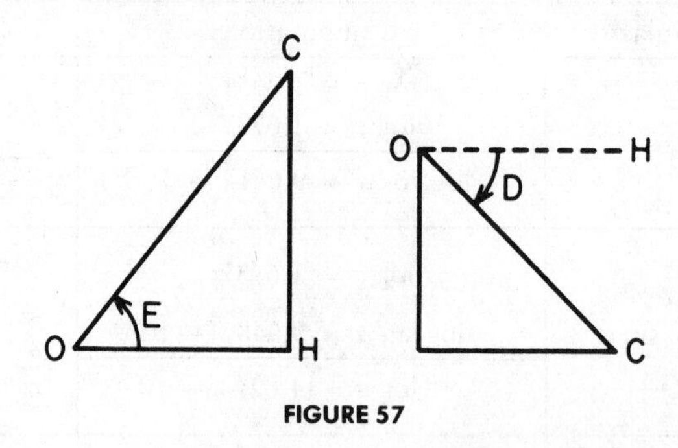

FIGURE 57

ILLUSTRATION. From the top of a lighthouse, 212 feet above a lake, a keeper sights a boat sailing directly towards him. He measures the angle of depression of the boat to be $6° 18'$; a half-hour later he measures it to be $13° 10'$. Find the distance the boat has sailed between the observations, and predict the time when it will arrive at the lighthouse.

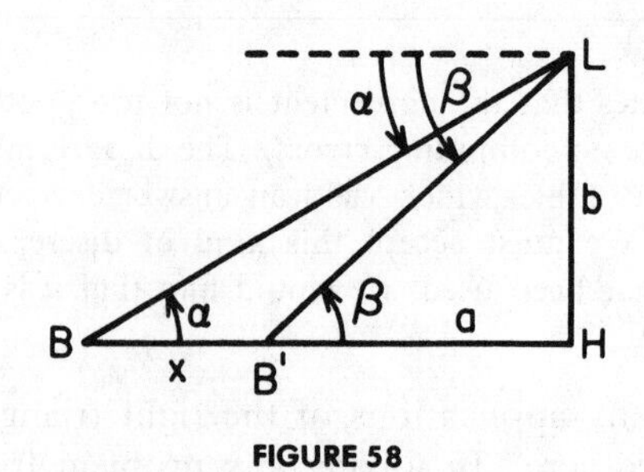

FIGURE 58

SOLUTION. Construct the relative diagram of Figure 58; i.e., not a scale drawing. B and B' represent the two positions of the boat; LH, the lighthouse; α, first angle of depression; β, second angle of depression. Denote the distances as shown in the figure.

In $\triangle B'LH$: $b = 212, \beta = 13^\circ 10'$		
$\dfrac{a}{b} = \cot \beta$ $a = b \cot \beta$	$\log b = 2.3263$ $\log \cot \beta = 0.6309$	
	$\log a = 2.9572$	$a = 906.2$
In $\triangle BLH$: $b = 212, \alpha = 6^\circ 18'$		
$y = x + a$ $= b \cot \alpha$	$\log b = 2.3263$ $\log \cot \alpha = 0.9570$	
	$\log y = 3.2833$	$y = 1920.0$
$x = y - a = 1920.0 - 906.2 = 1013.8$ ft.		
Since it took a half-hour to go $x = 1013.8$ ft., the average speed may be assumed to be $s = 2027.6$ ft./hr., or to four figures, $s = 2028$ ft./hr. The remaining distance to the lighthouse is $a = 906.2$ ft., so the predicted time is $\dfrac{a}{s}$.		
$t = \dfrac{a}{s}$	$\log a = 12.9572 - 10$ $\log s = 3.3071$	
	$\log t = 9.6501 - 10$	$t = .4468$ hr.
The boat should arrive approximately 27 minutes after the second observation.		

A **vector** is a quantity which has magnitude and direction. It is usually represented by an arrow, the length of which represents the magnitude and the head of which indicates its direction. The **sum** of two vectors is called the **resultant** and, in general, is the diagonal of the parallelogram of which the two vectors are the adjacent sides. The two vectors are called the **components** of the resultant. Vectors are used to represent forces, velocities, accelerations, and other physical quantities.

ILLUSTRATION. A force of 467 pounds is acting at an angle of 67°10′ with the horizontal. What are the horizontal and vertical components?

SOLUTION. Construct Figure 59.

OR = vector of the force = c

$b = OA$ = horizontal component

$a = OB$ = vertical component

FIGURE 59

In $\triangle OAR$: $c = 467, \alpha = 67° 10'$		
$a = c \sin \alpha$	$\log c = 2.6693$ $\log \sin \alpha = 9.9646 - 10$	
	$\log a = 2.6339$	$a = 430.4$ lbs.
$b = c \cos \alpha$	$\log c = 2.6693$ $\log \cos \alpha = 9.5889 - 10$	
	$\log b = 2.2582$	$b = 181.2$ lbs.

75. OBLIQUE TRIANGLES

An **oblique triangle** is a triangle in which there is no right triangle. In order to solve the triangle we need to know three parts of which at least one is a side. There are four distinct cases.

CASE I. *Given two angles and a side.*

CASE II. *Given two side and an angle opposite one of them.*

CASE III. *Given two sides and the included angle.*

CASE IV. *Given three sides.*

To solve the triangle we employ certain laws of the trigonometric functions which we shall discuss in the next sections. We shall also find occasion to use the elementary geometric fact

$$\textbf{A} \qquad \alpha + \beta + \gamma = 180°$$

76. LAWS OF COSINES, SINES, AND TANGENTS

We shall first state the laws and focus the resulting formulas; then a proof will be developed for each of the laws.

The Law of Cosines. *In any triangle, the square of any side is equal to the sum of the squares of the other sides minus twice their product times the cosine of their included angle.*

$$\textbf{B} \begin{cases} a^2 = b^2 + c^2 - 2bc \cos \alpha \\ b^2 = a^2 + c^2 - 2ac \cos \beta \\ c^2 = a^2 + b^2 - 2ab \cos \gamma \end{cases}$$

We can solve these formulas for the cosines of the angles.

$$\textbf{C} \begin{cases} \cos \alpha = \dfrac{b^2 + c^2 - a^2}{2bc} \\ \cos \beta = \dfrac{a^2 + c^2 - b^2}{2ac} \\ \cos \gamma = \dfrac{a^2 + b^2 - c^2}{2ab} \end{cases}$$

The Law of Sines. *In any triangle, any two sides are proportional to the sines of the opposite angles.*

$$\textbf{D}\qquad \frac{a}{\sin\alpha}=\frac{b}{\sin\beta}=\frac{c}{\sin\gamma}$$

The Law of Tangents. *In any triangle, the difference of any two sides divided by their sum equals the tangent of one-half the difference of the opposite angles divided by the tangent of one-half their sum.*

$$\textbf{E}\quad\begin{cases}\dfrac{a-b}{a+b}=\dfrac{\tan\frac{1}{2}(\alpha-\beta)}{\tan\frac{1}{2}(\alpha+\beta)}\\[2ex]\dfrac{a-c}{a+c}=\dfrac{\tan\frac{1}{2}(\alpha-\gamma)}{\tan\frac{1}{2}(\alpha+\gamma)}\\[2ex]\dfrac{b-c}{b+c}=\dfrac{\tan\frac{1}{2}(\beta-\gamma)}{\tan\frac{1}{2}(\beta+\gamma)}\end{cases}$$

The difference of any two sides may be taken in any order, but the difference of the corresponding angles must be in the same order; i.e., if we use $c - b$, then we must use $\gamma - \beta$. Always take the order in which the difference of the sides will be positive.

PROOFS. Let a be any side and let β be an acute angle of the triangle. Drop a perpendicular from C to AB, or AB extended. If α is acute, we obtain Figure 60; if α is obtuse, we obtain Figure 61. Let

$$AD = m \qquad BD = n \qquad DC = h$$

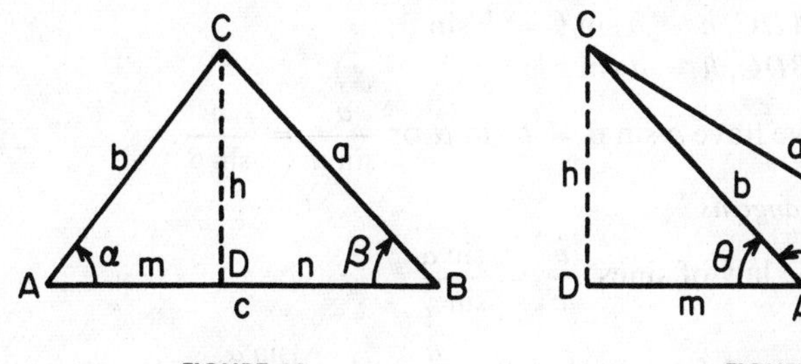

FIGURE 60 FIGURE 61

In either figure we have the following properties, by the Pythagorean theorem.

In rt. $\triangle ADC$, $\quad b^2 = h^2 + m^2 \qquad h^2 = b^2 - m^2 \qquad (1)$

In rt. $\triangle BDC$, $\quad a^2 = h^2 + n^2 \qquad (2)$

Substituting (1) into (2), $\quad a^2 = b^2 - m^2 + n^2 \qquad (3)$

The Law of Cosines

I. If α is acute, $n = c - m$.
In $\triangle ADC$, $m = b \cos \alpha$.
Substituting into (3) we obtain

$$\begin{aligned} a^2 &= b^2 - m^2 + (c - m)^2 \\ &= b^2 - m^2 + c^2 - 2cm + m^2 \\ &= b^2 + c^2 - 2bc \cos \alpha \end{aligned}$$

II. If α is obtuse, $n = c + m$.
In $\triangle ADC$, $m = b \cos \theta$.
Since $\alpha = 180° - \theta$, $\cos \alpha = -\cos \theta$.
$\therefore m = -b \cos \alpha$.
Substituting in (3),

$$\begin{aligned} a^2 &= b^2 - m^2 + (c + m)^2 \\ &= b^2 - m^2 + c^2 + 2cm + m^2 \\ &= b^2 + c^2 - 2bc \cos \alpha \end{aligned}$$

The Law of Sines

I. If α is acute:
In rt. $\triangle ADC$, $h = b \sin \alpha$.
In rt. $\triangle BDC$, $h = a \sin \beta$.
By substitution, $a \sin \beta = b \sin \alpha$.
Dividing both sides by $\sin \alpha \sin \beta$, $\dfrac{a}{\sin \alpha} = \dfrac{b}{\sin \beta}$.

II. If α is obtuse:
In rt. $\triangle ADC$, $h = b \sin \theta = b \sin \alpha$.
In rt. $\triangle BDC$, $h = a \sin \beta$.
Again, we have $a \sin \beta = b \sin \alpha$ or $\dfrac{a}{\sin \alpha} = \dfrac{b}{\sin \beta}$.

The Law of Tangents

From the law of sines, $\dfrac{a}{c} = \dfrac{\sin \alpha}{\sin \gamma}$.

Subtracting 1 from both sides, $\dfrac{a}{c} - 1 = \dfrac{\sin \alpha}{\sin \gamma} - 1$

or $$\frac{a - c}{c} = \frac{\sin \alpha - \sin \gamma}{\sin \gamma} \tag{1}$$

Adding 1 to both sides, $$\frac{a + c}{c} = \frac{\sin \alpha + \sin \gamma}{\sin \gamma} \tag{2}$$

Dividing each side of (1) by the corresponding side of (2),

$$\frac{a - c}{a + c} = \frac{\sin \alpha - \sin \gamma}{\sin \alpha + \sin \gamma}$$

Using (30) and (29) of Section 33,

$$\frac{a - c}{a + c} = \frac{2 \cos \frac{1}{2}(\alpha + \gamma) \sin \frac{1}{2}(\alpha - \gamma)}{2 \sin \frac{1}{2}(\alpha + \gamma) \cos \frac{1}{2}(\alpha - \gamma)} = \frac{\tan \frac{1}{2}(\alpha - \gamma)}{\tan \frac{1}{2}(\alpha + \gamma)}$$

77. THE HALF-ANGLE FORMULAS

A set of formulas known as the half-angle formulas is frequently used in solving a triangle. Let us first define two quantities:

s *denotes one-half the perimeter of the triangle;*
r *denotes the radius of the inscribed circle.*

Then in any triangle ABC, we have

$$s = \tfrac{1}{2}(a + b + c) \qquad r = \sqrt{\frac{(s - a)(s - b)(s - c)}{s}}$$

The first of these two formulas is the direct result of the definition. The second requires a derivation. The easiest proof is the result of the formula for the area of a triangle. Let K denote the area of $\triangle ABC$. See Figure 62.

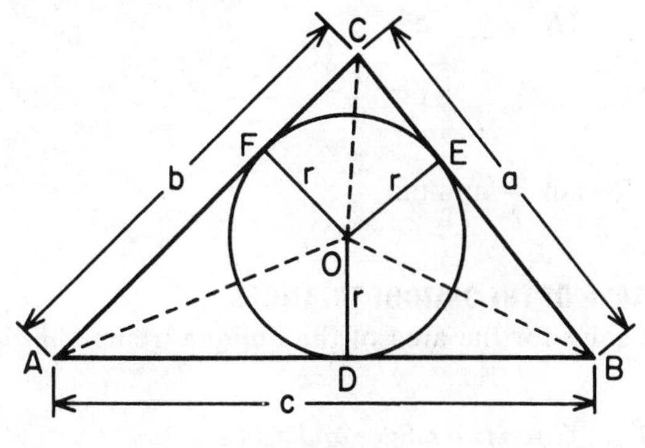

FIGURE 62

Then from elementary geometry we have

$$\begin{aligned} K &= \text{area of } \triangle AOB + \text{area of } \triangle BOC + \text{Area of } \triangle OAC \\ &= \tfrac{1}{2}rc + \tfrac{1}{2}ra + \tfrac{1}{2}rb = r \cdot \tfrac{1}{2}(a + b + c) \\ &= rs \end{aligned}$$

It may be recalled from elementary geometry (we shall also prove it using trigonometry, see Section 122) that

$$K = \sqrt{s(s-a)(s-b)(s-c)}$$

Thus $$rs = \sqrt{s(s-a)(s-b)(s-c)}$$

or dividing by s,

$$r = \sqrt{\frac{s(s-a)(s-b)(s-c)}{s^2}} = \sqrt{\frac{(s-a)(s-b)(s-c)}{s}}$$

Tangents of the Half-Angles. In any triangle ABC,

F $$\tan\frac{\alpha}{2} = \frac{r}{s-a} \qquad \tan\frac{\beta}{2} = \frac{r}{s-b} \qquad \tan\frac{\gamma}{2} = \frac{r}{s-c}$$

PROOF. From $\triangle ADO$ in Figure 62,

$$\tan\frac{\alpha}{2} = \frac{\overline{OD}}{\overline{AD}} = \frac{r}{\overline{AD}}$$

The perimeter of $\triangle ABC$ is

$$\begin{aligned} 2s &= (\overline{AD} + \overline{AF}) + (\overline{BD} + \overline{BE}) + (\overline{CE} + \overline{CF}) \\ &= 2\overline{AD} + 2\overline{BE} + 2\overline{CE} \end{aligned}$$

since $\overline{AD} = \overline{AF}$, $\overline{BD} = \overline{BE}$, and $\overline{CE} = \overline{CF}$. Thus,

$$s = \overline{AD} + (\overline{BE} + \overline{CE}) = \overline{AD} + a$$

or $$\overline{AD} = s - a$$

Therefore $$\tan\frac{\alpha}{2} = \frac{r}{s-a}$$

similarly, for $\tan\frac{\beta}{2}$ and $\tan\frac{\gamma}{2}$.

78. THE AREA OF THE OBLIQUE TRIANGLE

We can solve for the area of the oblique triangle in the following cases.

CASE I. *Given two angles and a side.*
CASE III. *Given two sides and the included angle.*
CASE IV. *Given three sides.*

To find the area for Case II of Section 75, it is necessary to obtain one of the other elements of the triangle first, and then use the formulas we shall now derive.

The area in terms of two sides and the included angle is given by

$$K = \tfrac{1}{2}bc \sin \alpha \qquad K = \tfrac{1}{2}ac \sin \beta \qquad K = \tfrac{1}{2}ab \sin \gamma \tag{1}$$

PROOF. Construct Figure 63.

Let b and c be the given sides;
α, the included angle;
h, the altitude.

In $\triangle ACD$, $h = b \sin \alpha$.
Then $K = \frac{1}{2}hc = \frac{1}{2}bc \sin \alpha$ and similarly, for the other formulas.

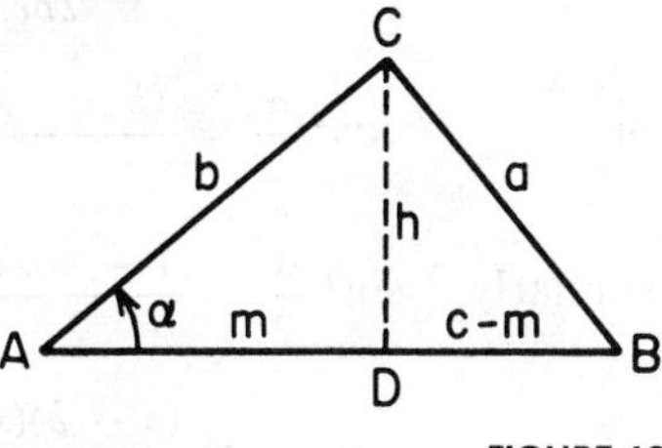

FIGURE 63

The area in terms of one side and the angles is given by

$$K = \frac{a^2 \sin \beta \sin \gamma}{2 \sin \alpha} = \frac{b^2 \sin \alpha \sin \gamma}{2 \sin \beta} = \frac{c^2 \sin \alpha \sin \beta}{2 \sin \gamma} \tag{2}$$

PROOF. From the law of sines, we have

$$\frac{c}{\sin \gamma} = \frac{b}{\sin \beta} \quad \text{or} \quad c = \frac{b \sin \gamma}{\sin \beta}$$

Substituting for c into (1), we get

$$K = \tfrac{1}{2}bc \sin \alpha = \tfrac{1}{2}b\left(\frac{b \sin \gamma}{\sin \beta}\right)\sin \alpha = \frac{b^2 \sin \alpha \sin \gamma}{2 \sin \beta}$$

similarly, for the other formulas.

The area in terms of the three sides is given by

$$K = \sqrt{s(s-a)(s-b)(s-c)} \tag{3}$$

where $$s = \tfrac{1}{2}(a + b + c)$$

PROOF. By Section 31,

$$\sin \alpha = 2 \sin \frac{\alpha}{2} \cos \frac{\alpha}{2}$$

so that in (1) we have

$$K = bc \sin \frac{\alpha}{2} \cos \frac{\alpha}{2}$$

By Section 31, $$2 \cos^2 \tfrac{1}{2}\alpha = 1 + \cos \alpha$$

By Section 76, $$\cos \alpha = \frac{b^2 + c^2 - a^2}{2bc}$$

Thus

$$2\cos^2\frac{\alpha}{2} = 1 + \frac{b^2 + c^2 - a^2}{2bc} = \frac{(b + c + a)(b + c - a)}{2bc}$$

or $$\cos^2\frac{\alpha}{2} = \frac{(b + c + a)(b + c - a)}{4bc} = \frac{s(s - a)}{bc}$$

similarly, $$\sin^2\frac{\alpha}{2} = \frac{1 - \cos\alpha}{2} = \frac{(a - b + c)(a + b - c)}{4bc}$$

$$= \frac{(s - b)(s - c)}{bc}$$

Substituting into the value for K, we have

$$K = bc\sqrt{\frac{s(s - a)}{bc}}\sqrt{\frac{(s - b)(s - c)}{bc}}$$

$$= \sqrt{s(s - a)(s - b)(s - c)}$$

79. THE AMBIGUOUS CASE

If two sides and an angle opposite one of them are given (Case II), there may be two triangles, one triangle, or no triangle. Consequently, Case II is called the ambiguous case. Let the given parts be a, b, and α; then the possibilities are shown in Figures 64–69.

These conditions, which are shown in the figures, may be summarized as follows.

$$\textbf{If } \alpha < 90^\circ \begin{cases} a < b\sin\alpha, \textbf{ no} \text{ solution} \\ a = b\sin\alpha, \textbf{ one} \text{ solution}, \beta = 90^\circ \\ a > b\sin\alpha, a < b, \textbf{ two} \text{ solutions} \\ a \geq b, \textbf{ one} \text{ solution} \end{cases}$$

$$\textbf{If } \alpha \geq 90^\circ \begin{cases} a > b, \textbf{ one} \text{ solution} \\ a \leq b, \textbf{ no} \text{ solution} \end{cases}$$

80. SOLUTIONS OF OBLIQUE TRIANGLES

The laws and formulas which we have been discussing in the previous sections are used to solve oblique triangles. We now submit a summary, regarding the cases cited in Section 75, which should be committed to memory.

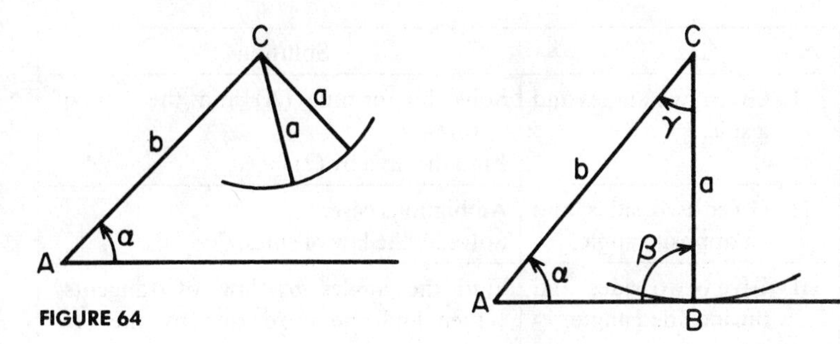

FIGURE 64

FIGURE 65

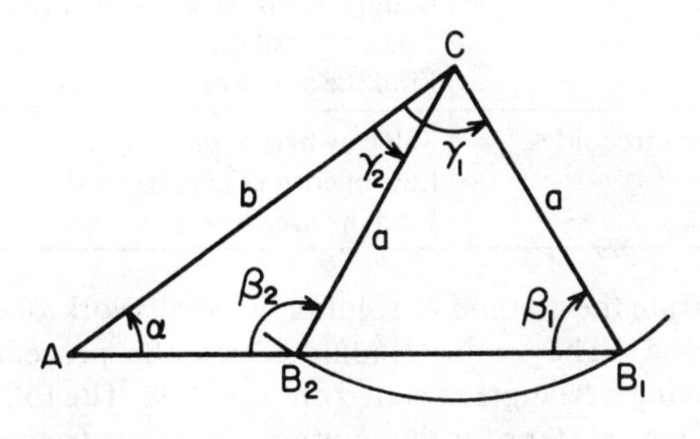

FIGURE 66

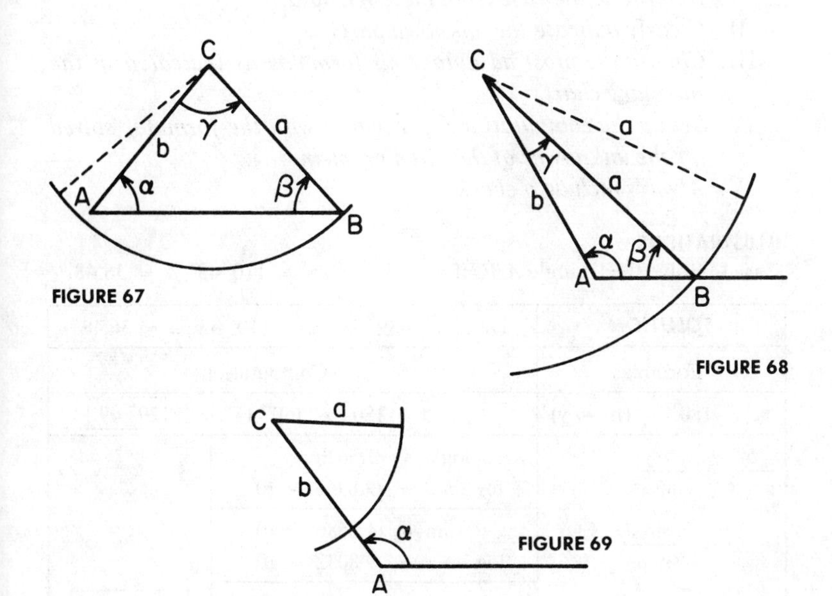

FIGURE 67

FIGURE 68

FIGURE 69

Cases	Solution
I. Given two angles and a side.	Solve by formula (A) and the law of sines. Find the area by (2).
II. Given two sides and an opposite angle.	Ambiguous case. Solve by the law of sines.
III. Given two sides and the included angle.	Find the angles by law of tangents; then find the third side by the law of sines. If only third side is required, use the law of cosines. Find the area by (1).
IV. Given three sides.	Solve by half-angle formulas. Law of cosines may be used. Find the area by (3).

To illustrate the method of solution we shall work one problem for each case. The student should review the procedure to be used in solving a triangle as stated on page 119. The following are additional logical steps for the solution of oblique triangles.

I. *Determine the case from the given data.*
II. *Clearly indicate the unknown parts.*
III. *Choose the most advantageous formulas as indicated in the summary chart.*
IV. *Set up the computational procedure with the formulas solved for the unknowns of the given problem.*
V. *Always include a check.*

ILLUSTRATIONS

Case I. Solve the triangle ABC if $\alpha = 39°28'$, $\gamma = 110°43'$, $a = 36.48$.

SOLUTION	Data: $\alpha = 39°28'$, $\gamma = 110°43'$, $a = 36.48$.	
Formulas	Computation	
$\beta = 180° - (\alpha + \gamma)$	$\beta = 180° - 150°11' =$	$\boxed{29°49'}$
$\dfrac{b}{\sin\beta} = \dfrac{a}{\sin\alpha}$	$\log a = 1.5620$ $\log \sin\beta = 9.6966 - 10$	
$b = \dfrac{a \sin\beta}{\sin\alpha}$	sum $= 11.2586 - 10$ $\log\sin\alpha = 9.8032 - 10$	
	$\log b = 1.4554$	$\boxed{b = 28.54}$

$\frac{c}{\sin\gamma} = \frac{a}{\sin\alpha}$	$\log a = 1.5620$ $\log\sin\gamma = 9.9710 - 10$	
$c = \frac{a\sin\gamma}{\sin\alpha}$	sum $= 11.5330 - 10$ $\log\sin\alpha = 9.8032 - 10$	
	$\log c = 1.7298$	$c = 53.68$
$K = \frac{a^2\sin\beta\sin\gamma}{2\sin\alpha}$	$2\log a = 3.1240$ $\log\sin\beta = 9.6966 - 10$ $\log\sin\gamma = 9.9710 - 10$	
	sum $= 12.7916 - 10$	
	$\log 2 = 0.3010$ $\log\sin\alpha = 9.8032 - 10$	
	sum $= 10.1042 - 10$	
	$\log K = 2.6874$	$K = 486.9$

CASE II. Solve the triangle ABC if $c = .513$, $b = .753$, $\gamma = 36^\circ 50'$.

SOLUTION	Data: $c = .513$, $b = .753$, $\gamma = 36^\circ 50'$.	
Formulas	Computation	
$\sin\beta = \frac{b\sin\gamma}{c}$	$\log b = 9.8768 - 10$ $\log\sin\gamma = 9.7778 - 10$	
	sum $= 19.6546 - 20$ $\log c = 9.7101 - 10$	$\beta_1 = 61^\circ 39'$
$\beta_2 = 180^\circ - \beta_1$	$\log\sin\beta = 9.9445 - 10$	$\beta_2 = 118^\circ 21'$
$\alpha = 180^\circ - (\gamma + \beta)$	$\alpha_1 = 180^\circ - 98^\circ 29' =$ $\alpha_2 = 180^\circ - 155^\circ 11' =$	$81^\circ 31'$ $24^\circ 49'$
$a_1 = \frac{c\sin\alpha_1}{\sin\gamma}$	$\log c = 9.7101 - 10$ $\log\sin\alpha_1 = 9.9952 - 10$	
	sum $= 19.7053 - 20$ $\log\sin\gamma = 9.7778 - 10$	
	$\log a_1 = 9.9275 - 10$	$a_1 = .8462$
$a_2 = \frac{c\sin\alpha_2}{\sin\gamma}$	$\log c = 9.7101 - 10$ $\log\sin\alpha_2 = 9.6229 - 10$	
	sum $= 19.3330 - 20$ $\log\sin\gamma = 9.7778 - 10$	
	$\log a_2 = 9.5552 - 10$	$a_2 = .3591$

$K_1 = \frac{1}{2} bc \sin \alpha$	$\log b = 9.8768 - 10$ $\log c = 9.7101 - 10$ $\text{colog } 2 = 9.6990 - 10$	
$S = \frac{1}{2} bc$	$\log S = 9.2859 - 10$ $\log \sin \alpha_1 = 9.9952 - 10$	
	$\log K_1 = 9.2811 - 10$	$K_1 = .1910$
$K_2 = \frac{1}{2} bc \sin \alpha_2$	$\log S = 9.2859 - 10$ $\log \sin \alpha_2 = 9.6229 - 10$	
	$\log K_2 = 8.9088 - 10$	$K_2 = .08106$

CASE III. Solve the triangle ABC if $b = .356$, $c = .938$, $\alpha = 62° 40'$.

SOLUTION. Note first that $\gamma = \frac{1}{2}(\gamma + \beta) + \frac{1}{2}(\gamma - \beta)$ and

$$\beta = \tfrac{1}{2}(\gamma + \beta) - \tfrac{1}{2}(\gamma - \beta).$$

Data: b = .356, $c = .938$, $\alpha = 62° 40'$.		
Formulas	Computation	
$\frac{1}{2}(\gamma + \beta) = \frac{1}{2}(180° - \alpha)$	$\frac{1}{2}(\gamma + \beta) = \frac{1}{2}(180° - 62°40') = 58°40'$	
$\tan \frac{1}{2}(\gamma - \beta) = \frac{(c - \beta) \tan \frac{1}{2}(\gamma + \beta)}{c + b}$	$\log (c - \beta) = 9.7649 - 10$ $\log \tan \frac{1}{2}(\gamma + \beta) = 0.2155$ $\text{colog } (c + b) = 9.8881 - 10$	
$c = 0.938$ $b = 0.356$ $c - b = 0.582$ $c + b = 1.294$	$\log \tan \frac{1}{2}(\gamma - \beta) = 9.8685 - 10$ $\frac{1}{2}(\gamma - \beta) = 36° 27'$	$\gamma = 95° 07'$ $\beta = 22° 13'$
$a = \frac{b \sin \alpha}{\sin \beta}$	$\log b = 9.5514 - 10$ $\log \sin \alpha = 9.9486 - 10$ $\text{colog} \sin \beta = 0.4224$	
	$\log a = 9.9224 - 10$	$a = 0.8364$

Check:

$\log a = 9.9224 - 10$	$\log b = 9.5514 - 10$	$\log c = 9.9722 - 10$
$\log \sin \alpha = 9.9486 - 10$	$\log \sin \beta = 9.5776 - 10$	$\log \sin \gamma = 9.9982 - 10$
Diff. $= 9.9738 - 10$	Diff. $= 9.9738 - 10$	Diff. $= 9.9740 - 10$

The finding of the area is left to the student.

CASE IV. Solve the triangle ABC if $a = 163.6, b = 397.5, c = 253.7.$

SOLUTION

Data: $a = 163.6, b = 397.5, c = 253.7.$		
Formulas	**Computation**	
$s = \frac{1}{2}(a + b + c)$	$a = 163.6$ $b = 397.5$ $c = 253.7$ $2s = 814.8$ $s = 407.4$ ←*check*→	$s - a = 243.8$ $s - b = 9.9$ $s - c = 153.7$ $s = 407.4$
$r^2 = \frac{(s-a)(s-b)(s-c)}{s}$	$\log(s - a) = 2.3870$ $\log(s - b) = 0.9956$ $\log(s - c) = 2.1867$ $\log N = 5.5693$ $\log s = 2.6100$ $2 \log r = 2.9593$ $\log r = 1.4796$	
$\tan \frac{1}{2}\alpha = \frac{r}{s-a}$	$\log r = 11.4796 - 10$ $\log(s - a) = 2.3870$ $\log \tan \frac{1}{2}\alpha = 9.0926 - 10$ $\frac{1}{2}\alpha = 7°3'$	$\alpha = 14°6'$
$\tan \frac{1}{2}\beta = \frac{r}{s-b}$	$\log r = 1.4796$ $\log(s - b) = 0.9956$ $\log \tan \frac{1}{2}\beta = 0.4840$ $\frac{1}{2}\beta = 71°50'$	$\beta = 143°40'$
$\tan \frac{1}{2}\gamma = \frac{r}{s-c}$	$\log r = 11.4796 - 10$ $\log(s - c) = 2.1867$ $\log \tan \frac{1}{2}\gamma = 9.2929 - 10$ $\frac{1}{2}\gamma = 11°6'$	$\gamma = 22°12'$
Check: $\alpha + \beta + \gamma = 180°$		*sum* $= 179°58'$
$K = \sqrt{s(s-a)(s-b)(s-c)}$ $N = (s-a)(s-b)(s-c)$	$\log s = 2.6100$ $\log N = 5.5693$ $2 \log K = 8.1793$ $\log K = 4.0896$	$K = 12290$

The use of four-place tables does not permit any greater accuracy than the check above indicates.

81. SOME APPLICATIONS

In this section we will consider a few applications.

Vectors. The resultant of two vectors is generally the diagonal of a parallelogram of which the two vectors form adjoining sides (see Section 74).

ILLUSTRATION. Two forces of 50 lbs. and 30 lbs. have an included angle of 60°. Find the magnitude and direction of their resultant.

SOLUTION. Construct a parallelogram and label it as in Figure 70. Since $\overline{AD}$ is parallel to $\overline{BC}$, we have

$$\angle ABC = \beta = 180^\circ - 60^\circ = 120^\circ$$

By the law of cosines

$$\begin{aligned} x^2 &= c^2 + a^2 - 2ac\cos\beta \\ &= 2500 + 900 - 2(50)(30)(-\tfrac{1}{2}) \\ &= 2500 + 900 + 1500 = 4900 \end{aligned}$$

$$\boxed{x = 70 \text{ lbs.}}$$

$$\cos\alpha = \frac{x^2 + c^2 - a^2}{2xc} = \frac{4900 + 2500 - 900}{2(70)(50)} = \frac{13}{14} = .9286$$

$$\boxed{\alpha = 21^\circ 47'}$$

Navigation. The subject of determining the positions and courses of ships at sea or airplanes in flight is called **navigation**. As a rough check, or if the distances are not too great, we may consider the surface of the earth as flat and apply the theory of **plane sailing**. In this case the problems are reduced to the solution of triangles. As in any subject matter, we must first become familiar with the terminology.

The direction from a given point O to a point B in the horizontal plane is described by *the angle through which ON* (the north direction) *must be rotated clockwise in order to coincide with OB*; this angle is called the **azimuth** of B from O. See Figure 71.

The **heading** of an airplane is the direction in which the *nose of the airplane is pointing*. The point on the ground directly (vertically) under the airplane describes a path called the **track**, and its *direction* is called the **course** of the airplane. The airplane is in motion with respect to the air, which in turn may be in motion because of wind. The speed of the airplane with respect to the air is called the **airspeed** and is in the direction of the heading. Since the wind has a speed in a certain direction, there is a resultant of the two speeds with respect to the ground; this is called the

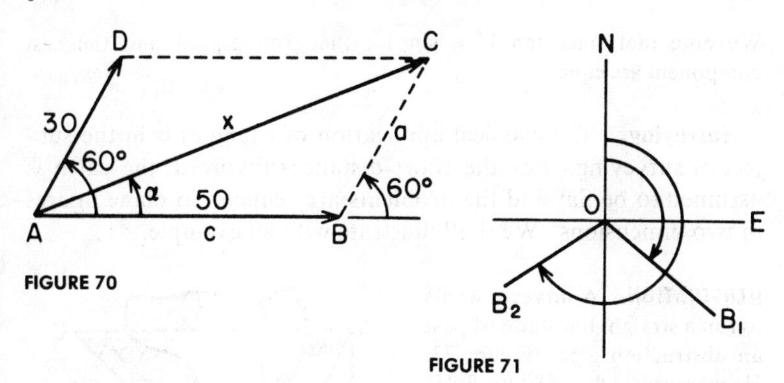

FIGURE 70

FIGURE 71

groundspeed. If we represent these speeds by vectors and there is an angle between them, this angle is called the **drift angle** and depends upon the *heading* as well as the *airspeed*. This is the familiar concept that the wind blows the airplane off its heading by an amount equal to the drift angle. This is also true of sailboats in water.

ILLUSTRATION. An airplane is heading 5° north of directly east with an airspeed of exactly 400 miles per hour. There is a wind blowing directly from the north with a speed of 41.85 miles per hour. Find the groundspeed, the drift angle, and the course of the airplane.

SOLUTION. Construct Figure 72.

$OP \equiv$ airspeed $= 400$

$OW \equiv$ wind velocity $= 41.85 = PG$

$OG \equiv$ resultant groundspeed $= x$

$\delta \equiv$ drift angle

In rt. $\triangle OEP$,

$EP = 400 \sin 5^\circ$

$= 400(.0872) = 34.88$

$OE = 400 \cos 5^\circ$

$= 400(.9962) = 398.48$

FIGURE 72

In rt. $\triangle OEG$,

$EG = 41.85 - EP = 41.85 - 34.88 = 6.97$

$$\tan(\delta - 5^\circ) = \frac{EG}{\text{OE}} = \frac{6.97}{398.48} = .0175$$

$$\delta - 5^\circ = 1^\circ \qquad \delta = 6^\circ$$

$$x = \frac{EG}{\sin(\delta - 5^\circ)} = \frac{6.97}{.0175} = 398.3$$

Groundspeed $= 398.3$ mi./hr.; drift angle $= \delta = 6^\circ$.
Course $= \delta - 5^\circ = 1^\circ$ south of east.

We note that since $\tan 1^\circ = \sin 1^\circ$, the groundspeed and the east component are equal.

Surveying. The classical application of triangles is in the subject of surveying. For the short distances involved, the earth is assumed to be flat and the problems are reduced to plane figures in two dimensions. We shall illustrate with an example.

ILLUSTRATION. A surveyor wants to run a straight line due east past an obstruction. See Figure 73. He measures AB = 780 ft., S 25° 20′ E (25° 20′ east of south), and then runs BC in the direction N 45° 50′ E. Find the length of BC so that C is due east of A, and find the length of AC.

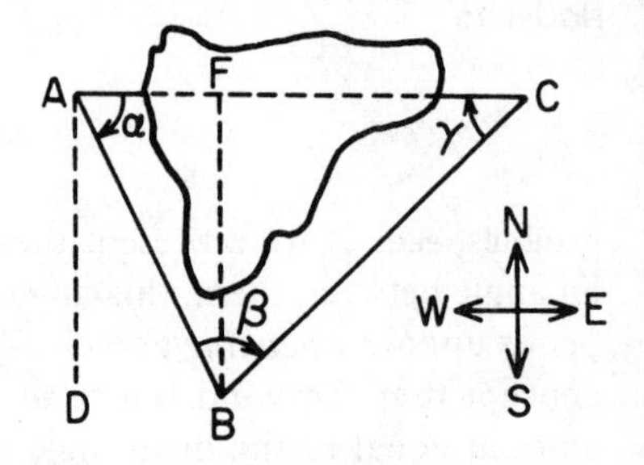

FIGURE 73

SOLUTION

Data: $\angle DAB = 25^\circ 20'$, AB = 780 ft., $\angle FBC = 45^\circ 50'$.		
$\alpha = 90^\circ - \angle DAB = 90^\circ - 25^\circ 20' = 64^\circ 40'$ $\gamma = 90^\circ - \angle FBC = 90^\circ - 45^\circ 50' = 44^\circ 10'$ $\beta = 180^\circ - (\alpha + \gamma) = 180^\circ - 108^\circ 50' = 71^\circ 10'$		
In $\triangle ABC$,	$\log AB = 2.8921$ $\log \sin \alpha = 9.9561 - 10$	
$BC = \dfrac{AB \sin \alpha}{\sin \gamma}$	sum = $12.8482 - 10$ $\log \sin \gamma = 9.8431 - 10$	
	$\log BC = 3.0051$	BC = 1012 ft.
	$\log AB = 2.8921$ $\log \sin \beta = 9.9761 - 10$	
$AC = \dfrac{AB \sin \beta}{\sin \gamma}$	sum = $12.8682 - 10$ $\log \sin \gamma = 9.8431 - 10$	
	$\log AC = 3.0251$	AC = 1060 ft.

82. AREA OF THE SEGMENT OF A CIRCLE

The area of a segment of a circle is the area between the arc and the chord formed by the given angle. This area can be found by subtracting the area of the triangle formed by the two radii and the chord from the area of the sector. See Figure 74. If the angle is measured in radians, the area of the sector (see Section 7) is

$$\tfrac{1}{2}r^2\theta$$

and the area of the triangle (see Section 78) is

$$\tfrac{1}{2}r^2 \sin \theta$$

Thus the area of a segment is

$$\tfrac{1}{2}r^2 (\theta - \sin \theta)$$

(θ measured in radians)

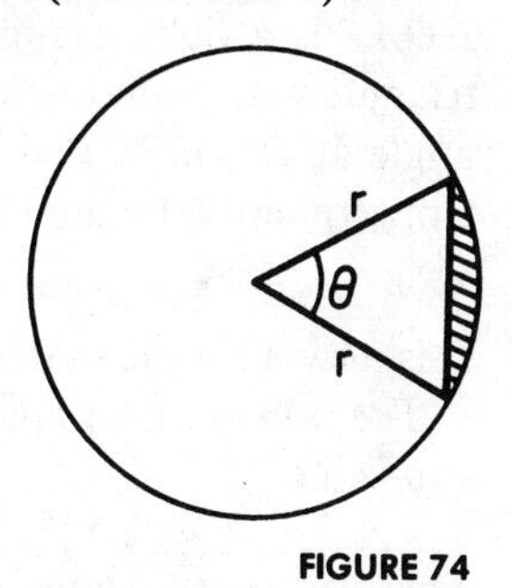

FIGURE 74

☆83. DIRECTION COSINES OF A LINE

We shall now discuss an application of trigonometry in three-dimensional geometry. First let us assume the notation associated with a rectangular coordinate system in three dimensions.

a) The coordinates of a point are denoted by $P(x,y,z)$.

b) The distance between two points is given by

$$d = \sqrt{(x_2 - x_1)^2 + (y_2 - y_1)^2 + (z_2 - z_1)^2}$$

c) The distance from a point to the origin is given by

$$\rho = \sqrt{x^2 + y^2 + z^2}$$

Consider the directed line OP, and let the angles formed by this line and the positive x-, y-, and z-axes be denoted by α, β, and γ. See Figure 75. These angles are called the **direction angles** of the

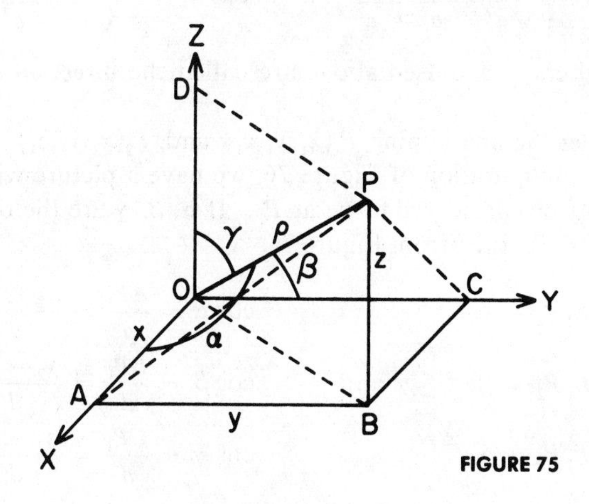

FIGURE 75

line, and the cosines of these angles are called the **direction cosines** of the line. A close inspection of the figure reveals that $\triangle OAP$ is a right triangle with right angle at A; $\triangle OCP$ is a right triangle with right angle at C; $\triangle ODP$ is a right triangle with right angle at D; and $OP = \rho$ is the hypotenuse of each right triangle. Consequently, by definition, we have

$$x = \rho \cos \alpha \qquad y = \rho \cos \beta \qquad z = \rho \cos \gamma \tag{1}$$

It is now a simple matter to prove that

The sum of the squares of the direction cosines of any line equals one, i.e.

$$\cos^2 \alpha + \cos^2 \beta + \cos^2 \gamma = 1 \tag{2}$$

The direction cosines of a line which do not pass through the origin are defined as *those of the parallel line through the origin.*

In practice, three numbers which are proportional to the direction cosines are frequently used. Let a, b, c be three numbers such that

$$\cos \alpha = ka \qquad \cos \beta = kb \qquad \cos \gamma = kc$$

Then by formula (2) above, we have

$$k^2a^2 + k^2b^2 + k^2c^2 = 1 = k^2(a^2 + b^2 + c^2)$$

so that
$$k = \frac{1}{\pm\sqrt{a^2 + b^2 + c^2}}$$

We then have
$$\cos \alpha = \frac{a}{\pm\sqrt{a^2 + b^2 + c^2}}$$

$$\cos \beta = \frac{b}{\pm\sqrt{a^2 + b^2 + c^2}} \qquad \cos \gamma = \frac{c}{\pm\sqrt{a^2 + b^2 + c^2}}$$

The numbers a, b, c used above are called the **direction numbers** of a line.

Consider the line joining $P_1(x_1,y_1,z_1)$ and $P_2(x_2,y_2,z_2)$. If we draw the configuration of Figure 76, we have a picture where the origin may be considered to be at P_1. If α, β, γ are the direction angles of P_1P_2, then from Figure 76,

$$\angle AP_1P_2 = \alpha \qquad\qquad \cos \alpha = \frac{AP_1}{d} = \frac{x_2 - x_1}{d}$$

$$\angle BP_1P_2 = \beta \qquad \text{and} \qquad \cos \beta = \frac{BP_1}{d} = \frac{y_2 - y_1}{d}$$

$$\angle CP_1P_2 = \gamma \qquad\qquad \cos \gamma = \frac{CP_1}{d} = \frac{z_2 - z_1}{d}$$

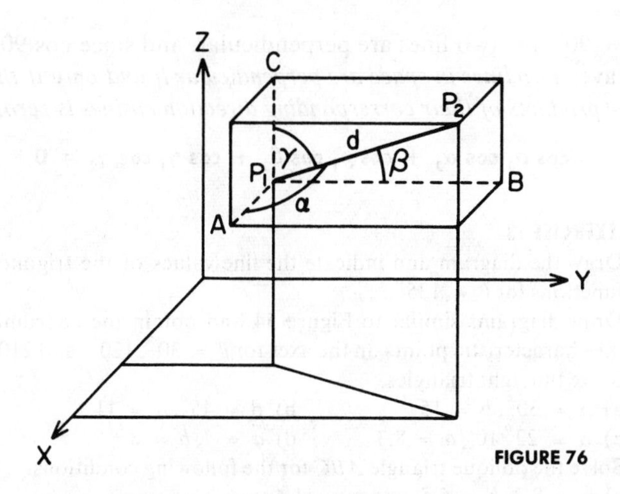

FIGURE 76

Now since $d = \sqrt{(x_2 - x_1)^2 + (y_2 - y_1)^2 + (z_2 - z_1)^2}$, we have the direction cosines of the line joining P_1 and P_2 proportional to $x_2 - x_1$, $y_2 - y_1$, and $z_2 - z_1$, and these may be taken as the direction numbers of P_1P_2.

The angle between two lines in space can be given in terms of the direction cosines. Consider the lines OP_1 and OP_2 shown in Figure 77, with $\angle P_1OP_2 = \theta$. By the law of cosines (see Section 76), we have

$$\cos\theta = \frac{\rho_1{}^2 + \rho_2{}^2 - d^2}{2\rho_1\rho_2}$$

FIGURE 77

Now

$$\begin{aligned}
\rho_1{}^2 &= x_1{}^2 + y_1{}^2 + z_1{}^2 \\
\rho_2{}^2 &= x_2{}^2 + y_2{}^2 + z_2{}^2 \\
d^2 &= (x_2 - x_1)^2 + (y_2 - y_1)^2 + (z_2 - z_1)^2 \\
\therefore \quad \rho_1{}^2 + \rho_2{}^2 - d^2 &= 2(x_1x_2 + y_1y_2 + z_1z_2)
\end{aligned}$$

Furthermore by (1) above,

$$x_1 = \rho_1\cos\alpha_1 \qquad x_2 = \rho_2\cos\alpha_2 \qquad y_1 = \rho_1\cos\beta_1 \text{ etc.}$$

so that after simplifying we have

$$\mathbf{\cos\theta = \cos\alpha_1\cos\alpha_2 + \cos\beta_1\cos\beta_2 + \cos\gamma_1\cos\gamma_2} \qquad (1)$$

If $\theta = 90°$, the two lines are perpendicular, and since $\cos 90° = 0$ we have: *Two lines in space are perpendicular if and only if the sum of the products of their corresponding direction cosines is zero,* i.e.

$$\cos \alpha_1 \cos \alpha_2 + \cos \beta_1 \cos \beta_2 + \cos \gamma_1 \cos \gamma_2 = 0 \qquad (2)$$

84. EXERCISE 13

1. Draw the diagram and indicate the line values of the trigonometric functions for $\theta = 135°$.
2. Draw diagrams similar to Figure 54 and obtain the coordinates of the characteristic points on the axes for $\theta = 30°, 120°$, and $210°$.
3. Solve the right triangles.
   a) $\alpha = 30°, b = 15$ b) $\beta = 45°, a = 11$
   c) $\alpha = 22° 10', a = 8.3$ d) $a = 7, b = 8$
4. Solve the oblique triangle ABC for the following conditions.
   a) $a = 8.3, b = 5.5$, and $c = 11.4$
   b) $c = 25, \alpha = 30°$, and $\beta = 70°$
   c) $a = 52.8, \alpha = 110° 10'$, and $\gamma = 45° 20'$
   d) $b = 4, c = \sqrt{3}, \alpha = 30°$
   e) $c = 272.6, a = 392.4$, and $\gamma = 37° 10'$
5. Find the areas of the triangles in problems 3 and 4.
6. To find the distance between two inaccessible points C and D on one side of a river, a surveyor on the other side, picks two points A and B which are 16 yards apart, and measures $\angle CAD = 32° 19'$, $\angle DAB = 28°41'$, $\angle ABC = 39° 30'$, and $\angle CBD = 36° 20'$. Find the distance between C and D.
7. A shell is fired from a gun raised to an angle $37° 43'$ with the horizontal. The muzzle velocity is 2860 feet per second. Find the horizontal and vertical components of this velocity.
8. An airplane is flying a heading due west at an airspeed of 260 mi./hr. A wind is blowing directly from the north with a speed of 60 mi./hr. Find the drift angle, the groundspeed, and the course of the airplane.
9. Find the radius of the inscribed circle of the triangles of problem 4.
10. An aviator finds that after traveling 300 miles in an easternly direction and then 250 miles in a southernly direction he is 400 miles due east of his starting point. Without logarithms, find the direction in which he traveled on the first leg of his journey. Draw the figure.

chapter 8

Complex Numbers

85. INTRODUCTION

The reader is probably familiar with the complex number system. We shall, however, repeat the definitions and simple operations for easy reference. The **imaginary unit** is defined as that quantity which when squared is equal to -1, i.e.

$$i^2 = -1$$

The square root of any negative number is then expressed as a product of a real number and the imaginary unit, for example

$$\sqrt{-16} = i\sqrt{16} = 4i$$

Any number of the form bi is called a **pure imaginary number.** If we combine it with a real number by an algebraic sign, $a + bi$, we have defined a **complex number.** The number a is the **real part,** and the number bi is the **imaginary part**. It is easily seen that the real number system and the pure imaginary system are special cases of the complex number system.

Complex numbers that differ only in the signs of their imaginary part are called **conjugate complex numbers.** Thus, $a - bi$ is the conjugate of $a + bi$, and in turn $a + bi$ is the conjugate of $a - bi$.

Two complex numbers are said to be equal if and only if their real parts are equal and their imaginary parts are equal.

Thus $a + bi = x + yi$ if and only if $a = x$ and $b = y$

$a + bi = 0$ if and only if $a = b = 0$

OPERATIONS

I. *To add or subtract two complex numbers, add or subtract the real and imaginary parts separately.*

$$(a + bi) + (c + di) = (a + c) + (b + d)i$$
$$(a + bi) - (c + di) = (a - c) + (b - d)i$$

II. *To find the product of two complex numbers, multiply them together as two binomials and substitute* -1 *for* i^2.

$$(a + bi)(c + di) = ac + adi + bci + bdi^2$$
$$= (ac - bd) + (ad + bc)i$$

III. *To express the quotient of two complex numbers as a single complex number, multiply both numerator and denominator by the conjugate of the denominator.*

$$\frac{a + bi}{c + di} = \frac{(a + bi)(c - di)}{(c + di)(c - di)} = \frac{ac + bd}{c^2 + d^2} + \frac{bc - ad}{c^2 + d^2}i$$

As in the real number system, division by zero is *not* permitted in the complex number system. We note that the product of two conjugate complex numbers is a real number.

$$(a + bi)(a - bi) = a^2 - (bi)^2 = a^2 - b^2i^2 = a^2 + b^2$$

Division of complex numbers can be thought of as an operation similar to that of rationalizing the denominator.

ILLUSTRATIONS

1. $(4 - 6i) + (3 + 2i) = 7 - 4i$
2. $(3 + 2i) - (7 + 4i) = -4 - 2i$
3. $(-4 + 7i)(3 - 2i) = [-12 - (-14)] + (8 + 21)i = 2 + 29i$
4. $\frac{6 - i}{1 + 3i} = \frac{(6 - i)(1 - 3i)}{(1 + 3i)(1 - 3i)} = \frac{3}{10} - \frac{19}{10}i$

86. TRIGONOMETRIC FORM OF A COMPLEX NUMBER

The graphical representation of a complex number can be accomplished by locating the real numbers a and b on a rectangular coordinate system. Let us choose the x-axis as the real axis and the y-axis as the imaginary axis. The point, $P(a + bi)$, is then located as in Figure 78, on the top of page 143. If we join P to the origin and denote $\overline{OP}$ by r, and the angle this line segment makes with the positive x-axis by θ, we can then verify the following relations.

$$a = r\cos\theta \qquad b = r\sin\theta$$
$$r = \sqrt{a^2 + b^2} \qquad \tan\theta = \frac{b}{a}$$

Thus we have

$$a + bi = r(\cos\theta + i\sin\theta)$$

The right-hand side of the last equation is the **trigonometric form** of the complex number; r is called the **modulus** and is always positive; the angle θ is called the **argument** or **amplitude.**

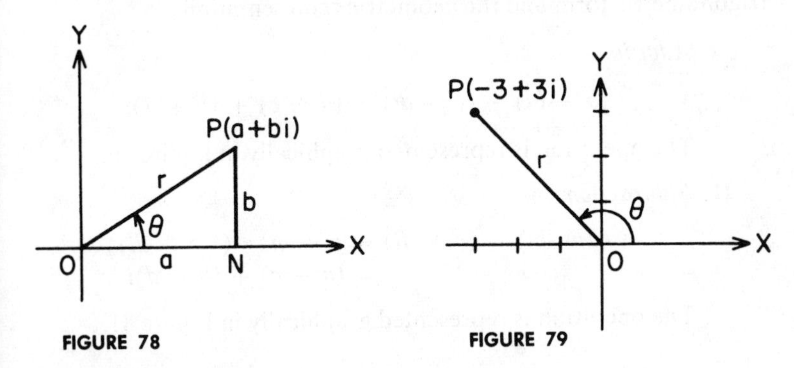

FIGURE 78

FIGURE 79

ILLUSTRATION. Change $-3 + 3i$ to its trigonometric form.
SOLUTION. See Figure 79.

$$r = \sqrt{(-3)^2 + (3)^2} = 3\sqrt{2}$$
$$\tan\theta = \frac{3}{-3} = -1$$

Since a is negative and b is positive, the angle is in the second quadrant;

$$\theta = 135^\circ$$
$$\therefore -3 + 3i = 3\sqrt{2}\,(\cos 135^\circ + i \sin 135^\circ)$$

Since we have associated a complex number with a rectangular coordinate system, the complex number is frequently written $x + yi$. The corresponding (x,y)-plane is then called the **complex plane,** and the figure is called the **Argand diagram.** A complex number is thus associated with a point whose coordinates are x and y; consequently, a complex number becomes an ordered pair (x,y). In advanced mathematics this is often given as the definition of a complex number. The trigonometric form

$$x + yi = r(\cos\theta + i \sin\theta)$$

can be abbreviated as r cis θ, arising from c (the cosine), i (the imaginary), and s (the sine). Let us also point out that because of the periodicity of the trigonometric function, we may write

$$x + yi = r[\cos(\theta + 2n\pi) + i \sin(\theta + 2n\pi)] = r \operatorname{cis}(\theta + 2n\pi)$$

This form is called the **polar form.**

87. OPERATIONS

The basic operations with complex numbers were given in Section 85. Let us now consider these operations in terms of the trigonometric form and the geometric representations.

I. *Addition*

$$(a + bi) + (c + di) = (a + c) + (b + d)i$$

The operation is represented graphically in Figure 80.

II. *Subtraction*

$$\begin{aligned}(a + bi) - (c + di) &= a + bi + (-c - di)\\ &= (a - c) + (b - d)i\end{aligned}$$

The operation is represented graphically in Figure 81.

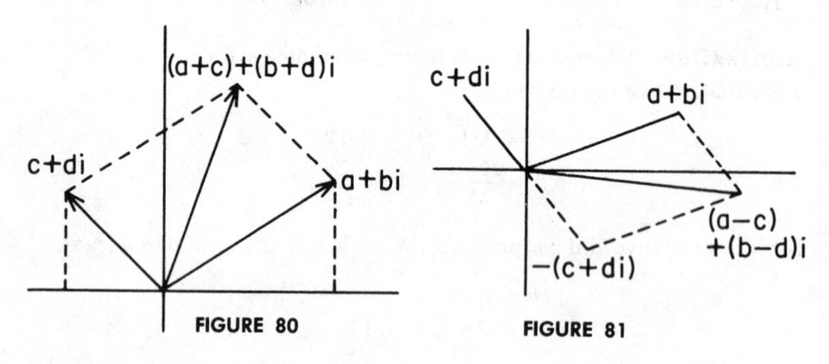

FIGURE 80

FIGURE 81

III. *Multiplication*

$$\begin{aligned}&r_1(\cos\theta_1 + i\sin\theta_1)\, r_2(\cos\theta_2 + i\sin\theta_2)\\ &\qquad = r_1 r_2[\cos(\theta_1 + \theta_2) + i\sin(\theta_1 + \theta_2)]\\ &\qquad = r_1 r_2 \operatorname{cis}(\theta_1 + \theta_2)\end{aligned}$$

This formula can be demonstrated by performing the multiplication and recalling the formulas for the sine and cosine of the sum of two arguments and the fact that $i^2 = -1$. Thus we see that the product of two complex numbers can be found by multiplying the moduli and finding the cosine and sine of the sum of the arguments. It is left to the reader to draw the corresponding Argand diagram.

ILLUSTRATION. Find the product of

$$z_1 = 2(\cos 30° + i\sin 30°) \text{ and } z_2 = 3(\cos 15° + i\sin 15°)$$

SOLUTION

$$\begin{aligned} z_1 z_2 &= (2)(3)\,[\cos(30^\circ + 15^\circ) + i \sin(30^\circ + 15^\circ)] \\ &= 6\,(\cos 45^\circ + i \sin 45^\circ) \\ &= 6\left(\frac{\sqrt{2}}{2} + \frac{\sqrt{2}}{2} i\right) = 3\sqrt{2} + 3\sqrt{2}\, i \end{aligned}$$

The product formula may be generalized to the product of n complex numbers.

$$z_1 z_2 z_3 \cdots z_n = r_1 r_2 \cdots r_n \operatorname{cis}(\theta_1 + \theta_2 + \cdots + \theta_n)$$

IV. *Division*

$$\frac{r_1(\cos\theta_1 + i \sin\theta_1)}{r_2(\cos\theta_2 + i\sin\theta_2)} = \frac{r_1}{r_2}[\cos(\theta_1 - \theta_2) + i \sin(\theta_1 - \theta_2)]$$
$$= \frac{r_1}{r_2} \operatorname{cis}(\theta_1 - \theta_2)$$

This formula has the obvious restriction that $r_2 \neq 0$. The proof is obtained by multiplying the numerator and denominator by $\cos\theta_2 - i\sin\theta_2$, noting that the denominator then becomes $r_2(\cos^2\theta_2 + \sin^2\theta_2)$, and applying the addition formulas to the resulting numerator.

ILLUSTRATION. Find the quotient of

$$z_1 = 6\,(\cos 45^\circ + i \sin 45^\circ) \text{ by } z_2 = 2\,(\cos 15^\circ + i \sin 15^\circ)$$

SOLUTION

$$\begin{aligned} \frac{z_1}{z_2} &= \frac{6\,(\cos 45^\circ + i\sin 45^\circ)}{2\,(\cos 15^\circ + i \sin 15^\circ)} \\ &= 3\,(\cos 30^\circ + i \sin 30^\circ) \\ &= \frac{3\sqrt{3}}{2} + \frac{3}{2} i \end{aligned}$$

88. POWERS OF COMPLEX NUMBERS

To obtain the powers of a complex number, we employ a famous theorem known as **De Moivre's theorem.** *If* n *is any positive integer, then*

$$[r(\cos\theta + i\sin\theta)]^n = r^n(\cos n\theta + i \sin n\theta)$$

This theorem can be considered as the direct result of the generalized product formula for the case in which all the complex numbers are equal, or it may be proved by mathematical induction.

PROOF. The theorem is easily verified for $n = 0$, 1, and 2 by actual multiplication. Let us then assume that the theorem is true for $n = k$.

$$z^k = r^k (\cos k\theta + i \sin k\theta)$$

Multiply both sides by z to obtain

$$\begin{aligned} z^k z = z^{k+1} &= [r^k (\cos k\theta + i \sin \theta)]\,[r(\cos\theta + i \sin\theta)] \\ &= r^k r \operatorname{cis}(k\theta + \theta) \\ &= r^{k+1} \operatorname{cis}(k+1)\theta \end{aligned}$$

and we obtain the same result as would have been obtained by letting $n = k + 1$ in the theorem which completes the proof by induction.

ILLUSTRATION. Find $(\sqrt{3} + i)^8$.
SOLUTION. Express the complex number in polar form:

$$r = \sqrt{3+1} = 2 \qquad \theta = \text{Arctan}\,\frac{1}{\sqrt{3}} = \frac{\pi}{6}$$

and
$$\sqrt{3} + i = 2\left(\cos\frac{\pi}{6} + i \sin\frac{\pi}{6}\right)$$

Now apply De Moivre's theorem.

$$\begin{aligned} (\sqrt{3} + i)^8 &= 2^8\left(\cos\frac{8\pi}{6} + i \sin\frac{8\pi}{6}\right) \\ &= 256\left(-\frac{1}{2} - \frac{i\sqrt{3}}{2}\right) \\ &= -128 - 128\sqrt{3}\,i \end{aligned}$$

89. ROOTS OF COMPLEX NUMBERS

Let us now consider the nth root of a complex number, i.e.

$$\sqrt[n]{a + bi} = [r \operatorname{cis}(\theta + 2k\pi)]^{1/n}$$

By definition we are seeking a complex number z such that $z^n = r \operatorname{cis}(\theta + 2k\pi)$, for then z would be the nth root of $a + bi$. Let $z = v \operatorname{cis} \theta$, then

$$z^n = (v \operatorname{cis}\phi)^n = v^n \operatorname{cis} n\phi = r \operatorname{cis}(\theta + 2k\pi)$$

The last equation is satisfied if

$$v^n = r \qquad \text{and} \qquad n\phi = \theta + 2k\pi$$

Consequently,
$$v = \sqrt[n]{r} \qquad \text{and} \qquad \phi = \frac{\theta}{n} + \frac{2k\pi}{n}$$

In the preceding discussion, k is an integer; i.e., $k = 0, \pm 1, \pm 2, \ldots$. However, we can show that the distinct roots of the original complex number can be found by limiting k to the values $0, 1, 2, \ldots, n - 1$. All other values will yield arguments whose terminal sides will coincide with one of those given by the restricted values of k and consequently, the trigonometric functions will be identical. We thus have the formula

$$\mathbf{z = \sqrt[n]{r}\,\text{cis}\left(\frac{\theta}{n} + \frac{2k\pi}{n}\right), \quad k = 0, 1, 2, 3, \ldots, n - 1}$$

The roots of a complex number have an interesting geometric property if we consider representing them as points on the circle whose radius is $\sqrt[n]{r}$. Locate the first root as the point of intersection between the terminal side of the angle $\frac{\theta}{n}$ (when placed in a standard position) and the circumference of this circle. The remaining roots can now be located by simply adding increments of

$$\Delta\theta = \frac{2\pi}{n}$$

to the angles, $\frac{\theta}{n}$, $\frac{\theta}{n} + \Delta\theta$, $\frac{\theta}{n} + 2\,\Delta\theta$, etc. See the illustrations below.

ILLUSTRATIONS

1. Find the cube roots of $4\sqrt{3} + 4i$.

SOLUTION. The polar form is

$$4\sqrt{3} + 4i = 8 \text{ cis}\left(\frac{\pi}{6} + 2k\pi\right)$$

The cube roots are then given by

$$[4\sqrt{3} + 4i]^{1/3} = 2 \text{ cis}\left(\frac{\pi}{18} + \frac{2k\pi}{3}\right)$$

$$k = 0, 1, 2$$

FIGURE 82

If we change the argument to degrees, we may write

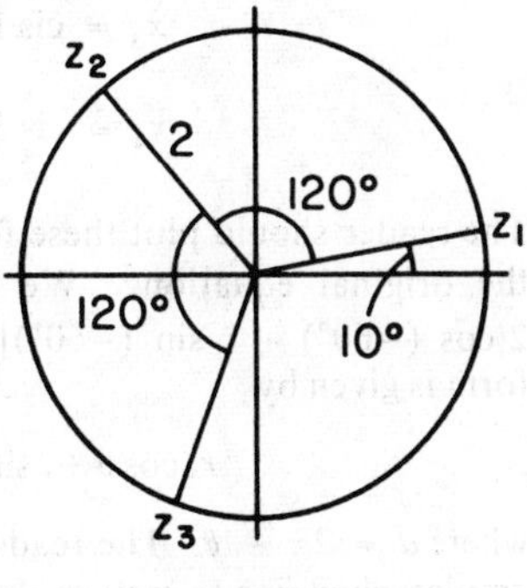

$$z_1 = 2 \text{ cis } 10^\circ \qquad z_2 = 2 \text{ cis } 130^\circ \qquad z_3 = 2 \text{ cis } 250^\circ$$

These three complex numbers may be plotted on a circle of radius 2. See Figure 82.

2. Find the cube roots of 1.

SOLUTION. The number 1 may be written in complex form as follows.

$$1 = 1 + 0i = 1(\cos 0^\circ + i \sin 0^\circ) = \text{cis } 2k\pi$$

The cube roots are given by

$$\text{cis } \frac{2k\pi}{3} = \text{cis}(120^\circ k), \ k = 0, 1, 2$$

In rectangular form we have

$$z_1 = 1 \qquad z_2 = -\frac{1}{2} + \frac{\sqrt{3}}{2} i \qquad z_3 = -\frac{1}{2} - \frac{\sqrt{3}}{2} i$$

[Solve the equation $x^3 - 1 = 0$ by algebraic methods and compare answers.]

3. Solve the equation $x^4 + x^2 + 1 = 0$.

SOLUTION. By using the quadratic formula, we obtain

$$x^2 = -\frac{1}{2} \pm \frac{\sqrt{3}}{2} i = \text{cis } 120^\circ, \text{cis } 240^\circ$$

We now find the square roots of these two complex numbers.

$$x_1 = \text{cis } 60^\circ = \frac{1}{2} + \frac{\sqrt{3}}{2} i$$

$$x_2 = \text{cis } 240^\circ = -\frac{1}{2} - \frac{\sqrt{3}}{2} i$$

$$x_3 = \text{cis } 120^\circ = -\frac{1}{2} + \frac{\sqrt{3}}{2} i$$

$$x_4 = \text{cis } 300^\circ = \frac{1}{2} - \frac{\sqrt{3}}{2} i$$

The reader should plot these four complex numbers and check them in the original equation. We note that $2(\cos 300^\circ + i \sin 300^\circ) = 2[\cos(-60^\circ) + i \sin(-60^\circ)] = 2 \cos 60^\circ - i \sin 60^\circ$. The general form is given by

$$r(\cos\theta + i \sin\theta) = r(\cos d - i \sin d)$$

where $d = 2\pi - \theta$. The reader can verify this by considering some examples, changing to rectangular form, or drawing the Argand diagram.

90. EXERCISE 14

1. Simplify the complex numbers, $i^2, i^3, i^4, i^7, i^{10}, i^{15}$.
2. Perform the indicated operations and write in the form $a + bi$.
   a) $3 - 6i + 7 - 8 + 2i + 4i - 1 + 3i$ b) $(\sqrt{-5})(\sqrt{-3})$
   c) $(3 - i)(4 + 2i)$ d) $(16 + 16i) \div (4 - 4i)$

3. Express the following complex numbers in their trigonometric form.
   a) $\sqrt{2} + \sqrt{2i}$ b) $-1 + i$ c) $5i$
   d) -1 e) $12 - 5i$ f) $3 - 3i$
4. Change the following complex numbers to their rectangular form.
   a) $2(\cos 30^\circ + i \sin 30^\circ)$ b) $3 \text{ cis } 45^\circ$
   c) $5\left(\cos \frac{\pi}{2} + i \sin \frac{\pi}{2}\right)$ d) $7 \text{ cis } \frac{3\pi}{4}$
   e) $-4(\cos 10^\circ + i \sin 10^\circ)$ f) $-5 \text{ cis } 300^\circ$
5. Perform the indicated operations graphically and check algebraically.
   a) $(2 + 3i) + (3 + i)$ b) $(2 - i) + (3 + 2i)$
   c) $(-1 + i) + (1 + i)$ d) $(3 + i) - (1 - i)$
   e) $(-2 + i) - (2 - 3i)$ f) $(4 + 2i) - (5 + i)$
6. Perform the indicated operation and write the result in the form $a + bi$.
   a) $3(\cos 18^\circ + i \sin 18^\circ) \times 2(\cos 72^\circ + i \sin 72^\circ)$
   b) $-2(\cos 15^\circ + i \sin 15^\circ) \times 5(\cos 30^\circ + i \sin 30^\circ)$
   c) $2(\cos 15^\circ + i \sin 15^\circ)^6$ d) $(1 - i\sqrt{3})^7$
   e) $12(\cos 142^\circ + i \sin 142^\circ) \div 6(\cos 97^\circ + i \sin 97^\circ)$
   f) $3 \text{ cis } 37^\circ \div 2 \text{ cis } 7^\circ$ g) $(3 + 2i) \div (1 - i)$
7. Solve the following equations.
   a) $z^3 + 1 = 0$ b) $z^5 + 4 + 4i = 0$
   c) $z^4 - i = 0$ d) $z^3 + i = 0$
   e) $x^4 - 8x^2 + 28 = 0$ f) $x^2 - 2x + 2 = 0$
8. Find the five fifth roots of unity.
9. Show that $\sqrt{2} - i\sqrt{2}$ is a fourth root of -16.
10. Show that $\text{cis } \frac{13\pi}{14}$ is a seventh root of i.

☆ 91. FURTHER PROPERTIES OF COMPLEX NUMBERS

The complex numbers have many interesting properties which can be demonstrated by using the definitions, properties of operations, method of representation, and conditions of equality. Before proceeding, recall the discussion of Section 85 and let us add the following notation. If z is a complex number, then $\bar{z}$ will be used to denote the conjugate; i.e., if

$$z = x + yi, \qquad \text{then} \qquad \bar{z} = x - yi$$

We shall also use the definition

$$|z| = r = \sqrt{x^2 + y^2}$$

In the Argand diagram $|z|$ represents the length of the line segment $\overline{OP}$, where P is (x,y). We shall illustrate some of the properties.

ILLUSTRATIONS

1. $z\bar{z} = |z|^2$.

SOLUTION. By writing in rectangular form, we have

$$z\bar{z} = (x + yi)(x - yi) = x^2 + y^2 = |z|^2$$

2. $\overline{z_1 + z_2} = \bar{z}_1 + \bar{z}_2$.

SOLUTION. In rectangular form we have

$$z_1 + z_2 = (x_1 + x_2) + (y_1 + y_2)i$$

The conjugate of this complex number is

$$(x_1 + x_2) - (y_1 + y_2)i = (x_1 - y_1 i) + (x_2 - y_2 i) = \bar{z}_1 + \bar{z}_2$$

3. $|z_1 z_2| = |z_1|\ |z_2|$.

SOLUTION. Let $z_1 = x_1 + y_1 i$ and $z_2 = x_2 + y_2 i$, then

$$z_1 z_2 = x_1 x_2 - y_1 y_2 + (x_1 y_2 + x_2 y_1)i$$

and

$$\begin{aligned} |z_1 z_2| &= [(x_1 x_2 - y_1 y_2)^2 + (x_1 y_2 + x_2 y_1)^2]^{1/2} \\ &= [x_1^2 x_2^2 - 2x_1 x_2 y_1 y_2 + y_1^2 y_2^2 + x_1^2 y_2^2 + 2x_1 x_2 y_1 y_2 + x_2^2 y_1^2]^{1/2} \\ &= [x_1^2(x_2^2 + y_2^2) + y_1^2(x_2^2 + y_2^2)]^{1/2} \\ &= [(x_1^2 + y_1^2)(x_2^2 + y_2^2)]^{1/2} = |z_1|\ |z_2| \end{aligned}$$

4. Let z_0 be a fixed complex number and R a positive constant. Show that if $|z + z_0| = R$, then z lies on a circle with radius R and center at $-z_0$.

SOLUTION. Let $z = x + yi$ and $z_0 = x_0 + y_0 i$, then

$$|z + z_0| = [(x + x_0)^2 + (y + y_0)^2]^{1/2} = R$$

By squaring both sides, we have

$$(x + x_0)^2 + (y + y_0)^2 = R^2$$

which is the equation of a circle of radius R and center, $c(-x_0, -y_0)$.

☆92. EXERCISE 15

1. Show the following.

a) $\overline{\bar{z} + i} = z - i$ b) $\bar{i} = -i$

c) $\left|\dfrac{z_1}{z_2}\right| = \dfrac{|z_1|}{|z_2|}$ d) $\overline{\left(\dfrac{z_1}{z_2 z_3}\right)} = \dfrac{\bar{z}_1}{\bar{z}_2 \bar{z}_3}$

e) $|\bar{z}| = |z|$ f) $\overline{(z^4)} = (\bar{z})^4$

2. Prove that if $\bar{z} = z$, then z is real.
3. Prove the conclusion of Illustration 4, page 150, with the same hypothesis, except that z satisfies the equation $z\bar{z} + \bar{z}_0 z + z_0 \bar{z} + z_0 \bar{z}_0 = R^2$ instead of $|z + z_0| = R$.
4. Let z_1 and z_2 be complex numbers with the property that $z_1 + z_2$ and $z_1 z_2$ are real numbers. Show either that z_1 and z_2 are real or that $z_1 = \bar{z}_2$.
5. Describe the geometric region of z for which the imaginary part of z^2 is positive.

chapter 9

Hyperbolic Functions

93. INTRODUCTION

In the preceding chapters we defined the trigonometric functions and discussed their properties and some of their applications. In our discussions we have associated the trigonometric functions with the unit circle, and for this reason these functions are often referred to as the **circular functions.** There is another set of functions which may be associated with a hyperbola and, consequently, these functions are called the **hyperbolic functions.** Their precise definitions, however, are stated in terms of the exponential function. In the following sections we shall state the definitions and discuss the properties of the hyperbolic functions. The reader will find that there is great similarity between the properties of the trigonometric and hyperbolic functions. However, there are differences, and we caution against falling into the trap of stretching the similarities too far.

94. DEFINITIONS

There are six hyperbolic functions.

hyperbolic sine abbreviated sinh
hyperbolic cosine abbreviated cosh
hyperbolic tangent abbreviated tanh
hyperbolic cotangent abbreviated coth
hyperbolic secant abbreviated sech
hyperbolic cosecant abbreviated csch

As in the case of the trigonometric functions, they are combined with an argument which we shall limit to the domain of real numbers. The definitions are given in terms of the exponential function on the top of page 153.

$$\sinh x = \frac{e^x - e^{-x}}{2} \qquad \coth x = \frac{e^x + e^{-x}}{e^x - e^{-x}} \quad x \neq 0$$

$$\cosh x = \frac{e^x + e^{-x}}{2} \qquad \operatorname{sech} x = \frac{2}{e^x + e^{-x}}$$

$$\tanh x = \frac{e^x - e^{-x}}{e^x + e^{-x}} \qquad \operatorname{csch} x = \frac{2}{e^x - e^{-x}} \quad x \neq 0$$

These functions occur in applied mathematics and engineering with a sufficient frequency that tables for their values, as x varies, have been published. Values for varying x can also be found from tables of the exponential function by performing the arithmetic indicated in the definitions.

ILLUSTRATION. Find a table of values for $\sinh x$.

SOLUTION

x	0	1	−1	1.5	−1.5	2	−2	2.5	− 2.5
e^x	1	2.72	0.37	4.48	0.22	7.39	0.14	12.18	0.08
e^{-x}	1	0.37	2.72	0.22	4.48	0.14	7.39	0.08	12.18
$e^x - e^{-x}$	0	2.35	−2.35	4.26	−4.26	7.25	−7.25	12.10	−12.10
$\sinh x$	0	1.18	−1.18	2.13	−2.13	3.63	−3.63	6.05	− 6.05

95. PROPERTIES OF THE HYPERBOLIC FUNCTIONS

There are many relations among the hyperbolic functions similar to the identities of the trigonometric functions. These properties are derived from the definitions. Note the similarities and differences between the forms for the circular and hyperbolic functions. We list the common properties.

(1) $\tanh x = \dfrac{\sinh x}{\cosh x}$ (2) $\coth x = \dfrac{\cosh x}{\sinh x}$

(3) $\operatorname{sech} x = \dfrac{1}{\cosh x}$ (4) $\operatorname{csch} x = \dfrac{1}{\sinh x}$

(5) $\cosh^2 x - \sinh^2 x = 1$ (6) $\operatorname{sech}^2 x = 1 - \tanh^2 x$

(7) $\operatorname{csch}^2 x = \coth^2 x - 1$ (8) $\cosh x + \sinh x = e^x$

(9) $\cosh x - \sinh x = e^{-x}$

(10) The hyperbolic functions of negative arguments can be changed to hyperbolic functions of positive arguments in the same manner as the trigonometric functions.

Each of these formulas can be proved directly from the definitions; for example,

$$\cosh^2 x - \sinh^2 x = \left(\frac{e^x + e^{-x}}{2}\right)^2 - \left(\frac{e^x - e^{-x}}{2}\right)^2$$

$$= \frac{1}{4}(e^{2x} + 2 + e^{-2x} - e^{2x} + 2 - e^{-2x})$$

$$= \frac{4}{4} = 1$$

The addition formulas are similar to those for the trigonometric functions and are proved in the same manner. They are listed in Exercise 16. We shall give the proof for the formula $\sinh\ (x + y)$ in the following illustration.

ILLUSTRATION. Prove $\sinh (x + y) = \sinh x \cosh y + \cosh x \sinh y$.

SOLUTION. Change the right side by using the definitions.

$$\sinh (x + y) = \frac{e^x - e^{-x}}{2} \cdot \frac{e^y + e^{-y}}{2} + \frac{e^x + e^{-x}}{2} \cdot \frac{e^y - e^{-y}}{2}$$

$$= \frac{1}{4}(e^{x+y} - e^{-x+y} + e^{x-y} - e^{-x-y} + e^{x+y} + e^{-x+y}$$
$$- e^{x-y} - e^{-x-y})$$

$$= \frac{1}{4}(2 e^{x+y} \quad 2 e^{-(x+y)})$$

$$= \sinh (x + y)$$

96. THE INVERSE HYPERBOLIC FUNCTIONS

The hyperbolic sine is the function y, defined by $y = \sinh x$. The **inverse hyperbolic sine** of x is the function y, defined by the equation $x = \sinh y$; the usual notation is $\sinh^{-1} x$. However, we can solve the equation $x = \sinh y$ for y by using the definition

$$x = \sinh y = \frac{e^y - e^{-y}}{2}$$

and multiplying by $2\,e^y$, we obtain

$$e^{2y} - 2\,x\,e^y - 1 = 0$$

which is a quadratic equation in the unknown e^y. Theoretically the equation has the two roots

$$e^y = x \pm \sqrt{1 + x^2}$$

however, as we saw in Chapter 6, the exponential e^y is positive for all values of y, and since

$$\sqrt{1 + x^2} > x$$

the solution $x - \sqrt{1 + x^2}$ will be a negative quantity for all x and thus, will be extraneous. We can now solve for y by taking the natural logarithms. We list them with their domain of definition for single-valued functions.

$$\sinh^{-1} x = \ln (x + \sqrt{x^2 + 1}) \qquad (\text{all } x)$$
$$\text{Cosh}^{-1} x = \ln (x + \sqrt{x^2 - 1}) \qquad (x \geq 1)$$
$$\tanh^{-1} x = \frac{1}{2} \ln \frac{1 + x}{1 - x} \qquad (x^2 < 1)$$
$$\coth^{-1} x = \frac{1}{2} \ln \left(\frac{x + 1}{x - 1}\right) \qquad (x^2 > 1)$$
$$\text{Sech}^{-1} x = \ln \left(\frac{1}{x} + \sqrt{\frac{1}{x^2} - 1}\right) \qquad (0 < x \leq 1)$$
$$\text{csch}^{-1} x = \ln \left(\frac{1}{x} + \sqrt{\frac{1}{x^2} + 1}\right) \qquad (\text{all } x \neq 0)$$

97. GRAPHS OF THE HYPERBOLIC FUNCTIONS

The hyperbolic functions may be graphed on a rectangular coordinate system by calculating a table of ordered pairs using the tables of the exponential function if necessary. We present a small table.

x	$\sinh x$	$\cosh x$	$\tanh x$
0	0	1	0
.1	.10	1.01	.10
.2	.20	1.02	.20
.3	.30	1.05	.29
.4	.41	1.08	.38
.5	.52	1.13	.46
1.0	1.18	1.54	.76
1.5	2.13	2.35	.91
2.0	3.63	3.76	.96
2.5	6.05	6.13	.99
3.0	10.02	10.07	1.00
3.5	16.54	16.57	1.00
4.0	27.29	27.31	1.00
5.0	74.20	74.21	1.00

The preceding table can be extended to negative values by the properties of the functions. The graphs for these three functions are shown in Figures 83–85.

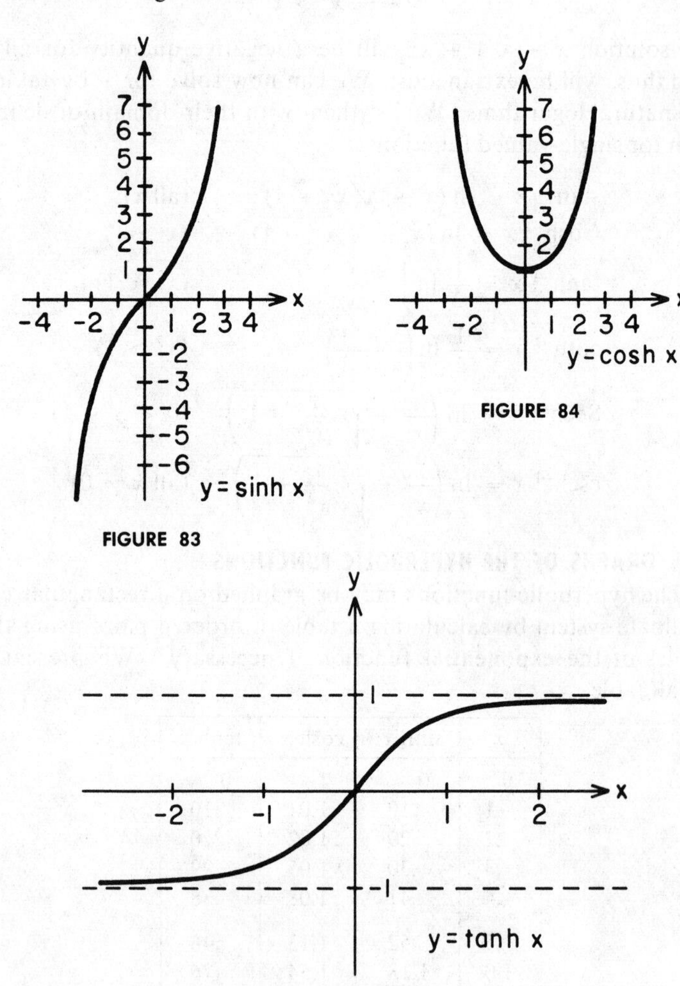

FIGURE 83

FIGURE 84

FIGURE 85

The graph of $y = \cosh x$ is given a special name, the **catenary**, which is the curve defined by a chain suspended from two points and hanging under its own weight. The general form of the catenary is $y = a \cosh \dfrac{x}{a}$.

98. ANALOGY BETWEEN TRIGONOMETRIC AND HYPERBOLIC FUNCTIONS

We mentioned (in Section 93) that the hyperbolic functions can be associated with a hyperbola. Let us develop this concept further, and show the analogy between the trigonometric (circular) functions and the hyperbolic functions.

The equation of the unit circle is $x^2 + y^2 = 1$, and we have seen that the coordinates of any point P on this circle can be given by $x = \cos\theta$ and $y = \sin\theta$. Consider a point P (see Figure 86). The area of the sector AOP is $\frac{1}{2}\theta$; that is, *the value of the argument θ is twice the area of the sector AOP.*

Now consider the equilateral hyperbola $x^2 - y^2 = 1$. Since $\cosh^2 u - \sinh^2 u = 1$, the coordinates of any point P on this hyperbola may be given by $x = \cosh u$ and $y = \sinh u$. The area of the sector AOP (see Figure 87) is given by the area of the triangle OMP minus the area bounded by the hyperbola, the x-axis, and the line PM; i.e.,

$$K = \text{area } \triangle OMP - \text{area "under" hyperbola from } A \text{ to } M$$

The area "under" the hyperbola is an exercise in the calculus which uses the principle of integration; we shall simply state the result here.*

$$\begin{aligned} K &= \frac{1}{2}xy - \frac{1}{2}\left[x\sqrt{x^2-1} - \ln\left(x + \sqrt{x^2-1}\right)\right] \\ &= \frac{1}{2}\ln\left(x + \sqrt{x^2-1}\right) \\ &= \frac{1}{2}\operatorname{Cosh}^{-1} x = \frac{u}{2} \end{aligned}$$

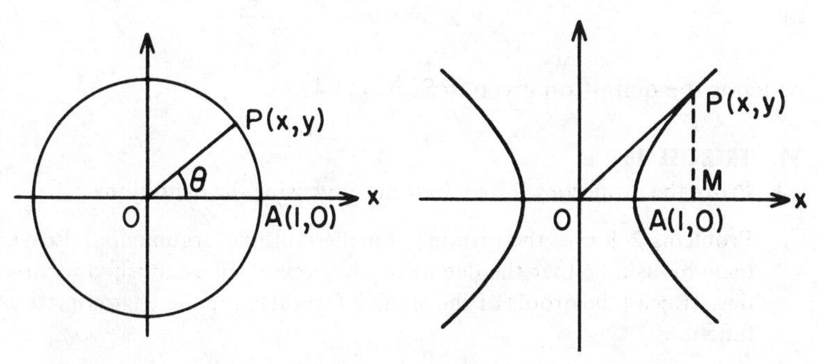

FIGURE 86

FIGURE 87

*See C. O. Oakley, *The Calculus* (New York: Barnes and Noble, 1957).

since $y = \sqrt{x^2 - 1}$; because $P(x,y)$ is on the hyperbola and $x = \cosh u$. Consequently, *the argument* u *is twice the area of the sector* AOP. This is the same statement we made for the circular function.

Although we defined the hyperbolic functions in terms of the exponential, we could start with the definitions

$$x = \cosh u \qquad y = \sinh u$$

provided that $P(x,y)$ satisfies the equation $x^2 - y^2 = 1$, which can be changed to $x^2 - 1 = y^2$. Then, from the above, we see that

$$u = \ln(x + \sqrt{x^2 - 1}) = \ln(x + y)$$

or

$$e^u = x + y$$

and by the definition of negative exponents,

$$e^{-u} = \frac{1}{x + y}$$

Now form the expression (and perform the indicated algebra).

$$\frac{1}{2}(e^u - e^{-u}) = \frac{1}{2}[x + y - (x + y)^{-1}]$$

$$= \frac{x^2 + 2xy + y^2 - 1}{2(x + y)}$$

[since $x^2 - 1 = y^2$]

$$= \frac{2xy + 2y^2}{2(x + y)} = \frac{2y(x + y)}{2(x + y)} = y$$

or

$$y = \sinh u = \frac{e^u - e^{-u}}{2}$$

which is the definition given in Section 94.

99. EXERCISE 16

1. Prove the properties 1–9 of Section 95 by using the definitions.

Problems 2–8 give the formulas for the multiple arguments. Prove them by using either the definitions or previously established identities. (Recall the proofs of the similar formulas for the trigonometric functions.)

2. $\sinh(x \pm y) = \sinh x \cosh y \pm \cosh x \sinh y$
3. $\cosh(x \pm y) = \cosh x \cosh y \pm \sinh x \sin y$

4. $\tanh(x \pm y) = \dfrac{\tanh x \pm \tanh y}{1 \pm \tanh x \tanh y}$

5. $\sinh 2x = 2 \sinh x \cosh x$

6. $\cosh 2x = \cosh^2 x + \sinh^2 x$

7. $\sinh \dfrac{x}{2} = \pm \sqrt{\dfrac{\cosh x - 1}{2}}$

8. $\cosh \dfrac{x}{2} = \pm \sqrt{\dfrac{\cosh x + 1}{2}}$

9. Develop formulas for the following.

 a) $\tanh 2x$ b) $\tanh \dfrac{x}{2}$

 c) $\coth 2x$ d) $\coth \dfrac{x}{2}$

10. Prove that $(\cosh x + \sinh x)^n = \cosh nx + \sinh nx$. (Hint: Show for $n = 2,3$; then use induction.)

11. Derive the logarithmic expressions for the following.

 a) $\text{Cosh}^{-1} x$ b) $\tanh^{-1} x$

12. Start with the definitions $x = \cosh u$ and $y = \sinh u$ and $x^2 - y^2 = 1$. Derive the exponential definition of $\cosh u$.

chapter 10

Series

100. INTRODUCTION

The reader may already have encountered the material in the next few sections in his study of algebra. It is, however, repeated here for the sake of continuity and review. If the material is familiar, we suggest going directly to Section 107.

101. SEQUENCE OF NUMBERS

It is frequently desired to order a group of objects in such a way that there is a first object, a second object, etc. For n objects we may represent the first object by a_1, the second by a_2, etc., so that we have the ordered set

$$a_1, a_2, a_3, \ldots, a_n \tag{1}$$

An ordered set of objects is called a **sequence**, and the individual objects are called **terms** or **elements** of the sequence. A term, a_i, where i* may equal 1, 2, 3, or any number up to n, is called a **general term**.

These concepts may be applied to numbers. Thus we say that a *sequence of numbers* is a set of numbers arranged in a definite order. A sequence may be represented in many ways, such as (1) above. Another representation is

$$a_i \quad (i = 1,2,3,\ldots,n) \tag{2}$$

which is read "a_i for i equal 1 to n." We may thus condense the writing of a sequence by giving the formula for the general term as a function of i, provided, of course, that the formula exists.

ILLUSTRATION. Write the first, second, and seventh terms of the sequence

$$a_i = 2i + 3 \qquad (i = 1,2,3,\ldots,7)$$

*In this case "i" does not represent the imaginary number.

SOLUTION

$$a_1 = 2(1) + 3 = 5$$
$$a_2 = 2(2) + 3 = 7$$
$$a_7 = 2(7) + 3 = 17$$

Another required concept is that of finding the sum of a given number of terms of a sequence. This may be indicated by two symbols.

$$S_n = a_1 + a_2 + a_3 + \cdots + a_n$$

$$\sum_{i=1}^{n} a_i = a_1 + a_2 + a_3 + \cdots + a_n$$

The use of capital sigma, Σ, is frequent in higher mathematics, $\sum_{i=1}^{n} a_i$ is read "the summation of a_i from $i = 1$ to n."

ILLUSTRATION. Find the sum of the first seven terms of $a_i = 2i + 3$ $(i = 1,2,\ldots,7)$.

SOLUTION $$\sum_{i=1}^{7} (2i + 3) = 5 + 7 + 9 + 11 + 13 + 15 + 17 = 77$$

102. ARITHMETIC PROGRESSION, A.P.

DEFINITION. *An* **arithmetic progression** *is a sequence of numbers in which each term, after the first, is obtained from the preceding one by adding to that term a fixed number called the* **common difference.**

Let: a denote the first term
d denote the common difference
n denote the number of terms
l denote the nth (last) term
s denote the sum of the first n terms

Then the A.P. is $a, a + d, a + 2d, \ldots, l$

The nth term is given by $\mathbf{l = a + (n - 1)d}$

The sum of the first n terms is given by

$$\sum_{i=1}^{n} a_i = s = \frac{n}{2}(a + l)$$
$$= \frac{n}{2}[2a + (n - 1)d]$$

Given any two terms of an A.P., the terms between them are called the **arithmetic means.**

If any three of the five elements a, d, n, l, and s are given, the other two may be found.

ILLUSTRATIONS

1. Given $a = 12$, $d = 7$, and $l = 75$, find n and s.

SOLUTION

$l = a + (n - 1)d$	$s = \frac{n}{2}(a + l)$
$75 = 12 + (n - 1)7$	$= \frac{10}{2}(12 + 75)$
$n - 1 = 9$	$= 5(87)$
$n = 10$	$= 435$

2. Given $a = 3$, $l = -47$, and $s = -242$, find n and d.

SOLUTION

$s = \frac{n}{2}(a + l)$	$l = a + (n - 1)d$
$2(-242) = n(3 - 47)$	$-47 = 3 + 10d$
$n = 11$	$d = -5$

3. Insert seven arithmetic means between -1 and 5.

SOLUTION. We have $a = -1$ and $l = 5$. If we insert seven numbers between two given numbers, we have $n = 7 + 2 = 9$ numbers. Now

$$l = a + (n - 1)d$$
$$5 = -1 + (9 - 1)d$$
$$d = \tfrac{3}{4}$$

Therefore, the A.P. is

$$-1, -\tfrac{1}{4}, \tfrac{1}{2}, \tfrac{5}{4}, 2, \tfrac{11}{4}, \tfrac{7}{2}, \tfrac{17}{4}, 5$$

103. GEOMETRIC PROGRESSION, G.P.

DEFINITION. *A* **geometric progression** *is a sequence of numbers in which each term, after the first, is obtained from the preceding one by multiplying that term by a fixed number called the* **common ratio.**

Let: a denote the first term
r denote the common ratio
n denote the number of terms
l denote the nth (last) term
s denote the sum of the first n terms

Then the G.P. is

$$a, ar, ar^2, ar^3, \ldots, l$$

The nth term is given by

$$l = ar^{n-1}$$

The sum of the first n terms is given by

$$\sum_{i=1}^{n} a_i = s = a\frac{1 - r^n}{1 - r}$$

$$= \frac{a - rl}{1 - r}, \; r \neq 1$$

Given any two terms of a G.P., the terms between them are called the **geometric means.**

If any three of the five elements a, r, l, n and s are given, the other two may be found.

ILLUSTRATIONS

1. Given $a = 2$, $r = 3$, and $n = 7$, find l and s.

SOLUTION

$l = ar^{n-1}$	$s = \frac{a - rl}{1 - r}$
$= 2(3)^{7-1}$	$= \frac{2 - 3(1458)}{1 - 3}$
$= 1458$	$= 2186$

2. Given $a = -2$, $n = 7$, and $l = -\frac{1}{32}$, find r and s.

SOLUTION

$l = ar^{n-1}$	$s = \frac{a - rl}{1 - r}$
$-\frac{1}{32} = (-2)r^6$	$= \frac{-2 - (\pm\frac{1}{2})(-\frac{1}{32})}{1 - (\pm\frac{1}{2})}$
$\frac{1}{64} = r^6$	$= -\frac{127}{32}, -\frac{43}{32}$
$\pm\frac{1}{2} = r$	

3. Insert five geometric means between 7 and 448.

SOLUTION. $a = 7, l = 448, n = 2 + 5 = 7.$

$$\begin{aligned} l &= ar^{n-1} \\ 448 &= 7(r)^6 \\ 64 &= r^6 \\ \pm 2 &= r \end{aligned}$$

There are two G.P.'s. (a) 7, 14, 28, 56, 112, 224, 448
(b) 7, −14, 28, −56, 112, −224, 448

104. INFINITE SEQUENCE

So far we have been discussing sequences of numbers which have a last term. It is also possible to have sequences of numbers which do not have a last term but continue indefinitely. If after each term of a sequence there exists another term, it is called an **infinite sequence.** Consider the sequence

$$a_i = \frac{1}{2^i} \qquad (i = 1,2,3,\ldots)$$

which can be written $\frac{1}{2}, \frac{1}{4}, \frac{1}{8}, \frac{1}{16}, \frac{1}{32}, \ldots$

We notice that the numbers are getting smaller but are always positive. In other words, they are getting closer and closer to zero. In such cases we say that the sequence approaches a limit.

DEFINITION. *A sequence* $a_i (i = 1,2,\ldots,n,\ldots)$ *is said to approach the constant* c *as a* **limit,** *if the value of* $|c - a_i|$ *becomes and remains less than any preassigned positive number, however small.*

If a sequence approaches a limit, it is said to be **convergent;** if it does *not* approach a limit, it is said to be **divergent.**

We are especially interested in an infinite geometric progression in which $|r| < 1$. Consider the sum

$$a + ar + ar^2 + ar^3 + \cdots + ar^n + \cdots$$

with $|r| < 1$. The sum of the first n terms is given by

$$S_n = a\,\frac{1 - r^n}{1 - r} = \frac{a}{1 - r} - \frac{ar^n}{1 - r}$$

Now, if we continue the sequence indefinitely, the term

$$\frac{ar^n}{1 - r}$$

becomes very small (since $|r| < 1$) and approaches zero as a limit. Thus we have the sum of the infinite geometric progression approaching the first term of the preceding equation, and we say

$$S = \frac{a}{1 - r}$$

for this progression.

ILLUSTRATIONS

1. Find the sum of $\frac{1}{2}, \frac{1}{4}, \frac{1}{8}, \frac{1}{16}, \frac{1}{32}, \ldots$

SOLUTION. $a = \frac{1}{2}$ and $r = \frac{1}{2}$.

$$S = \frac{a}{1 - r} = \frac{\frac{1}{2}}{1 - \frac{1}{2}} = 1$$

2. Find the sum of $\frac{4}{3} + \frac{4}{9} + \frac{4}{27} + \cdots$

SOLUTION. $a = \frac{4}{3}$ and $r = \frac{1}{3}$.

$$S = \frac{a}{1 - r} = \frac{\frac{4}{3}}{1 - \frac{1}{3}} = 2$$

A repeating decimal may be changed to an equivalent common fraction.

ILLUSTRATION. Change the repeating decimal $0.5363636\cdots$ into an equivalent common fraction.

SOLUTION. Write the decimal as the sum of an infinite geometric sequence.

$$0.5363636\cdots = .5 + .036 + .00036 + \cdots = .5 + a + \sum_{i=1}^{\infty} ar^i$$

Thus,

$$a = .036, \; r = .01 < 1$$

$$\therefore \qquad S = \frac{a}{1 - r} = \frac{.036}{1 - .01} = \frac{.036}{.99} = \frac{4}{110}$$

The decimal is then converted to

$$0.5363636\cdots = \frac{5}{10} + \frac{4}{110} = \frac{59}{110}$$

105. BINOMIAL THEOREM

The writing of a binomial to any integral power might necessitate long and tedious multiplications for large values of the exponent. To overcome this, mathematicians have proved a theorem which is most useful. The student can easily verify by multiplication that

$$(a+b)^1 = a+b$$
$$(a+b)^2 = a^2+2ab+b^2$$
$$(a+b)^3 = a^3+3a^2b+3ab^2+b^3$$
$$(a+b)^4 = a^4+4a^3b+6a^2b^2+4ab^3+b^4$$
$$(a+b)^5 = a^5+5a^4b+10a^3b^2+10a^2b^3+5ab^4+b^5$$

We seek an answer to the question, "What is the series of terms equal to $(a+b)^n$?" By noting the examples above for $n = 1,\ldots,5$, we can check the following characteristics.

(i) The first term is a^n.
(ii) The second term is $na^{n-1}b$.
(iii) The exponents of a decrease by 1 in succeeding terms, those of b increase by 1 in succeeding terms, and their sum in each term is n.
(iv) The coefficient of the "next" term is the coefficient of the preceding term multiplied by the exponent of a of that term and divided by the exponent of b of the "next" term.
(v) There are $n+1$ terms.
(vi) The coefficients of terms equidistant from the ends of the expansion are the same.

Assuming these characteristics to be true, we can write the **binomial formula**

$$(a+b)^n = a^n + na^{n-1}b + \frac{n(n-1)}{1\cdot 2}a^{n-2}b^2 + \frac{n(n-1)(n-2)}{1\cdot 2\cdot 3}a^{n-3}b^3 + \cdots + nab^{n-1} + b^n$$

It may be desired to find only a particular term of this expansion without having to write out the entire expansion. We shall therefore try to develop a formula for a general term, and the term usually considered is the rth term. We can easily check the following features.

(i) The exponent of b is 1 less than the number of the term, that is, $r-1$.
(ii) Since the sum of the exponents of a and b is n in each term, we have the exponent of a equal to $n-(r-1)$.
(iii) The denominator of the coefficient is $1\cdot 2\cdot 3\cdot\ \cdots(r-1)$.

(iv) The numerator has the same number of factors as the denominator, that is, $n(n - 1)(n - 2)\cdots(n - r + 2)$.

The rth term then becomes

$$\frac{n(n-1)(n-2)\cdots(n-r+2)}{1\cdot 2\cdot 3\cdot \cdots \cdot (r-1)}a^{n-r+1}b^{r-1}$$

ILLUSTRATIONS

1. Expand $(x + 2y)^4$.

SOLUTION

$$(x+2y)^4 = x^4 + 4x^3(2y) + \frac{4\cdot 3}{1\cdot 2}x^2(2y)^2 + \frac{4\cdot 3\cdot 2}{1\cdot 2\cdot 3}x(2y)^3 + \frac{4\cdot 3\cdot 2\cdot 1}{1\cdot 2\cdot 3\cdot 4}(2y)^4$$

$$= x^4 + 8x^3y + 24x^2y^2 + 32xy^3 + 16y^4$$

2. Find the first three terms and the 7th term of the expansion of $(x^{1/2} - y^2)^{14}$.

SOLUTION

$$(x^{1/2} - y^2)^{14} = (x^{1/2})^{14} + 14(x^{1/2})^{13}(-y^2) + \frac{14\cdot 13}{1\cdot 2}(x^{1/2})^{12}(-y^2)^2 + \cdots$$

$$= x^7 - 14x^{13/2}y^2 + 91x^6y^4 - \cdots$$

The 7th term is

$$\frac{14\cdot 13\cdot 12\cdot 11\cdot 10\cdot 9}{1\cdot 2\cdot 3\cdot 4\cdot 5\cdot 6}(x^{1/2})^{14-7+1}(-y^2)^{7-1} = 3003x^4y^{12}$$

106. FACTORIAL NOTATION

The denominators of the coefficients of the terms in the binomial expansion have a particularly interesting product; a special symbol is defined for such products.

The product of all the positive integers from 1 *to* n, *inclusive, is denoted by the symbol* n!.

$$1\cdot 2\cdot 3\cdot 4\cdot 5\cdots n = n!$$

It is read "*n* factorial" or less commonly "factorial *n*." Another symbol used for the same notation is $\underline{|n}$, but this is becoming obsolete. The values of factorials for $1 \leq n \leq 9$ are given in the table.

n	$n!$
1	1
2	2
3	6
4	24
5	120
6	720
7	5040
8	40320
9	362880

An interesting property of factorials is that

$$n! = n \cdot (n - 1)!$$

This permits us to *define* 0!, for by letting $n = 1$, we have

$$1! = 1 \cdot 0! \quad \text{or} \quad 0! = \frac{1}{1} = 1$$

Another property is that if $r < n$, then

$$\frac{n!}{r!} = n(n - 1) \cdots (r + 1)$$

107. SERIES

The sum of the terms of a sequence is called a **series.** In the preceding sections we have discussed the sums associated with arithmetic and geometric progressions, and in particular we found the sum of an infinite geometric progression. The binomial theorem gave us a formula for writing a particular series known as the binomial series.

An infinite series may *converge* or *diverge*; to establish the appropriate fact we consider the sequence of partial series and the definition given in Section 104. Let us consider the series

$$a_1 + a_2 + a_3 + a_4 + \cdots + a_n + \cdots$$

The partial sums are defined by

$$\begin{aligned} S_1 &= a_1 \\ S_2 &= a_1 + a_2 \\ S_3 &= a_1 + a_2 + a_3 \\ &\vdots \\ S_n &= a_1 + a_2 + \cdots + a_n \end{aligned}$$

If S_n approaches a limit S as n approaches infinity, then S is defined to be the sum of the infinite series, and the series is said to **converge to the sum S.** If S_n does not approach a limit, then the series is said to be **divergent.** The following are examples of divergent infinite geometric series.

$$1 + 1 + 1 + 1 + \cdots$$

and

$$1 - 1 + 1 - 1 + 1 - 1 + \cdots$$

We shall be interested in a particular class of series of the form

$$a_0 + a_1 x + a_2 x^2 + \cdots + a_n x^n + \cdots$$

where x is a variable and $a_0, a_1, \ldots, a_n, \ldots$ are constants. A series of this form is called a **power series in x**. If there exist values of x for which this series converges, then for each such value of x we can determine the sum, a number which we may denote by y. Thus a power series in x defines a function y over the domain of values of x for which the series converges; the domain is called the **interval of convergence.**

If we apply the binomial formula to a binomial with n equal to a number which is not a positive integer, we obtain a binomial series. Furthermore, if the binomial was of the form $a + bx$, we would have a formula for generating this special power series.

ILLUSTRATION. Write the series for $(1 + x)^{-1}$.

SOLUTION. Apply the binomial formula to obtain

$$\begin{aligned}(1 + x)^{-1} &= 1 - x + \frac{(-1)(-2)}{2} x^2 + \frac{(-1)(-2)(-3)}{3!} x^3 + \cdots \\ &= 1 - x + x^2 - x^3 + x^4 - \cdots\end{aligned}$$

In the study of the calculus, formulas are given for the generation of power series and methods are derived for determining the interval of convergence. One of these formulas generates the Maclaurin series to represent a given function.* This formula may be applied to the trigonometric, hyperbolic, and exponential functions. We list these series representations and their intervals of convergence.

(1) $\sin x = x - \dfrac{x^3}{3!} + \dfrac{x^5}{5!} + \cdots + (-1)^{n+1} \dfrac{x^{2n-1}}{(2n - 1)!} + \cdots$, all x

(2) $\cos x = 1 - \dfrac{x^2}{2!} + \dfrac{x^4}{4!} - \cdots + (-1)^{n+1} \dfrac{x^{2n-2}}{(2n - 2)!} + \cdots$, all x

(3) $\tan x = x + \dfrac{x^3}{3} + \dfrac{2x^5}{15} + \dfrac{17x^7}{315} + \cdots, x^2 < \dfrac{\pi^2}{4}$

(4) $e^x = 1 + x + \dfrac{x^2}{2!} + \cdots + \dfrac{x^n}{n!} + \cdots$, all x

(5) $\ln (1 + x) = x - \dfrac{x^2}{2} + \dfrac{x^3}{3} - \cdots + (-1)^{n+1} \dfrac{x^n}{n}$
$+ \cdots, -1 < x \leq 1$

*See C. O. Oakley, *The Calculus* (New York: Barnes and Noble, 1957).

$$(6)\ \text{Arcsin}\, x = x + \frac{1}{2}\cdot\frac{x^3}{3} + \frac{1\cdot 3}{2\cdot 4}\cdot\frac{x^5}{5} + \cdots$$
$$+ \frac{1\cdot 3\cdots(2n-3)}{2\cdot 4\cdots(2n-2)}\cdot\frac{x^{2n-1}}{(2n-1)},\ -1 < x < 1$$

$$(7)\ \text{Arctan}\, x = x - \frac{x^3}{3} + \frac{x^5}{5} - \cdots + (-1)^{n+1}\frac{x^{2n-1}}{2n-1} \pm \cdots$$
$$-1 \le x \le 1$$

$$(8)\ \sinh x = x + \frac{x^3}{3!} + \frac{x^5}{5!} + \cdots + \frac{x^{2n-1}}{(2n-1)!} + \cdots,\ \text{all } x$$

$$(9)\ \cosh x = 1 + \frac{x^2}{2!} + \frac{x^4}{4!} + \cdots + \frac{x^{2n}}{(2n)!} + \cdots,\ \text{all } x$$

$$(10)\ \sinh^{-1} x = x - \frac{1}{2}\cdot\frac{x^3}{3} + \frac{1\cdot 3}{2\cdot 4}\cdot\frac{x^5}{5} + \cdots,\ -1 < x < 1$$

The use of series provides us with a new representation of the functions discussed in this book. Most high speed computers use the series to find the values of these functions for given values of the argument x.

ILLUSTRATION. Find ln 0.98 to seven decimal places.

SOLUTION. We shall employ series (5).

$$\ln 0.98 = \ln(1 - .02) = (-.02) - \frac{(.02)^2}{2} - \frac{(.02)^3}{3} - \frac{(.02)^4}{4} - \cdots$$
$$= -.02 - .0002 - .00000267 - .00000004 - \cdots$$
$$= -.0202027$$

108. OPERATIONS WITH SERIES

We may add, subtract, multiply, and divide series to obtain a series which is usually convergent in the intersection of the intervals of the given series. By operating on known series we can obtain other series for some complicated functions or for simple functions, without using Maclaurin's formula.

ILLUSTRATIONS

1. By using the series for e^x find the series for cosh x.

SOLUTION. We employ series (4), Section 107.

$$e^x = 1 + x + \frac{x^2}{2!} + \frac{x^3}{3!} + \cdots + \frac{x^n}{n!} + \cdots$$

$$e^{-x} = 1 - x + \frac{x^2}{2!} - \frac{x^3}{3!} + \cdots + (-1)^n\frac{x^n}{n!} + \cdots$$

Add and divide by 2 to obtain

$$\frac{e^x + e^{-x}}{2} = 1 + \frac{x^2}{2!} + \frac{x^4}{4!} + \cdots + \frac{x^{2n}}{(2n)!} + \cdots = \cosh x$$

2. Find the first four nonzero terms of the series for $e^{\sin x}$.

SOLUTION. Employ the series of Section 107.

$$e^u = 1 + u + \frac{u^2}{2!} + \frac{u^3}{3!} + \frac{u^4}{4!} + \cdots$$

$$u = \sin x = x - \frac{x^3}{3!} + \frac{x^5}{5!} - \cdots$$

A direct substitution yields

$$e^{\sin x} = 1 + \left(x - \frac{x^3}{6} + \cdots\right) + \frac{1}{2}\left(x - \frac{x^3}{6} + \cdots\right)^2 + \frac{1}{6}\left(x - \frac{x^3}{6} + \cdots\right)^3 + \frac{1}{24}\left(x - \frac{x^3}{6} + \cdots\right)^4 + \cdots$$

$$= 1 + \left(x - \frac{x^3}{6} + \cdots\right) + \frac{1}{2}\left(x^2 - \frac{1}{3}x^4 + \cdots\right) + \frac{1}{6}\left(x^3 + \cdots\right) + \frac{1}{24}(x^4 + \cdots) + \cdots$$

$$= 1 + x + \frac{1}{2}x^2 - \frac{1}{8}x^4 + \ldots$$

NOTE. The division of a series by another is accomplished by "long hand" polynomial division.

3. Starting with the series of sin x and cos x, find the series for tan x.

SOLUTION

$$\tan x = \frac{\sin x}{\cos x} = \frac{x - \dfrac{x^3}{3!} + \dfrac{x^5}{5!} - \cdots}{1 - \dfrac{x^2}{2!} + \dfrac{x^4}{4!} - \cdots}$$

$$\begin{array}{r|l}
 & x + \frac{1}{3}x^3 + \frac{2}{15}x^5 \\
1 - \frac{x^2}{2} + \frac{x^4}{24} - \cdots & x - \frac{1}{6}x^3 + \frac{1}{120}x^5 - \cdots \\
 & x - \frac{1}{2}x^3 + \frac{1}{24}x^5 - \cdots \\
\hline
 & \frac{1}{3}x^3 - \frac{1}{30}x^5 - \cdots \\
 & \frac{1}{3}x^3 - \frac{1}{6}x^5 + \cdots \\
\hline
 & \frac{2}{15}x^5 + \cdots \\
 & \frac{2}{15}x^5 + \cdots \\
\hline
\end{array}$$

Thus,

$$\tan x = x + \frac{1}{3}x^3 + \frac{2}{15}x^5 + \cdots$$

109. EULER'S FORMULA

By employing series we can develop a well known formula and from it obtain some new expressions for the sine and cosine functions. Let us consider the series expansion of e^u. Let u be the pure imaginary number $i\theta$. By direct substitution and simplification, we obtain

$$\begin{aligned} e^{i\theta} &= 1 + (i\theta) + \frac{(i\theta)^2}{2!} + \frac{(i\theta)^3}{3!} + \frac{(i\theta)^4}{4!} + \frac{(i\theta)^5}{5!} + \cdots \\ &= \left(1 - \frac{\theta^2}{2!} + \frac{\theta^4}{4!} - \cdots\right) + i\left(\theta - \frac{\theta^3}{3!} + \frac{\theta^5}{5!} - \cdots\right) \\ &= \cos\theta + i \sin\theta \end{aligned}$$

The last line is obtained from series (1) and (2) of Section 107. This formula is known as **Euler's Formula.** By direct substitution we can also obtain

$$e^{-i\theta} = \cos\theta - i\sin\theta$$

If we add and subtract these expressions, we obtain

$$e^{i\theta} + e^{-i\theta} = 2\cos\theta \qquad \text{or} \qquad \frac{e^{i\theta} + e^{-i\theta}}{2} = \cos\theta$$

and

$$e^{i\theta} - e^{-i\theta} = 2i\sin\theta \qquad \text{or} \qquad \frac{e^{i\theta} - e^{-i\theta}}{2i} = \sin\theta$$

Thus we have obtained new expressions for the trigonometric functions which may be taken as definitions. Note the similarity between these forms and the definitions for the hyperbolic functions.

Euler's formula also gives us a new form for a complex number. Let us recall that

$$z = x + yi = r(\cos\theta + i\sin\theta)$$

If we now apply Euler's formula, we obtain

$$z = r e^{i\theta}$$

which is called the exponential form of a complex number. The conjugate complex number and the basic operations can be applied to this form; for example,

$$\bar{z} = r e^{-i\theta} \qquad \text{and} \qquad z_1 z_2 = r_1 r_2 e^{i(\theta_1 + \theta_2)}$$

Representations of this type permit us to extend the trigonometric, hyperbolic, and exponential functions to complex variables. This extension forms a complete topic in itself.

110. TRIGONOMETRIC SERIES

In the previous sections we stated that a function can be represented by a series expansion over the interval of convergence. It can also be demonstrated that a periodic function can, under certain conditions, be represented by a series of trigonometric terms; i.e.,

$$\begin{aligned} f(x) = a_0 &+ a_1 \sin\theta + a_2 \sin 2\theta + a_3 \sin 3\theta + \cdots \\ &+ b_1 \cos\theta + b_2 \cos 2\theta + b_3 \cos 3\theta + \cdots \end{aligned}$$

This is known as a trigonometric series. It was first announced by J. B. Fourier (1807), and is often referred to as the Fourier series. Formulas for the coefficients a_i and b_i are developed in more advanced mathematics courses.*

If the graph of the periodic function y is a periodic continuous curve, it can be approximated by a finite number of terms of the trigonometric series; i.e.,

$$y \doteq a_0 + \sum_{k=1}^{n} a_k \sin k\theta + \sum_{k=1}^{n} b_k \cos k\theta$$

If the curve is obtained from experimental data, we can calculate the trigonometric approximation by simple formulas; such an analysis is called **harmonic analysis**† and finds application in the study of the conduction of heat and wave motion.

111. EXERCISE 17

1. Find l and s in the following A.P.'s.
   a) $d = -3, a = 57,$ and $n = 6$
   b) $\frac{1}{3}, \frac{5}{6}, \frac{4}{3}, \ldots$ to 7 terms
2. Find l and s in the following G.P.'s.
   a) $\frac{1}{2}, \frac{1}{4}, \frac{1}{8}, \ldots$ to 8 terms
   b) $\sqrt{2}, \sqrt{6}, 3\sqrt{2}, \ldots$ to 7 terms
3. Insert seven arithmetic means between 2 and 6.
4. Insert five geometric means between $\frac{1}{2}$ and 2048.

*See J. M. H. Olmsted, *Advanced Calculus* (New York: Appleton-Century-Crofts, 1961).

†See Kaj L. Nielsen, *Methods in Numerical Analysis* (2nd ed.; New York: Macmillan, 1964).

5. Change the repeating decimal 0.3454545 . . . into an equivalent common fraction.
6. Find the first four terms in the expansions.
   a) $(3x - 2y)^{1/3}$
   b) $(x^{1/2} - 2y^{1/3})^{-2}$
   c) $(ax + by)^5$
7. Find the 7th term in the expansion of $(2x^{2/3} - 3y^{1/2})^{-1/2}$.
8. Use the series of Section 107 to find series expansions for the following functions (express the first 4 terms).
   a) $\cot x$ b) $\tanh x$
   c) $\sin^2 x$ d) $e^x \cos x$
   e) $e^x \sin x$ f) $\dfrac{1}{1 + e^x}$
   g) e h) $\sin 1$
9. Use the binomial theorem to find a series for $\dfrac{x^2}{(1 + x^2)^2}$.
*10. Find series expansions for the following.
   a) $\sin iu$ b) $\cos iu$
   c) $\sinh iu$ d) $\cosh iu$
   Inspect these series and determine relations between the trigonometric functions and hyperbolic functions.

chapter 11

Spherical Triangles

112. INTRODUCTION

A very interesting application of the trigonometric functions to three-dimensional geometry is in the study of the surface of a sphere. Since the earth can be approximated by a sphere and the heavenly bodies can be studied by projecting them on a fictitious sphere of infinite radius,* spherical trigonometry has been studied extensively by mathematicians and astronomers. We shall assume that the details of solid geometry associated with the sphere are familiar to the reader, and simply list some of the definitions and properties which will be needed.

113. REVIEW OF SOLID GEOMETRY

A plane is a surface such that a straight line joining any two points of the surface lies entirely in the surface. A plane is determined by

a) a line and a point not on the line,
b) two intersecting lines,
c) two parallel lines,
d) three points not in a straight line.

If two planes intersect, their intersection is a straight line, and four space angles are formed, which are called **dihedral angles.** The line of intersection of the two planes is called the **edge** of the dihedral angle, and each of the planes is called a **face**. Two lines, one in each face of the dihedral angle and perpendicular to the edge at a common point, form a plane angle which is the measure of the dihedral angle. See Figure 88, on the following page.

If three planes intersect, they will have only one point of intersection in common. The three planes thus form a space corner, and the angle is called a **trihedral angle.** The point of intersection

*This is called the celestial sphere; see Section 130.

is called the *vertex*, and there are three **face angles** formed by pairs of lines of intersection of the planes. The faces of a trihedral angle form dihedral angles. Figure 89 shows the trihedral angle *O-ABC*.

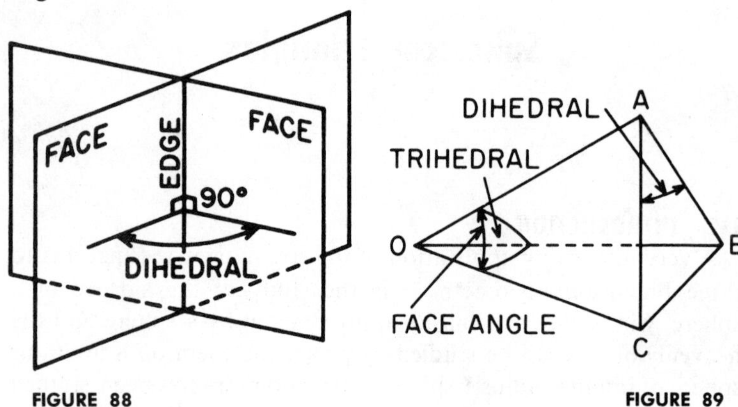

FIGURE 88 FIGURE 89

We shall now state some properties associated with solid angles and planes.

PROPERTY I. *If two angles which are not in the same plane have their sides parallel each to each and extended in the same direction from their vertices, the angles are equal.*

PROPERTY II. *If a plane is perpendicular to two intersecting planes, it is perpendicular to their line of intersection. The converse is also true.*

PROPERTY III. *The sum of two face angles of a trihedral angle is greater than the third face angle.*

114. THE SPHERE

Let us summarize the definitions and some properties. A **spherical surface** is a curved surface every point of which is at an equal distance from a fixed point on the surface. The fixed point is called the center, and a line joining the center with any point on the surface is called a radius. In spherical trigonometry the word sphere is used to mean the spherical surface.

If a plane intersects a sphere, the intersection is always a circle. If the intersecting plane passes through the center of the sphere, the intersection is called a **great circle**; if not, it is called a **small circle.** The **poles** of a circle of a sphere are the two ends of the diameter of the sphere which is perpendicular to the plane of the circle. The **polar distance** is the great circle arc between the given circle and its poles; in the case of a great circle, it is 90°; in the case of a small circle, the polar distance is the shorter of the two great circle arcs. See Figure 90.

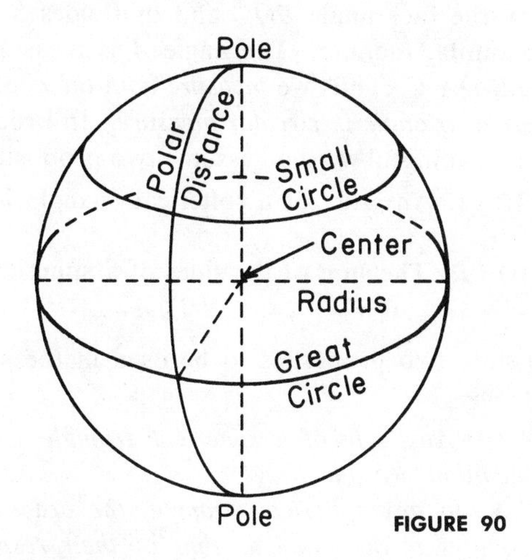

FIGURE 90

If two circles on a sphere intersect, the angle between them is defined to be the angle between their tangents. If the circles are great circles, the angle between them is called a **spherical angle** and the point of intersection is known as the *vertex*. A spherical angle has the same measure as the arc of the great circle drawn so that its pole is the vertex of the angle and included between the sides of the angle. In Figure 91, the arc BC has the same measure as the spherical angle BAC. If two great circles intersect at the ends of their diameter, i.e. have the same diameter, the spherical surface included between them is called a **lune** and the two angles are equal. The spherical angle and lune are shown in Figure 91.

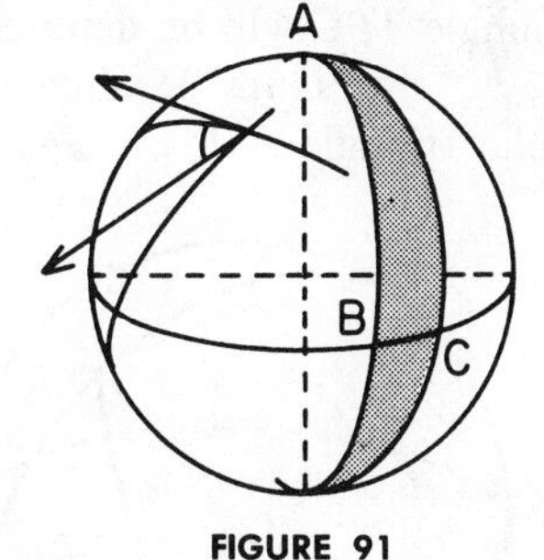

FIGURE 91

115. THE SPHERICAL TRIANGLE

The faces of a trihedral angle, whose vertex is at the center of a sphere, intersect the sphere in arcs which are great circles. The surface of the sphere bounded by these arcs is called a **spherical**

triangle and the arcs are *sides*. The three spherical angles formed by the intersections of these arcs form the vertices of the triangle. Figure 92 shows a spherical triangle with vertices at A, B, and C and sides a, b, and c. It should be clear from the figure that side a is measured by the face angle BOC and that sides b and c are measured in a similar manner. The angle A is measured by the dihedral angle B-OA-C. Thus we *measure both the angles and the sides of a spherical triangle in circular measure.* In order to make our discussion meaningful we shall assume two propositions.

PROPOSITION I. Any side of a spherical triangle is less than 180°.

PROPOSITION II. The sum of the sides of a spherical triangle is less than 360°; i.e., $a + b + c < 360°$.

We shall also state two properties to be used in the solution of spherical triangles.

PROPERTY IV. *Any side of a spherical triangle is less than the sum of the other two sides.*

PROPERTY V. *In any spherical triangle, the order of magnitude of the angles is the same as that of their respective opposite sides.*

116. POLAR TRIANGLES

If arcs of great circles are drawn with the vertices of a spherical triangle as poles, the arcs form a second triangle which is called the **polar triangle** of the first. The angles of the polar triangle ABC will be denoted by A', B', and C' and the sides, by a', b', c'. Figure 93 shows the spherical triangle ABC and its polar triangle $A'B'C'$. We note that the arc $A'C' = b'$ lies on

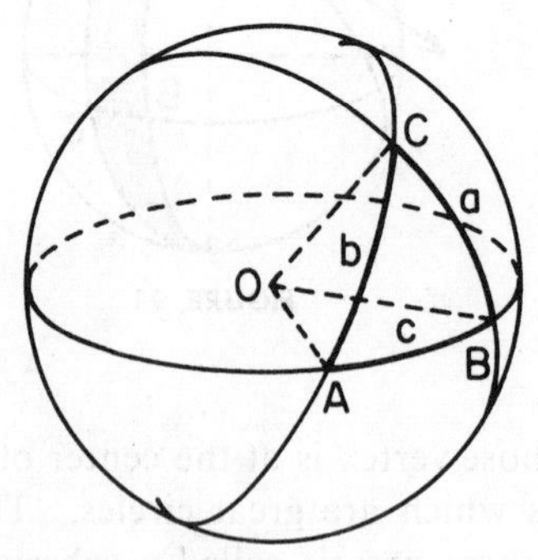

FIGURE 92

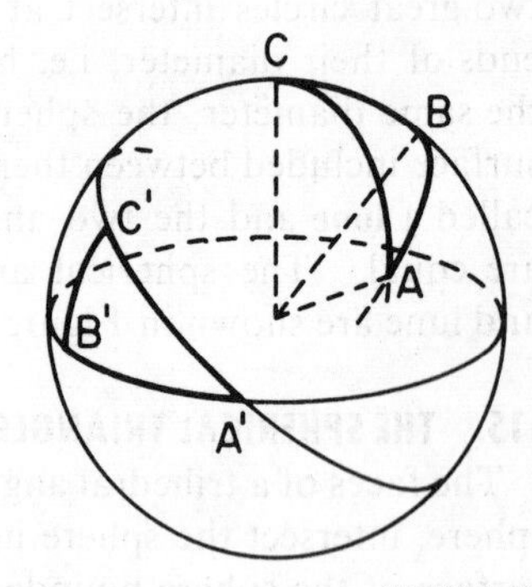

FIGURE 93

the great circle which has B as a pole; $A'B' = c'$, on the circle with C as a pole; and $B'C' = a'$, on the circle with A as a pole. We state two more properties without proof.

PROPERTY VI. *If $A'B'C'$ is the polar triangle of a spherical triangle ABC, then ABC is the polar triangle of $A'B'C'$.*

PROPERTY VII. *In two mutually polar triangles, an angle of one is the supplement of the side opposite the corresponding angle of the other.*

The last property can be stated in formula form.

the angles:	$A = 180° - a'$	or	$A' = 180° - a$
	$B = 180° - b'$	or	$B' = 180° - b$
	$C = 180° - c'$	or	$C' = 180° - c$
the sides:	$a = 180° - A'$	or	$a' = 180° - A$
	$b = 180° - B'$	or	$b' = 180° - B$
	$c = 180° - C'$	or	$c' = 180° - C$

Since we have assumed that any side of a spherical triangle is less than 180°, it follows that *any angle of a spherical triangle is less than* 180°. This can be demonstrated by using Property VII, for since $a' < 180°$ and $A = 180° - a'$, then $A < 180°$; similarly, for the other angles. We now have a theorem which may surprise some readers.

THEOREM. *The sum of the angles of a spherical triangle is greater than* 180° *but less than* 540°.

PROOF. By Property VII, we have $A = 180° - a'$, $B = 180° - b'$, and $C = 180° - c'$. If we add these, we obtain

$$A + B + C = 540° - (a' + b' + c') < 540°$$

By Proposition II, we have

$$a' + b' + c' < 360°$$

so that $A + B + C > 540° - 360° = 180°$

117. THE LAW OF SINES

In a spherical triangle the sines of the angles are proportional to the sines of their opposite sides.

$$\frac{\sin A}{\sin a} = \frac{\sin B}{\sin b} = \frac{\sin C}{\sin c}$$

PROOF. A spherical triangle ABC and its associated trihedral are shown in Figure 94. If planes are passed through C, perpendicular to OB and OA, they will intersect on a line CR which is perpendicular to the plane OAB, thus forming the right triangles PCR and QCR whose right angles are at R. Furthermore, by Property I, $\angle CPR = \angle B$ and $\angle CQR = \angle A$. Triangles OCP and OCQ are also right triangles with right angles at P and Q, respectively. From these plane right triangles, we have

$$\sin a = \frac{CP}{CO} \qquad \sin B = \frac{CR}{CP}$$

so that

$$\sin a \sin B = \frac{CP}{CO} \cdot \frac{CR}{CP} = \frac{CR}{CO}$$

Similarly

$$\sin b = \frac{CQ}{CO} \qquad \sin A = \frac{CR}{CQ}$$

so that

$$\sin b \sin A = \frac{CQ}{CO} \cdot \frac{CR}{CQ} = \frac{CR}{CO}$$

Consequently, $\dfrac{CR}{CO} = \sin a \sin B = \sin b \sin A$

or

$$\frac{\sin B}{\sin b} = \frac{\sin A}{\sin a}$$

By constructing planes through A perpendicular to OC and OB, we can prove similarly that

$$\frac{\sin B}{\sin b} = \frac{\sin C}{\sin c}$$

thus completing the proof of the law of sines.

118. THE LAW OF COSINES FOR SIDES

The cosine of a side of a spherical triangle is equal to the product of the cosines of the other two sides plus the product of the sines of those two sides multiplied by the cosine of their included angle.

$$\cos a = \cos b \cos c + \sin b \sin c \cos A$$
$$\cos b = \cos a \cos c + \sin a \sin c \cos B$$
$$\cos c = \cos a \cos b + \sin a \sin b \cos C$$

PROOF. The spherical triangle ABC and its associated trihedral are shown in Figure 95. Construct a plane tangent to the sphere at

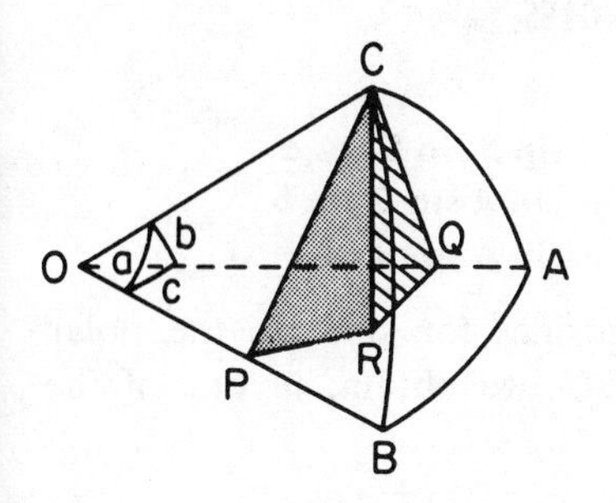

FIGURE 94

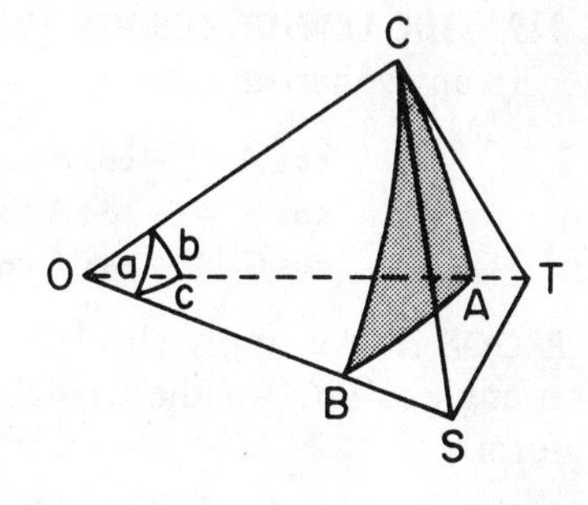

FIGURE 95

C so that the triangles SOC and TOC are right triangles with right angles at C and $\angle SCT = \angle C$. For ease of notation we let

$$ST = x \qquad OS = m \qquad OT = n$$
$$CS = y \qquad CT = z \qquad OC = w$$

The law of cosines for plane triangles applied to triangles SOT and SCT yields

$$x^2 = m^2 + n^2 - 2mn \cos c$$

and
$$x^2 = y^2 + z^2 - 2yz \cos C$$

If we subtract the second equation from the first, we get

$$0 = (m^2 - y^2) + (n^2 - z^2) - 2mn \cos c + 2yz \cos C$$

Since SOC and TOC are right triangles, we have

$$m^2 - y^2 = w^2 = n^2 - z^2$$

and the above equation reduces to

$$0 = 2w^2 - 2mn \cos c + 2yz \cos C$$

A division by $2mn$ yields

$$0 = \frac{w}{m} \cdot \frac{w}{n} - \cos c + \frac{y}{m} \cdot \frac{z}{n} \cos C$$

and again, since SOC and TOC are right triangles, we have

$$0 = \cos a \cos b - \cos c + \sin a \sin b \cos C$$

or
$$\cos c = \cos a \cos b + \sin a \sin b \cos C$$

The other forms of the cosine law for sides can be proved by constructing planes tangent to the sphere at A and B.

119. THE LAW OF COSINES FOR ANGLES

In any spherical triangle

$$\cos A = -\cos B \cos C + \sin B \sin C \cos a$$
$$\cos B = -\cos A \cos C + \sin A \sin C \cos b$$
$$\cos C = -\cos A \cos B + \sin A \sin B \cos c$$

PROOF. If we apply the law of cosines for sides to the polar triangle $A'B'C'$ of the triangle ABC, we obtain, in one of the forms,

$$\cos c' = \cos a' \cos b' + \sin a' \sin b' \cos C'$$

Property VII enables us to replace the sides and angles of this equation by the supplements of the corresponding angles and sides of triangle ABC. Recall also the properties of the trigonometric functions of supplementary angles. Consequently, the equation can be transformed into

$$-\cos C = \cos A \cos B - \sin A \sin B \cos c$$

or

$$\cos C = -\cos A \cos B + \sin A \sin B \cos c$$

The other forms of the cosine law for angles can be derived in a similar manner.

120. THE RIGHT SPHERICAL TRIANGLE

In a spherical triangle we may have one, two, or even three right angles. If we remember that the angles of a spherical triangle are measured the same as the dihedral angles of the associated trihedral, we can obtain the parts of a triangle with two or three right angles. We state the result as a property.

PROPERTY VIII. *If a spherical triangle has two right angles, the sides opposite these angles are quadrants (90° arcs) and the third angle has the same measure as its opposite side.*

If all three angles in a spherical triangle are right angles, the measure of each side is 90°.

Let us now consider the right spherical triangle with one right angle, and let the angle $C = 90°$ so that $\sin C = 1$ and $\cos C = 0$. If we apply this condition to the laws of sines and cosines for a spherical triangle (see Sections 117–119), we obtain, by direct substitution,

$$\frac{\sin A}{\sin a} = \frac{1}{\sin c} \quad \text{or} \quad \sin a = \sin c \sin A \qquad (1)$$

$$\frac{\sin B}{\sin b} = \frac{1}{\sin c} \qquad \text{or} \qquad \sin b = \sin c \sin B \tag{2}$$

$$\cos c = \cos a \cos b \tag{3}$$
$$\cos A = \sin B \cos a \tag{4}$$
$$\cos B = \sin A \cos b \tag{5}$$
$$0 = -\cos A \cos B + \sin A \sin B \cos c \qquad \text{or}$$
$$\cos c = \cot A \cot B \tag{6}$$

A substitution of (3) into the other two formulas for the cosine law for sides yields

$$\cos a = \cos^2 b \cos a + \sin b \sin c \cos A$$

or
$$\cos A = \frac{\cos a(1 - \cos^2 b)}{\sin b \sin c} = \frac{\cos a \sin b}{\sin c}$$

and, since $\cos a = \cos c/\cos b$, we further obtain

$$\cos A = \frac{\cos c \sin b}{\sin c \cos b} = \cot c \tan b \tag{7}$$

Similarly, $\cos b = \cos^2 a \cos b + \sin a \sin c \cos B$

or
$$\cos B = \frac{\cos b\,(1 - \cos^2 a)}{\sin a \sin c} = \frac{\cos b \sin a}{\sin c} = \cot c \tan a \tag{8}$$

Formulas (1) and (2) may be solved for $\sin A$ and $\sin B$ and used to write

$$\cot A = \frac{\cos A}{\sin A} = \frac{\cos a \sin b}{\sin a}$$

$$\text{or} \qquad \sin b = \tan a \cot A \tag{9}$$

Similarly,
$$\cot B = \frac{\cos B}{\sin B} = \frac{\cos b \sin a}{\sin b}$$

$$\text{or} \qquad \sin a = \tan b \cot B \tag{10}$$

These ten formulas will be used to find the parts of a right spherical triangle. Let us regroup them in summary form.

sin *a* = sin *c* sin *A*	**cos *A* = sin *B* cos *a***
sin *a* = tan *b* cot *B*	**cos *A* = tan *b* cot *c***
sin *b* = sin *c* sin *B*	**cos *B* = sin *A* cos *b***
sin *b* = tan *a* cot *A*	**cos *B* = tan *a* cot *c***
cos *c* = cos *a* cos *b*	**cos *c* = cot *A* cot *B***

121. NAPIER'S RULES

The preceding formulas are difficult to memorize; in order to facilitate their use, John Napier established two rules. Since we have taken $C = 90°$ in the right spherical triangle, we may consider a, b, A, c, and B to be the unknown parts. If we take them in this order (see Figure 96), we note that they have been listed in a counterclockwise manner. Instead of considering the hypotenuse and the two angles, let us consider their complements.

$$\bar{A} = 90° - A \qquad \bar{c} = 90° - c \qquad \bar{B} = 90° - B$$

and arrange them in the order $\bar{c}$, $\bar{B}$, a, b, and $\bar{A}$. These are known as **Napier's parts** of the triangle; the order is a clockwise cycle. The circular feature, illustrated in Figure 97, is a great aid in the application of the formulas. Considering any three parts, we see that either the three are together in the cycle, or one of them is separated from the other two. If they are all together, the middle one of the three is called the *middle* part and the two on each side are said to be *adjacent*. If one is separated from the other two, this one is called the *middle* part and the other two are said to be *opposite*.

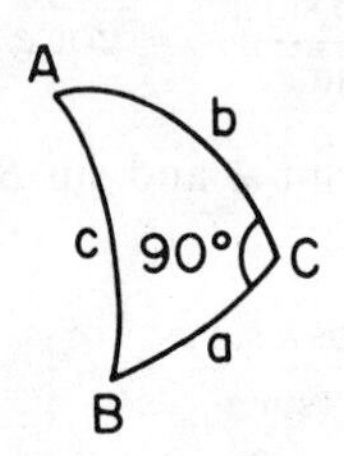

FIGURE 96

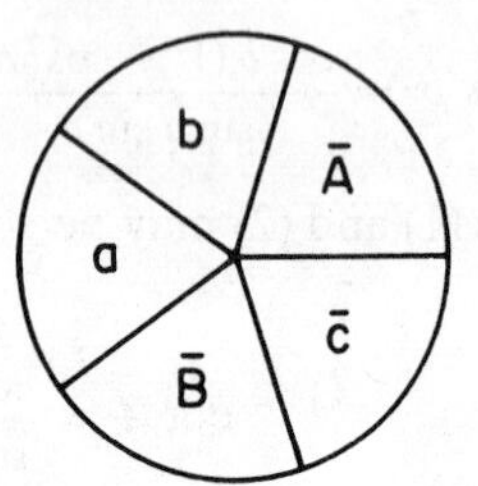

FIGURE 97

ILLUSTRATIONS

1. In the three parts a, b, and $\bar{A}$, the side b is the middle part, and a and $\bar{A}$ are adjacent parts.

2. In the three parts a, $\bar{A}$, and $\bar{c}$, the side a is the middle part, and $\bar{A}$ and $\bar{c}$ are opposite parts.

We now state the rules.

RULE I. *The sine of a middle part is equal to the product of the cosines of the opposite parts.*

RULE II. *The sine of a middle part is equal to the product of the tangents of the adjacent parts.*

ILLUSTRATIONS

1. Apply Napier's rules to the three parts, a, b, and $\overline{A}$ to obtain one of the formulas listed on the bottom of page 183.

SOLUTION. The side b is a middle part with a and $\overline{A}$ adjacent parts. We therefore apply Rule II to obtain

$$\sin b = \tan a \tan \overline{A} = \tan a \cot A$$

2. Apply Napier's rules to the three parts, a, b, and $\overline{c}$.

SOLUTION. The side $\overline{c}$ is a middle part with a and b opposite parts. Apply Rule I to obtain

$$\sin \overline{c} = \cos a \cos b$$

or
$$\sin \overline{c} = \sin (90° - c) = \cos c = \cos a \cos b$$

122. RULES OF QUADRANTS

Since the magnitude of an angle or side of a spherical triangle may be between 0° and 180°, it may represent either a first- or second-quadrant plane angle. When we use cosine, tangent, or cotangent functions to compute an unknown part; the sign of the function determines the quadrant in which the part lies and consequently, its magnitude. When, however, the sine function is used, there may be an ambiguity since the sine is positive in both quadrants. In order to determine whether one or both of the two angles found by means of a sine function is a solution of a triangle, we use the following rules.

RULE I. *In a right spherical triangle an oblique angle and the side opposite are in the same quadrant.*

The truth of this statement may be seen from the equation

$$\cos B = \sin A \cos b$$

where B represents either of the angles which is not the right angle. Since $A < 180°$, $\sin A > 0$; therefore, $\cos B$ and $\cos b$ have the same sign, and the angle B and the side b are in the same quadrant.

RULE II. *When the hypotenuse of a right spherical triangle is less than 90°, the legs are in the same quadrant; when the hypotenuse is greater than 90°, the legs are in different quadrants.*

The truth of this rule can be seen from the equation

$$\cos c = \cos a \cos b$$

When $c < 90°$, $\cos c$ is positive, and $\cos a$ and $\cos b$ are either both positive or both negative; in either case, a and b are in the same quadrant. On the other hand, when $c > 90°$, $\cos c$ is negative and either $\cos a$ is positive and $\cos b$ is negative, or the converse; in either case, a and b are in different quadrants.

RULE III. *When the two given parts are a leg and its opposite angle, there are always two solutions.*

Let a and A be the given parts. Consider the line AA' which is intersected by a great circle arc BC perpendicular to one side of the lune thus forming two right spherical triangles ABC and $A'BC$. (See Figure 98.) Since A and A' are vertices of a lune, they

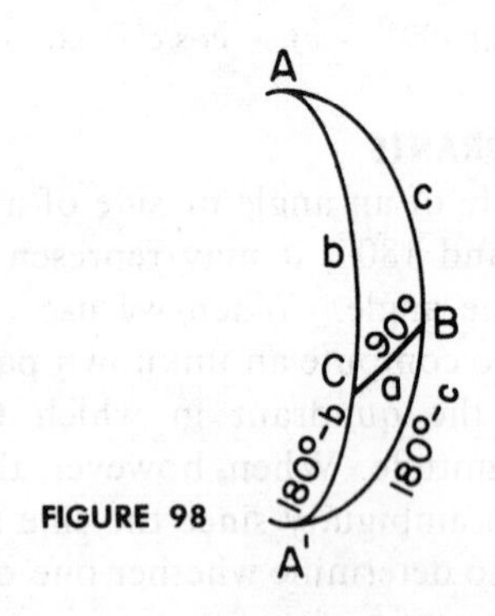

FIGURE 98

are equal, and the arc a is the opposite side in each triangle. Thus, unless BC divides AA' in half, we have two different triangles containing the angle A and the side a.

123. SOLUTION OF THE RIGHT SPHERICAL TRIANGLE

When any two parts (other than the right angle) of a right spherical triangle are given, the remaining three parts can be found. A systematic procedure should be used to perform the computation. The formulas for computation are obtained by applying Napier's rules. We also suggest the following.

a) Compute all unknown parts directly from the given data, not from previously computed parts.
b) Check the resulting solutions by a formula that requires the use of all three solutions at the same time.

In the following illustrations, we will abbreviate "log sin" as "l sin," "log cos" as "l cos," etc.

ILLUSTRATIONS

1. Solve the right spherical triangle given that $c = 110°30'$ and $B = 33°20'$.

SOLUTION

Given: $c = 110°30'$ and $B = 33°20'$.		
To find b: $\sin b = \cos \bar{c} \cos \bar{B}$ or $\sin b = \sin c \sin B$	$\text{l} \sin c = 9.9716{-}10$ $\text{l} \sin B = 9.7400{-}10$ $\text{l} \sin b = 9.7116{-}10$	$b = 30°59'$ $b = 149°1'$ [by Rule II]
To find a: $\sin \bar{B} = \tan \bar{c} \tan a$ or $\tan a = \tan c \cos B$	$\text{l} \tan c = 0.4273$ (n) $\text{l} \cos B = 9.9219{-}10$ $\text{l} \tan a = 0.3492$ (n)	$a = 114°7'$
To find A: $\sin \bar{c} = \tan \bar{A} \tan \bar{B}$ or $\cot A = \cos c \tan B$	$\text{l} \cos c = 9.5443{-}10$ (n) $\text{l} \tan B = 9.8180{-}10$ $\text{l} \cot A = 9.3623{-}10$ (n)	$A = 102°58'$
Check: $\sin b = \tan a \tan \bar{A}$ or $\sin b = \tan a \cot A$	$\text{l} \tan a = 0.3492$ (n) $\text{l} \cot A = 9.3623{-}10$ (n) $\text{l} \sin b = 9.7115{-}10$	check against $\text{l} \sin b$ above

NOTE: Rule II was applied after side a was found. The notation (n) indicates that the functional value is negative. The check is made against the logarithms which assumes that the antilogs have been correctly computed.

2. Solve the right spherical triangle if $A = 40°$ and $c = 115°$.

SOLUTION

Given: $A = 40°$ and $c = 115°$.		
To find a: $\sin a = \sin A \sin c$	$\text{l} \sin A = 9.8081{-}10$ $\text{l} \sin c = 9.9573{-}10$ $\text{l} \sin a = 9.7654{-}10$	$a = 35°38'$ ~~$a = 144°22'$~~ [by Rule I]

To find B: $\cot B = \cos c \tan A$	$\text{l} \cos c = 9.6259\text{–}10$ (n) $\text{l} \tan A = 9.9238\text{–}10$ $\text{l} \cot B = 9.5497\text{–}10$ (n)	$B = 109°32'$
To find b: $\tan b = \cos A \tan c$	$\text{l} \cos A = 9.8843\text{–}10$ $\text{l} \tan c = 0.3313$ (n) $\text{l} \tan b = 0.2156$ (n)	$b = 121°20'$
Check: $\sin a = \tan b \cot B$	$\text{l} \tan b = 0.2156$ (n) $\text{l} \cot B = 9.5497\text{–}10$ (n) $\text{l} \sin a = 9.7653\text{–}10$	$9.7654\text{–}10$

3. Solve the right spherical triangle if $a = 45°$ and $A = 60°$.

SOLUTION. The formulas furnish the following.

$$\sin c = \frac{\sin a}{\sin A} = \frac{\sqrt{2}/2}{\sqrt{3}/2} = \frac{\sqrt{6}}{3} = \frac{2.44949}{3} = 0.8165$$

$$\sin b = \tan a \cot A = (1)\left(\frac{\sqrt{3}}{3}\right) = \frac{1.73205}{3} = 0.5774$$

$$\sin B = \frac{\cos A}{\cos a} = \frac{1/2}{\sqrt{2}/2} = \frac{1}{\sqrt{2}} = \frac{\sqrt{2}}{2}$$

Check: $\sin b = \sin c \sin B$

$$\frac{\sqrt{3}}{3} = \frac{\sqrt{6}}{3} \cdot \frac{\sqrt{2}}{2} = \frac{\sqrt{12}}{6} = \frac{\sqrt{3}}{3}$$

The corresponding angles are:

$$c_1 = 54°44' \qquad b_1 = 35°16' \qquad B_1 = 45°$$
$$c_2 = 125°16' \qquad b_2 = 144°44' \qquad B_2 = 135°$$

124. THE ISOSCELES AND QUADRANTAL TRIANGLES

We shall now consider two special triangles. An *isosceles triangle* is one with two equal sides; for example, $a = b$. If a great circle is passed through C and the midpoint M of the arc AB, the triangle ABC is divided into two symmetrical right triangles with right angles at M. If any two of the distinct parts (a, c, A, C) are given, we can solve the triangle ABC by solving one of the resulting right triangles.

ILLUSTRATION. Solve the isosceles spherical triangle if $a = b = 40°30'$ and $A = 78°$.

SOLUTION. Consider the right spherical triangle ACM with arc $AC = b$ as the hypotenuse and angle A given. Then arc $AM = \frac{1}{2}c$ and $\angle ACM = \frac{1}{2}C$.

Given: $b = c^* = 40°30', A = 78°$.		
To find $\frac{1}{2}c = b^*$		
	$\text{l} \cos A = 9.3179\text{–}10$	
	$\text{l} \tan c^* = 9.9315\text{–}10$	$\frac{1}{2}c = 10°4'$
$\tan b^* = \cos A \tan c^*$	$\text{l} \tan b^* = 9.2494\text{–}10$	$c = 20°8'$
To find $\frac{1}{2}C = B^*$		
	$\text{l} \cos c^* = 9.8810\text{–}10$	
$\cot B^* = \cos c^* \tan A$	$\text{l} \tan A = 0.6725$	$\frac{1}{2}C = 15°37'$
	$\text{l} \cot B^* = 0.5535$	$C = 31°14'$
Check:		
	$\text{l} \sin c^* = 9.8125\text{–}10$	
$\sin b^* = \sin c^* \sin B^*$	$\text{l} \sin B^* = 9.4300\text{–}10$	
	$\text{l} \sin b^* = 9.2425\text{–}10$	$\frac{1}{2}c = 10°4'$

A **quadrantal triangle** is a spherical triangle with one *side* equal to 90°. The polar triangle of a quadrantal triangle is a right spherical triangle since the angle corresponding to the 90° side is $C = 180° - c = 180° - 90° = 90°$. The quadrantal triangle can be solved by first solving the corresponding polar triangle and then converting the values to the original triangle through the application of Property VII.

125. EXERCISE 18

1. Given c and B, use Napier's rules to write the formulas for finding a, b, and A.
2. Use Napier's rules to write three formulas, each involving A and B.
3. Given $A = 62°35'$ and $c = 71°17'$, solve for the remaining parts of the right spherical triangle.
4. Draw the right spherical triangle ABC with the right angle at C and the associated trihedral $O\text{-}ABC$ with $OA = OB = OC = 1$. Construct a plane AMN through A perpendicular to OB thus forming three plane right angles, $\angle ONA$, $\angle AMN$, and $\angle OMA$. Use the right triangle definitions of the trigonometric function to derive the formula $\sin b = \sin c \sin B$.

5. Find the unknown parts of the right spherical triangle ABC for each of the following.

   a) $c = 61°, B = 57°30'$ b) $a = 32°16', b = 125°25'$
   c) $a = 45°, B = 120°$ d) $b = 46°40', B = 57°30'$

6. Given $a = 90°, b = 65°30', B = 45°$, solve for A, C, and c.

126. FORMULAS

In order to solve the oblique spherical triangle, we use the addition formulas discussed in Sections 27-33 and a group of formulas which may be derived from previous laws. The common notation for half the perimeter of a triangle will also be used.

$$s = \tfrac{1}{2}(a + b + c)$$

The law of cosines for sides can be rewritten by solving for the cosine of the angle.

$$\cos A = \frac{\cos a - \cos b \cos c}{\sin b \sin c}$$

If we subtract this expression from 1, and make substitutions by applying the formulas indicated, we obtain

$$1 - \cos A = 1 - \frac{\cos a - \cos b \cos c}{\sin b \sin c}$$

$$= \frac{\sin b \sin c + \cos b \cos c - \cos a}{\sin b \sin c}$$

[(10), §27] $$= \frac{\cos (b - c) - \cos a}{\sin b \sin c}$$

[(32), §33] $$= \frac{2 \sin \frac{1}{2}(a + b - c) \sin \frac{1}{2}(a - b + c)}{\sin b \sin c}$$

The left-hand side can be changed by applying formula (18), § 31; and since

$$s - a = \tfrac{1}{2}(b + c - a)$$
$$s - b = \tfrac{1}{2}(a + c - b)$$
$$s - c = \tfrac{1}{2}(a + b - c)$$

we arrive at the expression

$$2 \sin^2 \tfrac{1}{2} A = \frac{2 \sin (s - b) \sin (s - c)}{\sin b \sin c}$$

or $$\sin \tfrac{1}{2} A = \sqrt{\frac{\sin (s - b) \sin (s - c)}{\sin b \sin c}}$$

If we add $\cos A$ to 1 and make the same type of substitutions by applying the addition formulas, we obtain

$$1 + \cos A = 1 + \frac{\cos a - \cos b \cos c}{\sin b \sin c}$$
$$= \frac{\cos a - (\cos b \cos c - \sin b \sin c)}{\sin b \sin c}$$
$$= \frac{\cos a - \cos (b + c)}{\sin b \sin c}$$

which can be changed to $\cos \frac{1}{2} A = \sqrt{\dfrac{\sin s \sin (s - a)}{\sin b \sin c}}$

Consequently, we have

$$\tan \tfrac{1}{2} A = \frac{\sin \frac{1}{2} A}{\cos \frac{1}{2} A}$$
$$= \sqrt{\frac{\sin (s - b) \sin (s - c)}{\sin (s - a) \sin s}}$$

If we multiply by $1 = \sqrt{\dfrac{\sin (s - a)}{\sin (s - a)}}$

and let $p = \sqrt{\dfrac{\sin (s - a) \sin (s - b) \sin (s - c)}{\sin c}}$

$$\tan \tfrac{1}{2} A = \frac{p}{\sin (s - a)}$$

Similar formulas can be derived in terms of angles B and C by starting with the expressions for $\cos B$ and $\cos C$,

$$\tan \tfrac{1}{2} B = \frac{p}{\sin (s - b)} \qquad \tan \tfrac{1}{2} C = \frac{p}{\sin (s - c)}$$

These are known as the half-angle formulas.

Formulas (11) and (12) of Section 28 may be written

$$\sin \tfrac{1}{2} (A + B) = \sin \tfrac{1}{2} A \cos \tfrac{1}{2} B + \cos \tfrac{1}{2} A \sin \tfrac{1}{2} B$$
$$\sin \tfrac{1}{2} (A - B) = \sin \tfrac{1}{2} A \cos \tfrac{1}{2} B - \cos \tfrac{1}{2} A \sin \tfrac{1}{2} B$$

Divide the second equation by the first, and then divide the numerator and denominator of the right-hand side by $\sin \frac{1}{2}A \sin \frac{1}{2}B$ to obtain

$$\frac{\sin \frac{1}{2}(A - B)}{\sin \frac{1}{2} (A + B)} = \frac{\cot \frac{1}{2} B - \cot \frac{1}{2} A}{\cot \frac{1}{2} B + \cot \frac{1}{2} A}$$

Since the cotangent is the reciprocal of the tangent, the half-angle formulas may be used to change the right-hand side of this equation to

$$\frac{\sin \frac{1}{2}(A - B)}{\sin \frac{1}{2}(A + B)} = \frac{\sin (s - b) - \sin (s - a)}{\sin (s - b) + \sin (s - a)}$$

[(29), (30), § 33]
$$= \frac{2 \cos \frac{1}{2} c \sin \frac{1}{2}(a - b)}{2 \sin \frac{1}{2} c \cos \frac{1}{2}(a - b)}$$

$$= \frac{\tan \frac{1}{2}(a - b)}{\tan \frac{1}{2} c}$$

If we start with the expressions for $\cos \frac{1}{2}(A - B)$ and $\cos \frac{1}{2}(A + B)$ and proceed in the same manner as above, we obtain

$$\frac{\cos \frac{1}{2}(A - B)}{\cos \frac{1}{2}(A + B)} = \frac{\cot \frac{1}{2} A \cot \frac{1}{2} B + 1}{\cot \frac{1}{2} A \cot \frac{1}{2} B - 1}$$

$$= \frac{\dfrac{\sin (s - a) \sin (s - b)}{p^2} + 1}{\dfrac{\sin (s - a) \sin (s - b)}{p^2} - 1}$$

$$= \frac{\sin s + \sin (s - c)}{\sin s - \sin (s - c)}$$

$$= \frac{\tan \frac{1}{2}(a + b)}{\tan \frac{1}{2} c}$$

If we write these formulas in terms of the parts of the polar triangle $A'B'C'$ and apply Property VII so that

$$\begin{aligned}
\tfrac{1}{2}(A' - B') &= \tfrac{1}{2}(180^\circ - a - 180^\circ + b) = -\tfrac{1}{2}(a - b) \\
\tfrac{1}{2}(A' + B') &= 180^\circ - \tfrac{1}{2}(a + b) \\
\tfrac{1}{2} C' &= 90^\circ - \tfrac{1}{2} c \\
\tfrac{1}{2}(a' - b') &= -\tfrac{1}{2}(A - B) \\
\tfrac{1}{2}(a' + b') &= 180^\circ - \tfrac{1}{2}(A + B) \\
\tfrac{1}{2} c' &= 90^\circ - \tfrac{1}{2} C
\end{aligned}$$

we then obtain
$$\frac{\sin \frac{1}{2}(a - b)}{\sin \frac{1}{2}(a + b)} = \frac{\tan \frac{1}{2}(A - B)}{\cot \frac{1}{2} C}$$

$$\frac{\cos \frac{1}{2}(a - b)}{\cos \frac{1}{2}(a + b)} = \frac{\tan \frac{1}{2}(A + B)}{\cot \frac{1}{2} C}$$

These formulas are known as **Napier's analogies.**

We shall derive one more set of formulas, known as Gauss's formulas, which are useful in checking the solution of triangles since each involves all the sides and all the angles of the spherical triangle. Let us again start with the equation

$$\sin \tfrac{1}{2}(A - B) = \sin \tfrac{1}{2}A \cos \tfrac{1}{2}B - \cos \tfrac{1}{2}A \sin \tfrac{1}{2}B$$

and alter the right-hand side by applying the formulas of this section and Sections 27–33 to obtain

$$\begin{aligned}
\sin \tfrac{1}{2}(A - B) &= \sqrt{\frac{\sin^2 (s - b) \sin (s - c) \sin s}{\sin a \sin b \sin^2 c}} \\
&\qquad - \sqrt{\frac{\sin s \sin^2 (s - a) \sin (s - c)}{\sin a \sin b \sin^2 c}} \\
&= \frac{\sin (s - b) - \sin (s - a)}{\sin c} \sqrt{\frac{\sin s \sin (s - c)}{\sin a \sin b}} \\
&= \frac{2 \sin \frac{1}{2}(a - b) \cos \frac{1}{2}c}{2 \sin \frac{1}{2}c \cos \frac{1}{2}c} \cos \tfrac{1}{2}C
\end{aligned}$$

$$\boldsymbol{\sin \tfrac{1}{2}(A - B) = \frac{\sin \frac{1}{2}(a - b)}{\sin \frac{1}{2}c} \cos \tfrac{1}{2}C}$$

There are three other forms of Gauss's formula which are listed in the summary below.

Consider the Napier analogy

$$\frac{\cos \frac{1}{2}(A - B)}{\cos \frac{1}{2}(A + B)} = \frac{\tan \frac{1}{2}(a + b)}{\tan \frac{1}{2}c}$$

Since the angles and sides must be less than 180°, we see that $A - B < 180^\circ$ and $\frac{1}{2}c < 90^\circ$. Therefore, $\cos \frac{1}{2}(A - B)$ and $\tan \frac{1}{2}c$ will always be positive, and then $\cos \frac{1}{2}(A + B)$ and $\tan \frac{1}{2}(a + b)$ must have the same sign, and the angles will be in the same quadrant. This fact is stated as a rule.

RULE OF QUADRANTS. *In any spherical triangle, one-half the sum of two angles is in the same quadrant as one-half the sum of the opposite sides.*

The above formulas are summarized for easy reference.

The Law of Sines

$$\frac{\sin A}{\sin a} = \frac{\sin B}{\sin b} = \frac{\sin C}{\sin c} \tag{1}$$

The Cosine Law

$$
\begin{aligned}
\cos a &= \cos b \cos c + \sin b \sin c \cos A \\
\cos b &= \cos a \cos c + \sin a \sin c \cos B \\
\cos c &= \cos a \cos b + \sin a \sin b \cos C
\end{aligned} \tag{2}
$$

The Half-Angle Formulas

$$\tan \tfrac{1}{2} A = \frac{p}{\sin (s - a)} \qquad \tan \tfrac{1}{2} B = \frac{p}{\sin (s - b)} \tag{3}$$

$$\tan \tfrac{1}{2} C = \frac{p}{\sin (s - c)}$$

where

$$p = \sqrt{\frac{\sin (s - a) \sin (s - b) \sin (s - c)}{\sin s}}, \quad s = \tfrac{1}{2}(a + b + c)$$

The Four Analogies of Napier

$$\frac{\sin \frac{1}{2}(A - B)}{\sin \frac{1}{2}(A + B)} = \frac{\tan \frac{1}{2}(a - b)}{\tan \frac{1}{2} c} \tag{4}$$

$$\frac{\cos \frac{1}{2}(A - B)}{\cos \frac{1}{2}(A + B)} = \frac{\tan \frac{1}{2}(a + b)}{\tan \frac{1}{2} c} \tag{5}$$

$$\frac{\sin \frac{1}{2}(a - b)}{\sin \frac{1}{2}(a + b)} = \frac{\tan \frac{1}{2}(A - B)}{\cot \frac{1}{2} C} \tag{6}$$

$$\frac{\cos \frac{1}{2}(a - b)}{\cos \frac{1}{2}(a + b)} = \frac{\tan \frac{1}{2}(A + B)}{\cot \frac{1}{2} C} \tag{7}$$

A Formula of Gauss

$$
\begin{aligned}
\sin \tfrac{1}{2}(A - B) &= \frac{\sin \frac{1}{2}(a - b)}{\sin \frac{1}{2} c} \cos \tfrac{1}{2} C \\
\sin \tfrac{1}{2}(A + B) &= \frac{\cos \frac{1}{2}(a - b)}{\cos \frac{1}{2} c} \cos \tfrac{1}{2} C \\
\cos \tfrac{1}{2}(A - B) &= \frac{\sin \frac{1}{2}(a + b)}{\sin \frac{1}{2} c} \sin \tfrac{1}{2} C \\
\cos \tfrac{1}{2}(A + B) &= \frac{\cos \frac{1}{2}(a + b)}{\cos \frac{1}{2} c} \sin \tfrac{1}{2} C
\end{aligned} \tag{8}
$$

127. THE OBLIQUE SPHERICAL TRIANGLE

A spherical triangle can be determined when three of its six parts are given. However, if the given parts are two sides and an angle opposite one of these sides, two triangles may exist. The triangles are determined by the proper use of the formulas derived in the last section. We will consider six cases classified according to the parts which are given.

Case I. *Given two sides and the included angle.*

Suppose that a, b, and C are given.

1. Apply Formula (6) to find $\frac{1}{2}(A - B)$.
   Apply Formula (7) to find $\frac{1}{2}(A + B)$.
   Add the results to obtain A and subtract to find B.
2. Apply Formula (4) or (5) to find c.
3. Check the solution by the law of sines.

NOTE: If it is desired to find only c, the law of cosines [Formula (2)] may be used.

Case II. *Given two angles and the included side.*

Let $\tilde{A}$, B, and c be the given parts.

1. Find $\frac{1}{2}(a - b)$ by Formula (4).
2. Find $\frac{1}{2}(a + b)$ by Formula (5).
3. Add to obtain a and subtract to obtain b.
4. Find C by Formula (6) or (7).
5. Check the solution by the law of sines.

Case III. *Given two sides and an angle opposite one of them.*

This is the ambiguous case. Let the given parts be b, c, and C.

1. Find the angle B by the law of sines. We obtain two angles, the acute angle B_1, and $B_2 = 180° - B_1$. If both B_1 and B_2 satisfy the *rule of quadrants*, there are two solutions. If only one of the values of B_1 and B_2 satisfies the rule, there is only one solution.
2. Find the angle A by Formula (7).
3. Find the side a by Formula (5).
4. Check the solution by Formula (8).

Case IV. *Given two angles and a side opposite one of them.*

Let A, B, and a be the given parts.

1. Find the side b by the law of sines. This yields two solutions for b, and we determine the number of triangles by the *rule of quadrants*.
2. Find the angle C by Formula (6) or (7).
3. Find the side c by Formula (4) or (5).
4. Check the solution by Formula (8).

Case V. *Given three sides.*

1. Find the three angles by Formula (3).
2. Check the solution by the law of sines.

Case VI. *Given three angles.*

1. Find a', b', c' of the polar triangle

$$a' = 180^\circ - A \qquad b' = 180^\circ - B \qquad c' = 180^\circ - C$$

2. Solve for A', B', C' by method of Case V.
3. Find a, b, and c by

$$a = 180^\circ - A' \qquad b = 180^\circ - B' \qquad c = 180^\circ - C'$$

4. Check the solution by the law of sines.

ILLUSTRATIONS. For Cases I–V see pages 197–201.

Since we are using 4-place tables, the checks will not always be exact to the last digit. If precise calculations are desired, at least 5-place tables should be used, and the angles and sides should be measured to the nearest second or fractions of units which correspond to the tables used.

In the ambiguous cases, it is possible to have given values for which there is no solution. This will be the condition if log sin of an unknown part is greater than zero; for example, if

$$\text{l} \sin B = \text{l} \sin b + \text{l} \sin C - \text{l} \sin c > 0$$

Case VI. Given $A = 60^\circ$, $B = 130^\circ$, $C = 85^\circ$, find a, b, and c.

SOLUTION. The associated polar triangle has

$$\begin{aligned} a' &= 180^\circ - A = 180^\circ - 60^\circ = 120^\circ \\ b' &= 180^\circ - B = 180^\circ - 130^\circ = 50^\circ \\ c' &= 180^\circ - C = 180^\circ - 85^\circ = 95^\circ \end{aligned}$$

The angles A', B', and C' are now found by the method of Case V. For this illustration, the values were chosen so that the sides of the polar triangle would have the same values of the triangle used in the illustration for Case V; thus we have

$$A' = 125^\circ 34' \qquad B' = 46^\circ \qquad C' = 69^\circ 20'$$

We now obtain

$$\begin{aligned} a &= 180^\circ - A' = 180^\circ - 125^\circ 34' = 54^\circ 26' \\ b &= 180^\circ - B' = 180^\circ - 46^\circ = 134^\circ \\ c &= 180^\circ - C' = 180^\circ - 69^\circ 20' = 110^\circ 40' \end{aligned}$$

Case I	*Given:* $a = 135°40'$, $c = 60°$, $B = 142°20'$, find b, A, and C.	
To find A and C: $\tan\frac{A + C}{2} = \frac{\cot\frac{1}{2}B\cos\frac{1}{2}(a - c)}{\cos\frac{1}{2}(a + c)}$	$a = 135°40'$ $c = 60°$ $a + c = 195°40'$ $a - c = 75°40'$ $B = 142°20'$	$\frac{1}{2}(a + c) = 97°50'$ $(82°10')$ $\frac{1}{2}(a - c) = 37°50'$ $\frac{1}{2}B = 71°10'$
$\tan\frac{A - C}{2} = \frac{\cot\frac{1}{2}B\sin\frac{1}{2}(a - c)}{\sin\frac{1}{2}(a + c)}$	$\text{l}\cot\frac{1}{2}B = 9.5329{-}10$ $\text{l}\cos\frac{1}{2}(a - c) = 9.8975{-}10$ $\text{col}\cos\frac{1}{2}(a + c) = 0.8655$ (n) $\text{l}\tan\frac{1}{2}(A + C) = 0.2959$ (n) $\text{l}\cot\frac{1}{2}B = 9.5329{-}10$ $\text{l}\sin\frac{1}{2}(a - c) = 9.7877{-}10$ $\text{col}\sin\frac{1}{2}(a + c) = 0.0041$ $\text{l}\tan\frac{1}{2}(A - C) = 9.3247{-}10$	$\frac{1}{2}(A + C) = 116°50'$ $\frac{1}{2}(A - C) = 11°56'$ $A = 128°46'$ $C = 104°54'$
To find b: $\tan\frac{1}{2}b = \frac{\tan\frac{1}{2}(a - c)\sin\frac{1}{2}(A + C)}{\sin\frac{1}{2}(A - C)}$	$\text{l}\tan\frac{1}{2}(a - c) = 9.8902{-}10$ $\text{l}\sin\frac{1}{2}(A + C) = 9.9505{-}10$ $\text{col}\sin\frac{1}{2}(A - C) = 0.6845$ $\text{l}\tan\frac{1}{2}b = 0.5252$	$\frac{1}{2}b = 73°23'$ $b = 146°46'$
Check: $\frac{\sin A}{\sin a} = \frac{\sin B}{\sin b} = \frac{\sin C}{\sin c}$	$\text{l}\sin A = 9.8919{-}10$ $\text{l}\sin a = 9.8444{-}10$ Diff. $= 0.0475$ $\text{l}\sin B = 9.7861{-}10$ $\text{l}\sin b = 9.7388{-}10$ Diff. $= 0.0473$	$\text{l}\sin C = 9.9851{-}10$ $\text{l}\sin c = 9.9375{-}10$ Diff. $= 0.0476$

Case II — *Given:* $A = 57°$, $B = 137°20'$, $c = 94°40'$, find a, b, and C.

To find a and b:

$$\tan \tfrac{1}{2}(b + a) = \frac{\cos \frac{1}{2}(B - A) \tan \frac{1}{2} c}{\cos \frac{1}{2}(B + A)}$$

$$\tan \tfrac{1}{2}(b - a) = \frac{\sin \frac{1}{2}(B - A) \tan \frac{1}{2} c}{\sin \frac{1}{2}(B + A)}$$

$B = 137°20'$		
$A = 57°$		
$B + A = 194°20'$	$\frac{1}{2}(B + A) = 97°10'$	$(82°50')$
$B - A = 80°20'$	$\frac{1}{2}(B - A) = 40°10'$	
$c = 94°40'$	$\frac{1}{2}c = 47°20'$	

$\text{l} \cos \frac{1}{2}(B - A) = 9.8832{-}10$	
$\text{l} \tan \frac{1}{2} c = 0.0354$	
$\text{col} \cos \frac{1}{2}(B + A) = 0.9039$ (n)	$\frac{1}{2}(b + a) = 98°33'$
$\text{l} \tan \frac{1}{2}(b + a) = 0.8225$ (n)	$\frac{1}{2}(b - a) = 35°12'$
$\text{l} \sin \frac{1}{2}(B - A) = 9.8096{-}10$	
$\text{l} \tan \frac{1}{2} c = 0.0354$	$b = 133°45'$
$\text{col} \sin \frac{1}{2}(B + A) = 0.0034$	$a = 63°21'$
$\text{l} \tan \frac{1}{2}(b - a) = 9.8484{-}10$	

To find C:

$$\cot \tfrac{1}{2} C = \frac{\tan \frac{1}{2}(B - A) \sin \frac{1}{2}(b + a)}{\sin \frac{1}{2}(b - a)}$$

$\text{l} \tan \frac{1}{2}(B - A) = 9.9264{-}10$	
$\text{l} \sin \frac{1}{2}(b + a) = 9.9951{-}10$	$\frac{1}{2} C = 34°38'$
$\text{col} \sin \frac{1}{2}(b - a) = 0.2393$	$C = 69°16'$
$\text{l} \cot \frac{1}{2} C = 0.1608$	

Check:

$$\frac{\sin A}{\sin a} = \frac{\sin B}{\sin b} = \frac{\sin C}{\sin c}$$

$\text{l} \sin A = 9.9236{-}10$	$\text{l} \sin B = 9.8311{-}10$	$\text{l} \sin C = 9.9709{-}10$
$\text{l} \sin a = 9.9512{-}10$	$\text{l} \sin b = 9.8588{-}10$	$\text{l} \sin c = 9.9986{-}10$
Diff. $= 9.9724{-}10$	Diff. $= 9.9723{-}10$	Diff. $= 9.9723{-}10$

Case III	*Given*: $b = 45°, c = 81°20', C = 83°40'$.	
	$c - b = 36°20'$ $c + b = 126°20'$	$\alpha = \frac{1}{2}(c - b) = 18°10'$ $\beta = \frac{1}{2}(c + b) = 63°10'$
To find B: $\sin B = \dfrac{\sin b \sin C}{\sin c}$	$\text{l} \sin b = 9.8495-10$ $\text{l} \sin C = 9.9973-10$ $\text{col} \sin c = 0.0050$ $\text{l} \sin B = 9.8518-10$	$B_1 = 45°18'$ ~~$B_2 = 134°42'$~~ (By Rule of Quadrants there is only one solution)
To find A: $\cot \frac{1}{2} A = \dfrac{\cos \beta \tan \gamma}{\cos \alpha}$ $\gamma = \frac{1}{2}(C + B) = 64°29'$	$\text{l} \cos \beta = 9.6546-10$ $\text{l} \tan \gamma = 0.3212$ $\text{col} \cos \alpha = 0.0222$ $\text{l} \cot \frac{1}{2} A = 9.9980-10$	$\frac{1}{2} A = 45°8'$ $A = 90°16'$
To find a: $\tan \frac{1}{2} a = \dfrac{\sin \gamma \tan \alpha}{\sin \delta}$ $\delta = \frac{1}{2}(C - B) = 19°11'$	$\text{l} \sin \gamma = 9.9554-10$ $\text{l} \tan \alpha = 9.5161-10$ $\text{col} \sin \delta = 0.4833$ $\text{l} \tan \frac{1}{2} a = 9.9548-10$	$\frac{1}{2} a = 42°1'$ $a = 84°2'$
Check: $\sin \frac{1}{2} a = \dfrac{\sin \alpha \cos \frac{1}{2} A}{\sin \delta}$	$\text{l} \sin \alpha = 9.4939-10$ $\text{l} \cos \frac{1}{2} A = 9.8485-10$ $\text{col} \sin \delta = 0.4833$ $\text{l} \sin \frac{1}{2} a = 9.8257-10$	$\text{l} \sin \frac{1}{2} a = 9.8257-10$

Case IV	*Given:* $A = 33°20'$, $B = 55°$, $a = 40°$, find b, c, and C.		
	$B - A = 21°40'$ $B + A = 88°20'$	$\alpha = \frac{1}{2}(B - A) = 10°50'$ $\beta = \frac{1}{2}(B + A) = 44°10'$	
<u>To find b:</u> $\sin b = \dfrac{\sin a \sin B}{\sin A}$	$\text{l} \sin a = 9.8081\text{–}10$ $\text{l} \sin B = 9.9134\text{–}10$ $\text{col} \sin A = 0.2600$ $\text{l} \sin b = 9.9815\text{–}10$	$b_1 = 73°23'$ $\gamma_1 = 16°42'$ $\delta_1 = 56°42'$	$b_2 = 106°37'$ $\gamma_2 = 33°18'$ $\delta_2 = 73°18'$
<u>To find C:</u> $\cot \frac{1}{2} C = \dfrac{\sin \delta \tan \alpha}{\sin \gamma}$ $\gamma = \frac{1}{2}(b - a)$ $\delta = \frac{1}{2}(b + a)$	$\text{l} \sin \delta_1 = 9.9221\text{–}10$ $\text{l} \tan \alpha = 9.2819\text{–}10$ $\text{sum} = 9.2040\text{–}10$ $\text{l} \sin \gamma_1 = 9.4584\text{–}10$ $\text{l} \cot \frac{1}{2} C_1 = 9.7456\text{–}10$	$\text{l} \sin \delta_2 = 9.9813\text{–}10$ $\text{l} \tan \alpha = 9.2819\text{–}10$ $\text{sum} = 9.2632\text{–}10$ $\text{l} \sin \gamma_2 = 9.7396\text{–}10$ $\text{l} \cot \frac{1}{2} C_2 = 9.5236\text{–}10$	$\frac{1}{2} C_1 = 60°54'$ $C_1 = 121°48'$ $\frac{1}{2} C_2 = 71°32'$ $C_2 = 143°4'$
<u>To find c:</u> $\tan \frac{1}{2} c = \dfrac{\sin \beta \tan \gamma}{\sin \alpha}$	$\text{l} \sin \beta = 9.8431\text{–}10$ $\text{l} \tan \gamma_1 = 9.4771\text{–}10$ $\text{col} \sin \alpha = 0.7260$ $\text{l} \tan \frac{1}{2} c_1 = 0.0462$	$\text{l} \sin \beta = 9.8431\text{–}10$ $\text{l} \tan \gamma_2 = 9.8175\text{–}10$ $\text{col} \sin \alpha = 0.7260$ $\text{l} \tan \frac{1}{2} c_2 = 0.3866$	$\frac{1}{2} c_1 = 48°3'$ $c_1 = 96°6'$ $\frac{1}{2} c_2 = 67°41'$ $c_2 = 135°22'$
<u>Check:</u> $\sin \alpha = \dfrac{\sin \gamma \cos \frac{1}{2} C}{\sin \frac{1}{2} c}$	$\text{l} \sin \gamma_1 = 9.4584\text{–}10$ $\text{l} \cos \frac{1}{2} C_1 = 9.6869\text{–}10$ $\text{col} \sin \frac{1}{2} c_1 = 0.1286$ $\text{l} \sin \alpha = 9.2739\text{–}10$	$\text{l} \sin \gamma_2 = 9.7396\text{–}10$ $\text{l} \cos \frac{1}{2} C_2 = 9.5007\text{–}10$ $\text{col} \sin \frac{1}{2} c_2 = 0.0338$ $\text{l} \sin \alpha = 9.2741\text{–}10$	

Case V	*Given:* $a = 120°$, $b = 50°$, $c = 95°$, find A, B, and C.		
To find $\log p$: [See Formulas (3)]	$a = 120°$ $b = 50°$ $c = 95°$ $2s = 265°$ $s = 132°30'$	$s - a = 12°30'$ $s - b = 82°30'$ $s - c = 37°30'$ $s = 132°30'$ (*check*)	$\text{l} \sin (s - a) = 9.3353{-}10$ $\text{l} \sin (s - b) = 9.9963{-}10$ $\text{l} \sin (s - c) = 9.7844{-}10$ $\text{col} \sin s = 0.1324$ $2 \log p = 9.2484{-}10$
To find A: $\tan \frac{1}{2} A = \dfrac{p}{\sin (s - a)}$	$\log p = 9.6242{-}10$ $\text{l} \sin (s - a) = 9.3353{-}10$ $\text{l} \tan \frac{1}{2} A = 0.2889$		$\frac{1}{2} A = 62°47'$
To find B: $\tan \frac{1}{2} B = \dfrac{p}{\sin (s - b)}$	$\log p = 9.6242{-}10$ $\text{l} \sin (s - b) = 9.9963{-}10$ $\text{l} \tan \frac{1}{2} B = 9.6279{-}10$		$\frac{1}{2} B = 23°$ $B = 46°$
To find C: $\tan \frac{1}{2} C = \dfrac{p}{\sin (s - c)}$	$\log p = 9.6242{-}10$ $\text{l} \sin (s - c) = 9.7844{-}10$ $\text{l} \tan \frac{1}{2} C = 9.8398{-}10$		$\frac{1}{2} C = 34°40'$ $C = 69°20'$
Check: $\dfrac{\sin A}{\sin a} = \dfrac{\sin B}{\sin b} = \dfrac{\sin C}{\sin c}$	$\text{l} \sin A = 9.9103{-}10$ $\text{l} \sin a = 9.9375{-}10$ Diff. $= 9.9728{-}10$	$\text{l} \sin B = 9.8569{-}10$ $\text{l} \sin b = 9.8843{-}10$ Diff. $= 9.9726{-}10$	$\text{l} \sin C = 9.9711{-}10$ $\text{l} \sin c = 9.9983{-}10$ Diff. $= 9.9728{-}10$

128. EXERCISE 19

1. Solve the spherical triangles having the following parts.

a) $b = 110°40', c = 70°20', A = 130°$
b) $A = 62°, C = 135°20', b = 98°$
c) $a = 72°, b = 52°, A = 48°$
d) $B = 55°, C = 110°, b = 60°$
e) $a = 132°, b = 54°, c = 110°$
f) $A = 110°, B = 120°, C = 130°$

129. THE TERRESTRIAL TRIANGLE

For many practical problems the earth is considered to be a sphere. The diameter about which the earth rotates is called its **axis** and the two extremeties of this axis are the **north** and **south poles**. The great circle for these poles is called the **equator**.

A **meridian** is a great circle on the earth which passes through the north and south pole; it is thus perpendicular to the equator. The meridian of Greenwich, England is chosen as a reference, and any other meridian is identified by its angular distance (0° to 180°) measured along the equator to the east or west of the meridian of Greenwich. This angular distance is called the **longitude**, λ, of a point on the given meridian. Since the earth makes a complete revolution in twenty-four hours, it turns through 15° in one hour. The longitude therefore can be measured in units of time according to the conversion units.

1 hour = 15°   1 minute = 15′   1 second = 15″

In this notation "h, m, s" are used for hours, minutes, and seconds.

The **latitude** L of a point on the earth is the angular distance north or south of the equator. All points on the earth that have the same latitude lie on a small circle parallel to the equator which is called a **parallel of latitude**. A point on the earth is determined by its latitude and longitude.

The distance between two points $P_1(L_1,\lambda_1)$ and $P_2(L_2,\lambda_2)$ is the measure of the great circle arc between these points. This distance may be measured in circular measure (the **angular distance**), but more frequently the linear distance is desired. For this purpose, a **nautical mile** has been defined to be the length along a great circle on the earth of an arc whose angular distance is 1′. In recent years there has been much discussion concerning the exact relationship between the nautical mile and the ordinary (statute) mile. We shall use the international nautical mile equal to 1852 meters or 1.15078 miles.

The two points P_1 and P_2 (Figure 99) together with the north pole form a spherical triangle P_1P_2N. (The south pole could be used if desired.) The given parts of this triangle are the colatitudes, arc $NP_1 = 90° - L_1$, arc $NP_2 = 90° - L_2$, and the included angle, $P_1NP_2 = \lambda_1 - \lambda_2$. The arc P_1P_2 is the distance d. The angles at P_1 and P_2 are called the **initial courses** from P_1 to P_2 and P_2 to P_1. It is necessary to adopt some conventions. The longitude of a point *west* of Greenwich is *positive*; *east* of Greenwich, *negative*. When using the north pole as one vertex of the triangle, a southern latitude (south of the equator) is negative. With these conventions it is possible for $\lambda_1 - \lambda_2$ to be greater than 180°. This means that we are considering the larger spherical triangle formed by P_1P_2N; but to stay within our assumptions of the previous sections, we transform to the smaller triangle by using $360° - (\lambda_1 - \lambda_2)$. The triangle is called the **terrestrial triangle** and the given parts are those of Case I.

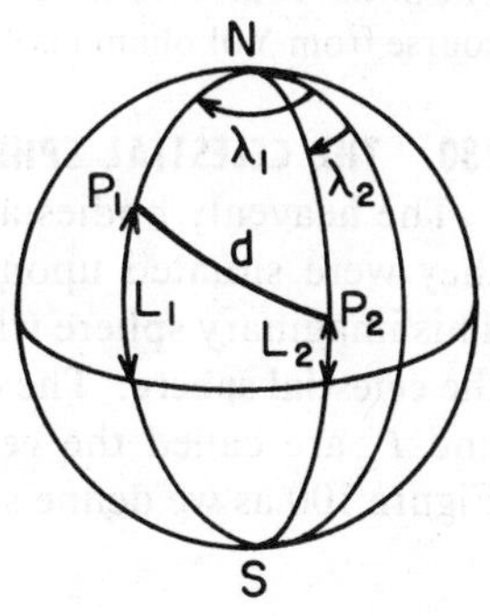

FIGURE 99

ILLUSTRATION.* Find the great circle course and distance between Vancouver, $L = 49°16' N$, $\lambda = 8^h\ 12^m\ 44^s$ W, and Yokohama, $L = 35°30'$ N, $\lambda = 9^h\ 18^m\ 40^s$ E.

SOLUTION. Let A represent Vancouver; B, Yokohama; and C, the north pole. The spherical triangle is

$$a = BC = 90° - 35°30' = 54°30'$$
$$b = AC = 90° - 49°16' = 40°44'$$
$$C = \lambda_1 - \lambda_2 = 8^h\,12^m\,44^s - (-9^h\,18^m\,40^s) = 17^h\,31^m\,24^s$$

Since this is greater than 12 hours, we use

$$C = 24^h - 17^h\,31^m\,24^s = 6^h\,28^m\,36^s = 97°9'$$

The triangle is solved by the method of Case I, obtaining

$$A = 60°36' \qquad B = 44°18' \qquad c = 68°4'$$

*The values have been rounded off to the nearest minute of circular measure to permit use of the tables in the back of this book. For practical purposes at least 5-place tables should be used.

The distance is $c = 4084' = 4084$ nautical miles $= 4700$ statute miles. The initial course from Vancouver is $60°36'$ west of north and the initial course from Yokohama is $44°18'$ east of north.

130. THE CELESTIAL SPHERE

The heavenly bodies appear to an observer on the earth as if they were situated upon the inner surface of a hollow sphere. This imaginary sphere which is concentric with the earth is called the **celestial sphere**. The extensions of the earth's axis to points P_n and P_s are called the **celestial north** and **south poles**. Refer to Figure 100 as we define some additional terminology.

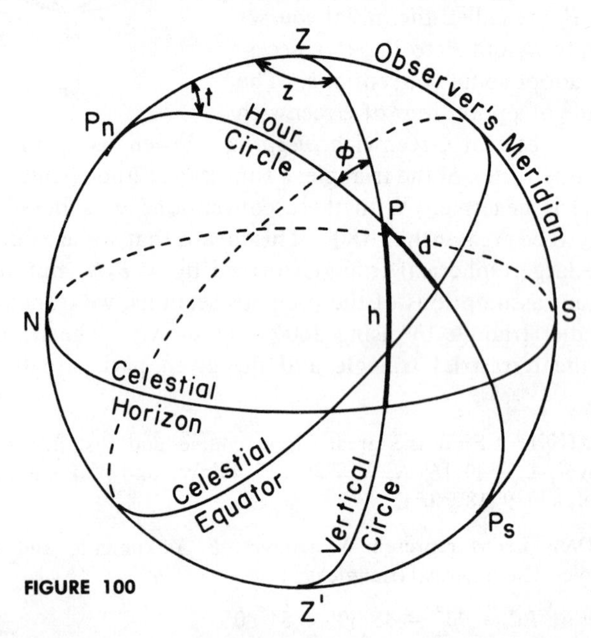

FIGURE 100

The **celestial equator** is the great circle on the celestial sphere whose poles are the celestial north and south poles. It is in the same plane as the earth's equator. The **zenith** of an observer is the point Z vertically above him on the celestial sphere. The diametrically opposite point Z' is called the **nadir**. The **horizon** of an observer is the great circle on the celestial sphere which has the zenith and nadir as its poles.

The **hour circle** of a point P on the celestial sphere is the great circle which passes through the poles and also through the point P. It corresponds to the meridian on the earth. The hour circle of the zenith is called the **observer's meridian**. The points at which the observer's meridian intersects the horizon are the **north point** and **south point**; an observer in the northern latitudes finds the north point below the celestial north pole. The **east** and **west points** are points on the horizon to the *right* and *left*, respectively, of an observer facing north.

The **vertical circle** of a point P on the celestial sphere is the great circle which passes through the zenith and nadir and the point P; it is perpendicular to the horizon.

The **hour angle** t of a celestial point P is the angle at the pole between the observer's meridian and the hour circle of P. It is measured positively through the west from 0 to 24 hours. The **sidereal day** is the time of one complete rotation of the star from the observer's meridian back to that meridian. The **sidereal hour** (sidereal day divided by 24) is less than the ordinary solar hour.*

The **declination** d of a celestial point P is the angular distance from P to the celestial equator measured along the hour circle of P. It is measured from $0°$ to $90°$ and corresponds to the latitude of a point on the earth.

The **altitude** h of a celestial point P is the angular distance from P to the horizon measured along the vertical circle of P. It is considered positive if P is above the horizon; negative, if P is below.†

The **azimuth** z of a celestial point P is the angle at the zenith between the observer's meridian and the vertical circle of P. It is measured from the north point around through the east from $0°$ to $360°$.

A point on the celestial sphere can be designated by its declination and hour angle, or by its altitude and azimuth. The declination of most stars is given in the *Nautical Almanac*. Unfortunately, both of these coordinate systems depend upon the time and place of the observation. Astronomers use a third coordinate system in which the astronomical bodies are located by means of declination and right ascension. The **right ascension** of a point on the celestial sphere is the angle at the pole between the hour circle of the point and a fixed hour circle determined by international agreement (like the Greenwich meridian). It is measured eastward from 0 hours to 24 hours.

The **position angle** ϕ of a heavenly body is the angle at the heavenly body between its hour circle and its vertical circle.

*Mean sidereal day = $23^h\ 56^m\ 4.1^s$ of mean solar time.
Mean solar day = $24^h\ 3^m\ 56.6'$ of mean sidereal time.

†The altitude can be determined with the aid of a sextant.

The **astronomical triangle** of a heavenly body is defined to be the spherical triangle formed by the celestial pole, the zenith Z, and the celestial point P, the projection of the heavenly body upon the celestial sphere. If the observer is at a northern latitude, the north pole P_n is used; south of the equator, P_s is used. The astronomical triangle for an observer north of the equator has the following parts.

$$\text{sides}\begin{cases}\text{arc } P_nZ = 90^\circ - L = \text{colatitude of observer} \\ \qquad\qquad\text{(codeclination of the zenith)} \\ \text{arc } P_nP = 90^\circ - d = \text{codeclination of star} \\ \text{arc } PZ = 90^\circ - h = \text{co-altitude of star}\end{cases}$$

$$\text{angles}\begin{cases}\text{angle } ZP_nP = t = \text{hour angle of star} \\ \text{angle } P_nZP = z = \text{azimuth of star} \\ \text{angle } ZPP_n = \phi = \text{position angle of star}\end{cases}$$

131. TYPICAL PROBLEMS

In this section we shall discuss some typical problems which may be solved through the use of the astronomical triangle.

I. *To find the time of day.* (Given latitude and date.)

If the latitude and date are known, the declination of the sun can be determined from the *Nautical Almanac* for the given date, and the altitude of the sun can be measured by means of a sextant. Thus the three sides of the astronomical triangle are known, and the problem can be solved by the method of Case V. The hour angle t of the sun represents the time since noon; it will be the observer's local time. If the sun is west of the meridian, t is the local time; if east of the meridian, the local time is $24^h - t$ since noon.

ILLUSTRATION. A navigator west of the meridian, at latitude 45°N, measures the altitude of the sun at 32°. The *Nautical Almanac* reports the declination of the sun on that day to be 5° S. Find the local time.

SOLUTION. $L = 45°\text{N}, d = 5°\ S = -5°, h = 32°.$

$$\begin{array}{ll} a = 90^\circ - L = 45^\circ & s - a = 54^\circ \\ b = 90^\circ - d = 95^\circ & s - b = 4^\circ \\ c = 90^\circ - h = 58^\circ & s - c = 41^\circ \\ \hline 2s = 198^\circ & \\ s = 99^\circ \longleftrightarrow & s = 99^\circ \end{array}$$

$\log \sin (s - a) = 9.9080{-}10$	
$\log \sin (s - b) = 8.8436{-}10$	
$\log \sin (s - c) = 9.8169{-}10$	
$\text{col} \sin s = 0.0054$	
$2 \log p = 8.5739{-}10$	$\frac{1}{2}t = 16°27'$
$\log p = 9.2869{-}10$	$t = 32°54'$
$\log \sin (s - c) = 9.8169{-}10$	$\doteq 2^h12^m$
$\log \tan t/2 = 9.4700{-}10$	

II. *To find the observer's longitude.* (Latitude, date known.)

This is an extension of the preceding problem. The local time t is determined and the Greenwich local time t' is found by radio. Then $t' - t$ is the angle between the Greenwich meridian and the observer's meridian and is, therefore, the longitude of the observer. In degrees, the longitude is $\lambda = 15\,(t' - t)°$.

III. *To find the time of sunrise* (*or sunset*).

This is a special case of Problem I, since at this time the altitude of the sun is 0° and the arc PZ is 90°. The spherical triangle has become a quadrantal triangle and can be solved by the method indicated in Section 124.

IV. *To find the observer's latitude.* (Local time known.)

If the local time is known and the altitude and declination of the sun can be measured and read from the almanac, then we know two sides and an angle opposite one of them in the astronomical triangle. This triangle can be solved by using the method of Case III to determine the third side, which is the colatitude of the observer.

V. *To find the altitude and azimuth of a star.*

If the declination and hour angle of a star and the observer's latitude are known, then the altitude and azimuth of the star can be calculated. In the astronomical triangle two sides and the included angle are known, and the triangle can be solved by the method of Case I.

Modern technology, in the form of radar and automatic computers, has eliminated the necessity of solving navigational and other problems associated with spherical triangles. However, the design of this equipment has its basis in the application of spherical trigonometry. The analysis of combat between airplanes, airplanes versus missiles, and missiles versus missiles uses the applications of spherical trigonometry and more advanced mathematics.

132. EXERCISE 20

1. Find the distance and initial course from Los Angeles (lat. 34° 3′ N, long. 118° 15′ W) to Honolulu (lat. 21° 18′ N, long. 157° 52′ W).
2. Find the solar time of a place on earth whose latitude is 35° N at a moment in the afternoon when the sun's declination is 20° 20′, if the sun's altitude is observed to be 50°.
3. Find the time of sunrise at a place with latitude 37° N on the day when the sun's declination is 10°.
4. The local time is $1^h\ 10^m$, the sun's declination is 8° 7′ and its altitude is 65°. Find the observer's latitude.

Appendix

Mathematical tables or tables which exhibit values that remain constant for specified arguments form a body of knowledge which is becoming more and more important. The subject of trigonometry provides an excellent introduction to the use of tables.

A.1. TRIGONOMETRIC TABLES*

Since the trigonometric functions of a given angle are constant, they may be computed and placed in a table. The values are usually expressed in decimal form, correct to a given number of places. There are many such tables; here, we shall use an abbreviated four-place table.

Before turning to the tables, let us first consider the matter of expressing a number correct to a certain number of places. Since, in general, these numbers are continuing decimals, we must "round off" the numbers to four places. This is accomplished by discarding all the digits to the right of the fourth place. If the digit in the 5th place is

(a) greater than half a unit in the 4th place, increase the digit in that place by 1;

(b) less than half a unit in the 4th place, leave the digit in that place unaltered;

(c) exactly half a unit in the 4th place,

  (i) increase an *odd* digit in the 4th place by 1,

  (ii) leave an *even* digit in the 4th place unaltered.

ILLUSTRATIONS. Express the following numbers to four places.

1. 64.347. Answer: 64.35.

2. 13.342. Answer: 13.34.

3. 7.2345. Answer: 7.234.

4. 7.2375. Answer: 7.238.

*For a complete discussion of more extensive tables see K. L. Nielsen and J. H. Vanlonkhuyzen, *Plane and Spherical Trigonometry* (New York: Barnes and Noble, 1954) pp. 8–12.

Table V, on page 254, is a four-place table of the trigonometric functions of the acute angles at every 10 minutes. The angles from 0° to 45° are labeled in the left-hand column, and the angles from 45° to 90° in the extreme right-hand column. The trigonometric functions are labeled at the top and the bottom. For the angles in the *left*-hand column we read the titles of the columns at the *top* of the page. For those in the extreme *right*-hand column we read the titles of the columns at the *bottom* of the page.

Let us now consider the following operations.

I. *Given the angle, find the trigonometric function.*

ILLUSTRATIONS

1. Find tan 13°40′.

SOLUTION. Find 13° in the extreme left column. Find 40′ under 13°, and opposite this, in the column headed by "tan" at the *top* of the page, read .2432. Thus tan 13°40′ = .2432.

2. Find cos 79°20′.

SOLUTION. Find 79° in the extreme right column. Find 20′ *above* 79° in this column, and opposite 20′, in the column headed by "cos" at the *bottom* of the page, read .1851. Thus cos 79°20′ = .1851.

II. *Given the trigonometric function, find the angle.*

ILLUSTRATIONS

1. Given $\sin\theta = .3283$, find θ.

SOLUTION. In the columns headed by "sin" search for .3283. This is found in the column headed by "sin" at the *top* of the page; therefore, look in the *left*-hand column and read 19°10′. Thus $\theta = 19°10'$.

2. Given $\cot\theta = .4950$, find θ.

SOLUTION. In the columns headed by "cot" search for .4950. This is found in the column headed by "cot" at the *bottom* of the page; therefore, look in the *right*-hand column for the angle and read 63°40′.

NOTE: In the *left* columns of the table the angle increases as one reads *downward*; in the **right** columns, as one reads *upward*.

III. *Interpolation.* If it is desired to obtain the trigonometric functions of an angle given to the nearest minute, it becomes necessary to interpolate for the values. This is done either by linear interpolation or by using tables of proportional parts. Let us discuss linear interpolation, which will suffice for our work. The difference between two entries in the table is called a **tabular difference**. Interpolation is accomplished by taking appropriate portions of this difference.

ILLUSTRATIONS

1. Find tan 49°33′.

SOLUTION. Since tan 49°33′ is between tan 49°30′ and tan 49°40′, we obtain these values and subtract one from the other.

$$\begin{array}{rcl} \tan 49^\circ 40' &=& 1.1778 \\ \tan 49^\circ 30' &=& 1.1708 \\ \hline \text{tabular difference} &=& .0070 \end{array}$$

The tabular difference of the function is now multiplied by the difference between the given angle and the smaller chosen angle, and then divided by the tabular difference of the angle. Thus $(.0070)(33 - 30)/10 = (.0070)(.3) = .0021$. This amount is now added to the tangent of the smaller angle (49°30′) and rounded off to four decimal places; the result is 1.1729. Thus

$$\tan 49^\circ 33' = 1.1729$$

The work may be arranged as follows.

$$\left.\begin{array}{rcl} \tan 49^\circ 30' &=& 1.1708 \\ \tan 49^\circ 33' &=& 1.1729 \\ \tan 49^\circ 40' &=& 1.1778 \end{array}\right] .0070(.3) = .0021$$

The difference between the angles in this table is a constant 10′; the division by 10 is a simple matter of inserting the decimal point.

2. Find cos 32°47′.

SOLUTION

$$\left.\begin{array}{rcl} \cos 32^\circ 40' &=& .8418 \\ \cos 32^\circ 47' &=& .8408 \\ \cos 32^\circ 50' &=& .8403 \end{array}\right] (-.0015)(.7) = -.00105$$

3. Given $\tan\theta = .8312$, find θ.

SOLUTION. Search the table in the column headed "tan" for the value closest to .8312. We find

$$\left.\begin{array}{rcl} \tan 39^\circ 40' &=& .8292 \\ \tan\theta &=& .8312 \\ \tan 39^\circ 50' &=& .8342 \end{array}\right] \tfrac{20}{50}(10) = 4$$

Subtract the value of the trigonometric function of the smaller angle from that of the larger angle and also from the given value. Form the fraction of these differences and multiply by 10 (the angular difference). The result rounded off to the nearest integer is the number of minutes to be added to the *smaller* angle. Thus

$$\theta = 39^\circ 40' + 4' = 39^\circ 44'$$

A.2. TRIGONOMETRIC FUNCTIONS FOR RADIAN MEASURE

If the argument is given in radian measure or is a pure number, we may use a table which has been calculated for this argument, or change the argument to degree measure. Table VI, on page 260, is a four-place table for radian measure.

ILLUSTRATIONS. Refer to Table VI.

1. Sin 0.53 = 0.5055.

2. Find tan 0.736.

SOLUTION.

$$\left.\begin{array}{ll} \tan 0.73 = & .8949 \\ & \underline{109} \\ \tan 0.736 = & .9058 \\ \tan 0.74 = & .9131 \end{array}\right] 182(.6) = 109.2$$

3. If $\cos x = .9020$, find x.

SOLUTION.

$$\left.\begin{array}{ll} \cos .44 = & .9048 \\ \cos x = & .9020 \\ \cos .45 = & .9004 \end{array}\right] \tfrac{28}{44}(10) \doteq 6$$

$$\therefore x = .446$$

A.3. COMMON LOGARITHMS*

It is easy to write down a table of numbers which are integral powers of 10 and the corresponding common logarithms.

Exponential Form	Logarithmic Form
...	...
$10^3 = 1000$	$\log 1000 = 3$
$10^2 = 100$	$\log 100 = 2$
$10^1 = 10$	$\log 10 = 1$
$10^0 = 1$	$\log 1 = 0$
$10^{-1} = 0.1$	$\log 0.1 = -1$
$10^{-2} = 0.01$	$\log 0.01 = -2$
$10^{-3} = 0.001$	$\log 0.001 = -3$
...	...

To find the logarithm of a number which is not an integral power of 10 can best be explained by considering an example such as $N = 110$. Since 110 is between 100 and 1000, it is natural to suppose, from the above table, that log 110 is between 2 and 3 or, in other words, the logarithm is 2 + (a proper fraction). Since

*See Section 66.

we can express the proper fraction in decimal form, we have, in general,

$$\log N = (\text{an integer}) + (0 \leq \text{decimal fraction} < 1)$$

The integral part is called the **characteristic**. The decimal fraction is called the **mantissa**. Thus

$$\log N = \text{characteristic} + \text{mantissa}$$

Since the mantissas may be nonrepeating infinite decimals, they are approximated to as many places as desired. The approximations have been tabulated in four-place, five-place, or higher-place tables which are called logarithmic tables. Thus the mantissas, or decimal parts, are found from tables, and the values in the tables are **always positive**.

The characteristic is determined according to the following two rules.

RULE I. *If the number* N *is greater than* 1, *the characteristic of its logarithm is one less than the number of digits to the left of the decimal point.*

RULE II. *If the number* N *is less than* 1, *the characteristic of its logarithm is negative; if the first digit which is not zero occurs in the* k*th decimal place, the characteristic is* $-k$.

Since the characteristic and mantissa are combined to give the complete logarithm,

$$\log N = \text{characteristic} + \text{mantissa}$$

and, further, since the mantissa is always positive, it is best to write a negative characteristic, $-k$, as $(10 - k) - 10$.

ILLUSTRATIONS

1. Write the logarithms given a mantissa of .3942 and characteristics 1, 0, -1, and -2.

SOLUTION

Characteristic	Logarithm
1	1.3942
0	0.3942
−1	9.3942 − 10
−2	8.3942 − 10

2. Find the characteristics of the logarithms of the numbers given in the left column.

SOLUTION

Number	Characteristic
197.3	2
81.72	1
6.291	0
0.3962	9 − 10
0.0815	8 − 10
0.000073	5 − 10

If the logarithm is given, the problem becomes one of finding the number corresponding to this logarithm. The number is called the **antilogarithm**. The characteristic of the given logarithm determines the position of the decimal point in the antilogarithm. In placing the decimal point we use the same two rules given for determining the characteristic of a number. However, we must remember it is a reverse problem.

ILLUSTRATION. The digits of an antilogarithm are 7329. Place the decimal point if the characteristic is 1, 2, 8 − 10.

SOLUTION

Characteristic	Number
1	73.29
2	732.9
8 − 10 = −2	0.07329

A.4. LOGARITHMIC TABLES AND THEIR USE

To find the mantissa of a logarithm, we shall use the Table of the Common Logarithms on page 246. This table is a four-place table, and, although it will not yield as accurate results as a table of more places, it will serve to illustrate the methods.* The numbers .04021, .4021, 402100.0 are said to have the *same sequence of digits*. The *mantissa* of the logarithm for each of these numbers is the *same*; the characteristics are, of course, different.

I. *To find the logarithm of a given number.*

ILLUSTRATIONS

1. Find log 32.4.

*For a five-place table see Kaj L. Nielsen, *Logarithmic and Trigonometric Tables* (2nd ed.; New York: Barnes and Noble, 1961), pp. 2–21.

SOLUTION. By Rule 1 the characteristic is 1. To find the mantissa turn to the table and locate the first two digits (32) in the left column headed by "*N*." In the "32" row and in the column headed by "4" (the third digit of the number) find 5105. This number is the mantissa. Thus log 32.4 = 1.5105.

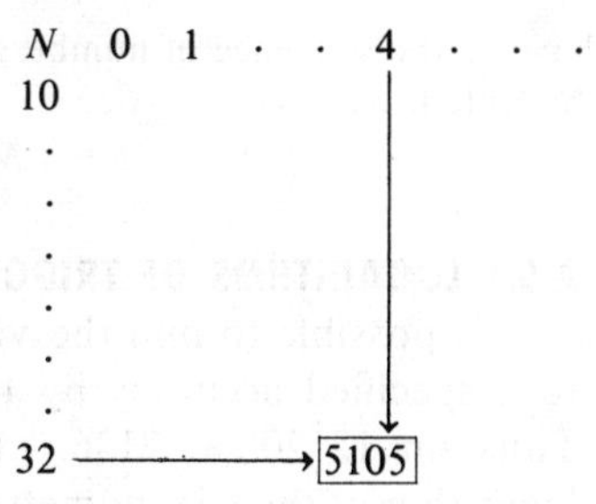

2. Find log .06732.

SOLUTION. By Rule 2 the characteristic is 8 − 10. To find the mantissa it is necessary to interpolate. We shall use simple linear interpolation as we did in Section A.1. Locate "67" in left column. In the "67" row and column headed by "3," find 8280, and in the column headed by "4," find 8287. The difference 8287 − 8280 = 7 is now multiplied by .2 (2 being the fourth significant digit of given number); (7)(.2) = 1.4, which is rounded off to the nearest whole number, 1. This number is added to the smaller mantissa; 8280 + 1 = 8281. The answer is given by

$$\log .06732 = 8.8281 - 10$$

II. *To find the antilogarithm of a given logarithm.*

ILLUSTRATIONS

1. Find N if $\log N = 7.6503 - 10$.

SOLUTION. First find the mantissa 6503 in the table; it appears in the column headed by "7" in the row which has "44" at the left under "*N*." Thus the sequence of digits is 447.

Now the characteristic is 7 − 10 = −3. Using Rule 2 for the characteristic, the first significant digit after the decimal point should occur in the *third* place. The answer is $N = .00447$.

2. Find N if $\log N = 1.5952$.

SOLUTION. First seek the mantissa in the table. We find 5944 and 5955 corresponding to the numbers 393 and 394. Consequently, our number is between these two numbers. Form the quotient.

$$\left(\frac{\text{given mantissa} - \text{smaller table mantissa}}{\text{larger table mantissa} - \text{smaller table mantissa}}\right) 10$$

$$10\left(\frac{5952 - 5944}{5955 - 5944}\right) = \frac{8}{11}(10) = \frac{80}{11} = 7 +$$

This number is rounded off to an integer which is "attached to" the smaller number N resulting from the table. Thus 7 is attached to 393 to give us the sequence of numbers 3937. The given characteristic is 1; so by Rule 1, we have

$$N = 39.37$$

A.5. LOGARITHMS OF TRIGONOMETRIC FUNCTIONS

It is possible to find the values of the trigonometric functions to a specified accuracy by the use of the trigonometric tables. Thus $\sin 12°20' = .2136$. It is therefore possible to find the logarithm of the trigonometric function.

$$\log \sin 12°20' = \log 0.2136 = 9.3296 - 10$$

This procedure, however, would necessitate first using the table of natural trigonometric functions to find $\sin 12°\ 20'$, and then using the logarithm table to find $\log 0.2136$. To eliminate this two-step procedure, tables which give the logarithms of trigonometric functions directly have been computed. An abridged four-place table is on page 248.

The table lists the angles from $0°$ to $45°$ in the left column at every 10 minutes. The corresponding logarithms are labeled at the *top* of the pages. The angles from $45°$ to $90°$ are given in reverse order in the right-hand column, and the heading of the columns is to be taken from the *bottom* of the pages. In order to aid the interpolation, columns of differences have been calculated. These tabular differences are given in the table in the columns headed by "d" to the right of the columns of "L sin" and "L cos"; "L tan" and "L cot" have a common difference, and the "cd" column between them refers to both.

The sines and cosines are less than one (except for $90°$ and $0°$), and therefore the logarithms of these trigonometric functions have a negative characteristic. Similarly, the logarithms of $\tan \theta$ for $\theta < 45°$ and $\cot \theta$ for θ between $45°$ and $90°$ have a negative characteristic. The first part of the characteristic for the logarithm of these functions is explicitly given in the table; it is, therefore, necessary to *subtract* 10 from the logarithm of the above-mentioned functions as given in the table. The characteristic for the logarithm of $\tan \theta$ for θ between $45°$ and $90°$ and $\cot \theta$ for $\theta < 45°$ as given in the table is the exact characteristic, and no subtraction is to be made in this case. No logarithms are given for the secant and cosecant; if these functions occur in any computation, express them by means of the reciprocal identities in terms of sine and cosine.

ILLUSTRATIONS

1. Find log sin 23°43′.

SOLUTION. Find 23°40′ in the *left* column of the table beginning on page 248; opposite this number look in the "L sin" column and read 9.6036. In the "d" column between 40′ and 50′ read 29; this is the tabular difference. Take the difference between the given minutes 43 and the tabulated minutes 40 and calculate,

$$\frac{43-40}{10}(29) = \frac{3}{10}(29) = 8.7$$

or 9 to the nearest integer. Add this to 9.6036 to get

$$\log \sin 23^\circ 43' = 9.6045 - 10$$

2. Find log cos 61°36′.

SOLUTION. Find 61°30′ in the right-hand column of the table, and opposite this number in the column headed by "L cos" at the *bottom* of the page read 9.6787. The tabular difference between 30′ and 40′ is given in the "d" column and reads 24. Calculate

$$\tfrac{6}{10}(24) = 14.4 \quad \text{or} \quad 14$$

This amount is to be subtracted from 9.6787 because the cosine is a decreasing function. Thus

$$\log \cos 61^\circ 36' = 9.6773 - 10$$

The student should recall that the sine and tangent functions are increasing functions and that the cosine and cotangent functions are decreasing functions.

ILLUSTRATIONS

1. Given log tan θ = 9.6932 − 10, find θ.

SOLUTION. In the "L tan" columns, the number just smaller than 9.6932 is 9.6914 and is found opposite 26°10′ in the left-hand column. The tabular difference is found from the "cd" column to be 32. Calculate

$$\frac{9.6932 - 9.6914}{32} = \frac{18}{32} = \frac{9}{16} = .56$$

This number is now multiplied by 10, which is the tabular difference of the angles, and rounded off to an integer; i.e.,

$$10(.56) = 5.6 = 6 \text{ to the nearest integer}$$

Since the tangent is an increasing function,

$$\theta = 26^\circ 10' + 6' = 26^\circ 16'$$

2. Given log cot θ = 9.8780 − 10, find θ.

SOLUTION. In the "L cot" column search for the number closest to 9.8780, and find 9.8797 and 9.8771 in the column headed "L cot" at the bottom of the page corresponding to angles 52°50′ and 53°0′. Calculate

$$\left.\begin{matrix} 9.8771 \\ 9.8780 \\ 9.8797 \end{matrix}\right] \tfrac{9}{26}(10) = [3]$$

to the nearest integer. Subtract 3′ from 53°0′, thus

$$\theta = 53^\circ 0' - 3' = 52^\circ 57'$$

A.6. COLOGARITHMS

The cologarithm of a number is defined as the logarithm of the reciprocal of the number.

$$\text{colog}\, N = \log \frac{1}{N} = \log 1 - \log N = 0 - \log N$$
$$= [10.0000 - 10] - \log N$$

ILLUSTRATION. Find colog 35.7.

SOLUTION

$$\begin{aligned} \text{colog}\, 35.7 &= \log 1 - \log 35.7 \\ &= 10.0000 - 10 \\ \text{minus} \quad & \quad\ \underline{1.5527} \\ &= 8.4473 - 10 \end{aligned}$$

To find a cologarithm mentally: Subtract each digit of the logarithm of the number, except the last one, from 9 and the last one from 10, and subtract 10 from the result. Thus, in the above illustration, $9 - 1 = 8$; $9 - 5 = 4$; $9 - 5 = 4$; $9 - 2 = 7$; $10 - 7 = 3$; giving $8.4473 - 10$.

Cologarithms are used in computation when it appears to be desirable to add all the logarithms instead of subtracting some of of them.

ILLUSTRATION

$$\log \frac{306}{(98.1)(1.52)} = \log 306 - \log 98.1 - \log 1.52$$
$$= \log 306 + \text{colog}\, 98.1 + \text{colog}\, 1.52$$

(1)	log 306 = 2.4857
(2)	colog 98.1 = 8.0083 − 10
(3)	colog 1.52 = 9.8182 − 10
(4)	sum = 20.3122 − 20

A.7. COMPUTATIONS USING LOGARITHMS

In carrying out computations using logarithms it is desirable to have a systematic form in which to display the work. We recommend the form given in the illustrations. The student should check each step carefully.

ILLUSTRATIONS

1. Find N if $N = \dfrac{5.367}{(12.93)(0.06321)}$.

SOLUTION. We employ logarithms and, using the three properties of logarithms, we get

$$\log N = \log 5.367 - [\log 12.93 + \log 0.06321]$$

(1)	$\log 5.367 = 0.7298$	
(2) (3)	$\log 12.93 = 1.1116$ $\log 0.06321 = 8.8008 - 10$	
(4)	$(2) + (3) = 9.9124 - 10$	
(5)	$(1) - (4) = 0.8174$	$N = 6.567$

NOTE: To subtract (4) from (1) we first change 0.7298 to 10.7298 − 10.

2. Find N if $N = \dfrac{(\sqrt[3]{0.9573})(3.21)^2}{98.32}$.

SOLUTION $\log N = \frac{1}{3}\log 0.9573 + 2\log 3.21 - \log 98.32$

(1) (2)	$\frac{1}{3}\log 0.9573 = \frac{1}{3}[9.9810 - 10]$ $2\log 3.21 = 2[0.5065]$	$= 9.9937 - 10$ $= 1.0130$
(3) (4)	$(1) + (2)$ $\log 98.32$	$= 11.0067 - 10$ $= 1.9927$
(5)	$(3) - (4)$	$= 9.0140 - 10$
	$N = 0.1033$	

NOTE: To find

$$\tfrac{1}{3}[9.9810 - 10] = \tfrac{1}{3}[29.9810 - 30] = 9.9937 - 10$$

always rearrange a negative characteristic so that *after* dividing the result will be x.xxxx minus an integer.

3. Find N if $N = \sqrt{\dfrac{(0.3592)^3}{673.5}}$.

SOLUTION $\log N = \frac{1}{2}[3 \log 0.3592 - \log 673.5]$

(1) (2)	$3 \log 0.3592 = 3[9.5553 - 10] = 28.6659 - 30$ $\log 673.5 = 2.8284$	
(3)	$(1) - (2) = 25.8375 - 30$	
(4)	$\frac{1}{2}(3) = \frac{1}{2}[15.8375 - 20] = 7.9188 - 10$	
	$N = 0.008294$	

NOTE: In going from (3) to (4) we made the change

$$25.8375 - 30 = 15.8375 - 20$$

This was done so that after dividing by 2 we would have x.xxxx − 10.

A.8. PROBLEMS

1. Find the values of each function by use of a table.

a) $\sin 67°10'$ b) $\cos 35°15'$ c) $\tan 43°18'$
d) $\cot 72°28'$ e) $\cos 57°14'$ f) $\tan 81°20'$
g) $\sin 13°53'$ h) $\cot 29°30'$ i) $\sin 45°31'$

2. Find the acute angle.

a) $\sin \alpha = .5398$ b) $\cos \alpha = .6841$ c) $\tan \alpha = .7720$
d) $\cos \alpha = .7000$ e) $\tan \alpha = 1.0631$ f) $\sin \alpha = .7380$
g) $\cot \alpha = .9004$ h) $\cos \alpha = .6000$ i) $\sin \alpha = .8650$

3. Find the values of the following functions.

a) $\sin 463°$ b) $\cos 157°$ c) $\tan 242°10'$
d) $\cot(-17°32')$ e) $\sin(-136°22')$ f) $\cos(-260°30')$

4. Find the values of the following functions.

a) $\sin .32$ b) $\cos 1.01$ c) $\tan .45$
d) $\sin 1.32$ e) $\tan 1.40$ f) $\cos 1$

5. Find the common logarithms of 2.953, 0.000431, and 172300.

6. Find the antilogs of 9.8243 − 10 and 2.2563.

7. Find $\log \tan 17°18'$ and $\log \sin 68°24'$.

8. Find θ if $\log \sin \theta = 9.9245 - 10$.

9. Evaluate to 4 significant figures by means of logarithms.

a) $(32.18)^2(0.009638)(\sqrt{0.3628})$
b) $(32.18) \sin 18°36' \csc 23°42'$
c) $(3.742) \sin 46°10' \tan 32°18'$

10. a) Find the values of $e^{.2}$, $e^{-.3}$, $e^{1.5}$, e^{-2}.
b) Find the values of ln 0.3, ln 1.4, ln 5.6.

Examinations

We now present a set of examinations in order to provide the reader with an opportunity to test his knowledge of the subject matter in this book. These examinations have been divided into two groups. The first part is a set of tests covering the specific subject matter of each chapter. The second part is a set of final examinations which covers the entire book.

Subject Matter Tests

TEST 1. (Covers the material in Chapter 1)
Given the points $P_1(5,5)$ and $P_2(-3,4)$:

1. Plot the points on a rectangular coordinate system.
2. Find the length of the line segment $P_1 P_2$.
3. Find the length of the radius vector for each point.
4. Draw a circle with its center at the origin and passing through P_1.
5. Find the coordinates of the point of intersection of the circle with the positive x-axis. Label this point A.
6. Find the circular coordinates of P_1.
7. Find the polar coordinates of P_1.
8. Write an expression for the set of points which lie inside the circle whose coordinates are greater than the coordinates of P_2, respectively.
9. Shade in the area covered by the set of points of problem 8.
10. Is $\angle P_2 O P_1$ a right angle?
11. Let B be the foot of the perpendicular from P_1 to the x-axis. Find the coordinates of B and the area of $\triangle O B P_1$.
12. Find the area and circumference of the circle.
13. Find the area of the sector $O A P_1$.
14. If a particle moves along the circle from A to P_1 in one second, what are its linear and angular velocities?
15. Write the equation of the circle.
16. Let C and D be the points of intersection of the perpendiculars from P_2 to the x- and y-axes, respectively. Find the lengths of $\overline{CD}$ and $\overline{BD}$.
17. Find the polar and circular coordinates of D.
18. Find the size of $\angle O P_1 A$.

19. Find the size of the negative angle from OA to OP_1 in both degree and radian measure.

20. The portion of the circle from A to P_1 can be described by the function $y = \sqrt{50 - x^2}, 5 \leq x \leq 5\sqrt{2}$. Find the range and inverse of this function. Indicate the inverse function on the diagram.

TEST 2. (Covers the material in Chapter 2)

1. Define the cosine function
   a) if the argument is a real number,
   b) if the argument is an angle,
   c) if the argument is an acute angle.
2. Complete the following table with exact values.

	$30°$	$-\frac{\pi}{4}$	π	$60°$	$-\frac{3\pi}{2}$	$-120°$
sin						
cos						
tan						
cot						

3. Evaluate the following expressions.
   a) $(\sin 270°)(\sin 30°)(\sin 765°)$
   b) $\left(\sin \frac{\pi}{3}\right)\left(\cos \frac{\pi}{4}\right)\left(\tan \frac{\pi}{6}\right)\left(\cot \frac{5\pi}{4}\right)$
   c) $\left(\cos \frac{3\pi}{2}\right)\left(\sec \frac{\pi}{3}\right)\left(\csc \frac{\pi}{4}\right)$
4. Sketch the angles given by $\left\{\frac{2\pi}{3} + n\pi\right\}$ and find the trigonometric values.
5. Find the following values.
   a) Arcsin $\frac{1}{2}$ b) Arctan $(-\sqrt{3})$
   c) tan arccos $\frac{3}{5}$ d) sin arccsc y

TEST 3. (Covers the material in Chapter 3)

Prove the following identities.

1. $\cos^4 x - \sin^4 x = \cos 2x$
2. $\sec^2 x + \csc^2 x = \sec^2 x \csc^2 x$
3. $\dfrac{\sin x}{\sin x + \sec y} = \dfrac{\cos y}{\cos y + \csc x}$
4. $\dfrac{\sin x + \cot x}{\cos x \sin x} - 1 = \cot^2 x + \sec x$

5. $\tan 3x = \dfrac{3\tan x - \tan^3 x}{1 - 3\tan^2 x}$

6. $\tan 2x = \dfrac{2}{\cot x - \tan x}$

7. $\dfrac{\cot x - \tan x}{\cot x + \tan x} = \cos 2x$

8. $\dfrac{\sin 4x - \sin 2x}{\cos 4x + \cos 2x} = \tan x$

9. $\sin 75^\circ = \dfrac{\sqrt{6} + \sqrt{2}}{4}$

10. $\sin 7^\circ 30' = \frac{1}{2}\sqrt{2 - \sqrt{2 + \sqrt{3}}}$, if $\cos 15^\circ = \frac{1}{2}\sqrt{2 + \sqrt{3}}$

TEST 4. (Covers the material in Chapter 4)

1. Sketch the graph of $y = 2\cos\dfrac{x}{2}$.
2. Sketch the graph of $y = \arcsin x$.
3. Sketch the graph of $y = 2\sin\left(x + \dfrac{\pi}{4}\right)$.
4. Sketch the graph of $y = \sin x + \cos x$.
5. Determine the phase angle of $y = \sqrt{3}\cos x + \sin x$.

TEST 5. (Covers the material in Chapter 5)

1. Find the complete solution set of the following.

   a) $\tan 3x = \sqrt{3}$ b) $\sin 2x = \cos 3x$

   c) $\sin\dfrac{x}{2} = \dfrac{1}{2}$ d) $\cot 2x = 1$

2. Solve the following equations for $0 \leq x < 2\pi$.

   a) $\sin^2 x = \frac{1}{2}\sin x$ b) $\tan(2x + 30^\circ) = 1$

   c) $\cos 2x + \sin x = 0$ d) $\frac{3}{2} - \frac{3}{2}\sin x - \cos^2 x = 0$

   e) $\sin \arcsin x = 2x - 1$

TEST 6. (Covers the material in Chapter 6)

1. Sketch the inverse function of y defined by $y = \log_2 x$.
2. Given: $\log 2 = 0.30103$, $\log 3 = 0.47712$, $\log 5 = 0.69897$

   Find:

   a) $\log 80$

   b) $\log\sqrt{45}$

3. a) Write the equivalent exponential form of

$$\log_8 2 = \frac{1}{3} \quad \text{and} \quad \log_5 \frac{1}{25} = -2$$

   b) Write the equivalent logarithmic form of

$$3^3 = 27 \quad \text{and} \quad \left(\tfrac{1}{2}\right)^{-3} = 8$$

4. If $\log 3 = 0.47712$ and $\log 5 = 0.69897$, find $\log_3 5$ to four decimal places.
5. Use the values given in problem 2 to solve for x in the equation

$$(2^{x+1})(3^x) = 5^{x+2}$$

TEST 7. (Covers the material in Chapter 7)

1. Locate the point $P\left(1, \frac{5\pi}{4}\right)$ on a circular coordinate system. Construct the line values of the six trigonometric functions of $\frac{5\pi}{4}$.
2. State the law of sines, law of cosines, and law of tangents for a $\triangle ABC$.
3. Find the area of the triangle with

$$b = 18 \qquad c = 11 \qquad \alpha = 30^\circ$$

4. A surveyor's notebook reads:

   AB: 32 ft. due E; *BC*: 13 ft. S 23° E; *CD*: 19 ft. E 28° S; *DE*: 27 ft. N 52° 30′ E; *EF*: 18 ft. N 44° 30′W.

   He desires to find the distance *BF*. Draw the geometric configuration and discuss the method of solution.
5. From a remote point 30 feet above the foot of a tower, the angle of depression of the base is arctan $\frac{3}{4}$ and the angle of elevation of the top is 60°. Find the height of the tower.

TEST 8. (Covers the material in Chapter 8)

1. Given the complex numbers

$$z_1 = 2 - 3i, \quad z_2 = 4 + 2i, \quad z_3 = -1 - i$$

   a) find $z_1 z_2 z_3$,
   b) find $z_2 \div z_1$,
   c) find the polar form of z_3,
   d) find $z_1 - z_2 + \bar{z}_3$.
2. Find the rectangular form of 28 cis 147° ÷ 7 cis 117°.
3. Find the five fifth roots of $1 + i$.
4. Find the roots of $x^4 - \sqrt{3}\,x^2 + 1 = 0$.
5. Find $(\sqrt{2} - i\sqrt{2})^{11}$ in rectangular form.

TEST 9. (Covers the material in Chapter 9)

1. Use the definitions to prove the following.

   a) $\tanh x = \dfrac{\sinh x}{\cosh x}$ b) $\cosh^2 x - \sinh^2 x = 1$

   c) $\cosh x + \sinh x = e^x$ d) $\sinh 2x = 2 \sinh x \cosh x$

2. Derive the logarithmic expression for $\sinh^{-1} x$.
3. Prove that $\sinh^{-1} x = \operatorname{Cosh}^{-1} \sqrt{x^2 + 1}$.
4. Prove that $\cosh(x \pm y) = \cosh x \cosh y \pm \sinh x \sinh y$.
5. Develop a formula for $(\cosh x + \sinh x)^3$.

TEST 10. (Covers the material in Chapter 10)

1. Find the first five terms of the expansion for $\sqrt{x - 2y}$.
2. Let $x = 1$ and $y = i$ in problem 1. Simplify the series and find the approximation for $\sqrt{1 - 2i}$.
3. Use the binomial series to approximate $\sqrt{98}$.
4. Develop Euler's Formula.
5. Find a series expansion for $e^x \sin x$.

TEST 11. (Covers the material in Chapter 11)

1. Use Napier's rules to write the formulas for finding a, b, and B when A and c are given.
2. Find b, A, and C of a spherical triangle if $a = 115°$, $c = 75°$, and $B = 160°$.
3. Indicate the method for finding the time of day when the latitude and date are given.

Final Examinations

EXAM 1

1. Reduce the following expressions to a single trigonometric function of the argument x.

   a) $\cos(90° - x)$

   b) $\dfrac{\sin x}{\cos(-x)}$

   c) $\cos 3x \cos 2x + \sin 3x \sin 2x$

   d) $\left(\sin \dfrac{x}{2} + \cos \dfrac{x}{2}\right)^2 - 1$

2. Prove the identities.

   a) $(\tan t - \sec t)^2 + \dfrac{2 \sin t}{\cos^2 t} - 1 = 2 \tan^2 t$

   b) $\left(\dfrac{\cos x}{1 - \sin x} + \dfrac{1 - \sin x}{\cos x}\right) \tan x = 2 \sin x \sec^2 x$

   c) $\sec x = \tan\left(\dfrac{x}{2} + \dfrac{\pi}{4}\right) - \tan x$

3. Find the solution set for the following equations.

   a) $\cos^2 x - \sin x \cos x = 0$

   b) $\sin x - 2 = 0$

   c) $3 \tan \dfrac{x}{2} = \cot \dfrac{x}{2}$

4. State the law of cosines for triangles and write the formulas for the angles in terms of the sides.
5. Find the four fourth roots of unity.

6. Evaluate.

a) $\sin\left(\arctan \frac{5}{12}\right)$

b) $\text{Arcsec}\,(-\sqrt{2})$

c) $\sin\left(\text{Arccos}\,\frac{3}{5} + \text{Arcsin}\,\frac{5}{13}\right)$

7. Given the two complex numbers $z_1 = \frac{1}{2} - \frac{\sqrt{3}}{3}i$ and $z_2 = 2 \text{ cis } 135^\circ$.

a) locate the two points in the Argand plane,

b) locate the point $z_3 = z_1 + z_2$,

c) find $z_4 = \frac{z_1}{z_2}$.

8. Sketch the graph of the inverse of e^x.

9. In $\triangle ABC$, let $\alpha = 45^\circ, \gamma = 30^\circ$, and $c = 5\sqrt{2}$; find a.

10. Use the definitions to prove the following.

a) $\sinh(x - y) = \sinh x \cosh y - \cosh x \sinh y$

b) $\sinh 3x = 3 \sinh x + 4 \sinh^3 x$

EXAM 2

1. If a force of 5 pounds pulls due east and another of 3 pounds pulls in a direction of 30° east of north, what is the magnitude of the resultant force?

2. Find the solutions for x and y if $3^{x+y} = 81$ and $81^{x-y} = 3$.

3. A wheel, 2 feet in diameter, makes 420 revolutions per minute. Find the linear velocity of a point on the rim in feet per seconds.

4. Given $\triangle ABC$ with

a) $\alpha = 45^\circ, \gamma = 30^\circ, \ c = 5\sqrt{2}$, find a.

b) $a = 6, b = 7, \ \gamma = 120^\circ$, find c.

c) $a = 3, b = 4, \gamma = 90^\circ$, find c.

5. Sketch the graph of $y = 3 \sin \frac{x}{3}$ for $0 \le x \le \pi$.

6. Sketch the graph of $\rho = 6 \cos \theta$.

7. Given the polar coordinates of a point P to be $\rho = 2$ and $\theta = 30^\circ$; find the circular and rectangular coordinates of P.

8. The function $y = \sqrt{25 - x^2}$ describes an arc of a circle. Find the inverse function for y in the domain $0 \le x \le 3$. Sketch the graph of both.

EXAM 3

1. Prove the following identities.

a) $\tan \frac{x}{2} + 2 \cot x = \cot \frac{x}{2}$

b) $\dfrac{\cos^2 x - \cot x}{\sin^2 x - \tan x} = \cot^2 x$

c) $\sin 2x \tan 2x = \dfrac{4 \tan^2 x}{1 - \tan^4 x}$

2. Solve for $0 \leq x < 2\pi$.
   a) $\sin 3x = \sin 2x$
   b) $\sin 2x + \cos 2x = 1$
   c) $\cos 2x + \sin x - 1 = 0$
3. Prove that $\text{Arctan}\, x = \text{Arcsin}\, \dfrac{x\sqrt{1 + x^2}}{1 + x^2}$.
4. Evaluate $\tan 2 \arcsin \frac{1}{2}$.
5. Find the three cubic roots of unity.
6. Sketch the graph of $y = \sin x + \cos x$.
7. Solve for x if $2^{x+1} = 3^x$.
8. Solve for x if $\text{Arcsin}\left(x^2 - \dfrac{1}{2}\right) = \dfrac{\pi}{6}$.
9. Find $(1 + i)^{14}$.
10. Find the series for $e^x \cos x$.

EXAM 4 (A True or False Test)

If statement is true, circle T; if false, circle F.

T F 1. $\sin 60° = \dfrac{1}{2}$.

T F 2. $\cos \dfrac{\pi}{3} = \sin \dfrac{\pi}{6}$.

T F 3. $\tan \dfrac{\pi}{4} = \sqrt{3}$.

T F 4. $\sin 243° = \sin 63°$.

T F 5. $\tan(-135°) = \tan 45°$.

T F 6. $\sin 2x = \cos^2 x - \sin^2 x$.

T F 7. $\cos \dfrac{2\pi}{3} = -\dfrac{1}{2}$.

T F 8. $\sin x \csc x = 1$.

T F 9. $\cos\left(2 \text{ Arcsin } \dfrac{1}{2}\right) = \dfrac{1}{2}$.

T F 10. $\sec^2 x - \tan^2 x = 1$.

T F 11. $\sin x = \sin(\pi + x)$.

T F 12. $\tan x = \tan(2\pi + x)$.

T F 13. $\cos x = \cos(-x)$.

T F 14. $\tan\left(\text{Arctan}\sqrt{3}\right) = \dfrac{\pi}{3}$.

T F 15. $\text{Arcsin}\, \dfrac{1}{2} + \text{Arccos}\, \dfrac{1}{2} = \dfrac{\pi}{2}$.

T F 16. $\tan 22.5° = \sqrt{2} - 1$.

T F 17. $\sin 30^\circ \cos 60^\circ + \cos 60^\circ \sin 30^\circ = 1$.
T F 18. $|z| = |\bar{z}|$ when z is a complex number.
T F 19. $\sin\left(\text{Arccos } \frac{3}{5}\right) = \frac{3}{5}$.
T F 20. $\sqrt[3]{1} = -\frac{1}{2} + \frac{\sqrt{3}}{2} i$.
T F 21. $\log 20 = 0.30103$.
T F 22. $\log 2x = \log 2 - \log x$.
T F 23. $\tan 675^\circ = 1$.
T F 24. $\sin x = \tan x$ if $x = 8\pi$.
T F 25. All the real numbers form the domain of the function, $\tan x$.

EXAM 5 (An Open Book Exam)

1. Find the series for $\frac{e^x}{\cos x}$. (First 5 terms)
2. Solve the spherical triangle ABC given $A = 36^\circ$, $B = 62^\circ$, and $a = 26^\circ$.
3. Find the distance and initial course from New York (long. 73°58′W, lat. 40°49′N) to Los Angeles (lat. 34°3′N, long. 118°15′W).

Challenging Problems

The following problems are more difficult than the problems designed to establish a concept or make the reader proficient in a technique. They can, however, be solved by applying the knowledge of mathematics obtained by the reader who has mastered this book. The first problem is directly related to trigonometry; some of the later ones require other mathematical concepts, but on the same level of development.

1. Prove the following identities.

a) $\left(1 + \tan \frac{x}{2} - \sec \frac{x}{2}\right)\left(1 + \tan \frac{x}{2} + \sec \frac{x}{2}\right) = \sin x \sec^2 \frac{x}{2}$

b) $(\cos x + \cos y)^2 + (\sin x - \sin y)^2 = 4 \cos^2 \frac{x + y}{2}$

c) $$\sin \frac{\alpha}{2} + \sin \frac{\beta}{2} + \sin \frac{\gamma}{2} - 1 = 4 \sin \frac{\pi - \alpha}{4} \sin \frac{\pi - \beta}{4} \sin \frac{\pi - \gamma}{4}, \text{ if } \alpha + \beta + \gamma = \pi.$$

2. Sketch the graph of

$$y = \frac{x \ln x - ax}{x - b}$$

if $a = \frac{e - 2}{e - 1}$ and $b = e - 1$.

3. A 20 foot ladder and a 30 foot ladder are placed against directly opposite walls with the foot of the ladders at the base of the opposite

wall. If they intersect 10 feet above the floor level, how far apart are the walls?

4. Given the square $ABCD$ with side a, draw four circular arcs with centers at A, B, C, and D, and radii equal to a. These arcs intersect at points P_1, P_2, P_3, and P_4. Find the area of the curved configuration $P_1P_2P_3P_4$.
5. Show that the diameter of a circle circumscribed about a triangle is equal to the ratio of any side to the sine of the opposite angle. Find the radius of a circle circumscribed about the triangle with sides $a = 4$, $b = 4\sqrt{2}$, and $c = 2\sqrt{6} + 2\sqrt{2}$.

Answers to the Exercises

EXERCISE 1

2. (a) 5. (b) $2\sqrt{5}$. (c) $5\sqrt{5}$. (d) $\sqrt{13}$. (e) 5. (f) 6.

4. (a) 2. (b) $\frac{\pi}{90}$. (c) $\frac{2\pi}{3}$. (d) $\frac{5\pi}{3}$. **5.** 64π ft. $\doteq$ 16.8 ft.

6. $\frac{\pi}{3}$ in. $\doteq$ 1.05 in. **7.** 60π ft./sec. $\doteq$ 188 ft./sec.

10. $x^2 + y^2 = 2$. **11.** Semicircle; domain: $-2 \le x \le 2$, range: $0 \le y \le 2$. **12.** $f^{-1} = \frac{1}{3}(x + 6)$; domain: $-6 \le x \le 3$, range: $0 \le y \le 3$. **13.** $\frac{\pi}{6}, \frac{\pi}{2}, \frac{5\pi}{6}, \frac{3\pi}{2}$, $-0.99484, 0.22980, 0.12217$, $-\frac{3\pi}{2}, \frac{\pi}{4}$, 0.00087. **14.** $90°, 60°, \frac{180}{7}°$, $114° \, 35' \, 29.6''$, $135°$.

15. In circular coordinates: (a) $\left(3, \frac{\pi}{2}\right)$. (b) $\left(3, \frac{3\pi}{4}\right)$. (c) $(3,\pi)$.

16. *merry* – *xmas*.

EXERCISE 2

1.

	sin	cos	tan	cot	sec	csc
30°	$\frac{1}{2}$	$\frac{\sqrt{3}}{2}$	$\frac{\sqrt{3}}{3}$	$\sqrt{3}$	$\frac{2\sqrt{3}}{3}$	2
210°	$-\frac{1}{2}$	$-\frac{\sqrt{3}}{2}$	$\frac{\sqrt{3}}{3}$	$\sqrt{3}$	$-\frac{2\sqrt{3}}{3}$	-2
45°	$\frac{\sqrt{2}}{2}$	$\frac{\sqrt{2}}{2}$	1	1	$\sqrt{2}$	$\sqrt{2}$
120°	$\frac{\sqrt{3}}{2}$	$-\frac{1}{2}$	$-\sqrt{3}$	$-\frac{\sqrt{3}}{3}$	-2	$\frac{2\sqrt{3}}{3}$
150°	$\frac{1}{2}$	$-\frac{\sqrt{3}}{2}$	$-\frac{\sqrt{3}}{3}$	$-\sqrt{3}$	$-\frac{2\sqrt{3}}{3}$	2
300°	$-\frac{\sqrt{3}}{2}$	$\frac{1}{2}$	$-\sqrt{3}$	$-\frac{\sqrt{3}}{3}$	2	$-\frac{2\sqrt{3}}{3}$

1. (continued)

	sin	cos	tan	cot	sec	csc
225°	$-\frac{\sqrt{2}}{2}$	$-\frac{\sqrt{2}}{2}$	1	1	$-\sqrt{2}$	$-\sqrt{2}$
330°	$-\frac{1}{2}$	$\frac{\sqrt{3}}{2}$	$-\frac{\sqrt{3}}{3}$	$-\sqrt{3}$	$\frac{2\sqrt{3}}{3}$	-2
135°	$\frac{\sqrt{2}}{2}$	$-\frac{\sqrt{2}}{2}$	-1	-1	$-\sqrt{2}$	$\sqrt{2}$
60°	$\frac{\sqrt{3}}{2}$	$\frac{1}{2}$	$\sqrt{3}$	$\frac{\sqrt{3}}{3}$	2	$\frac{2\sqrt{3}}{3}$
315°	$-\frac{\sqrt{2}}{2}$	$\frac{\sqrt{2}}{2}$	-1	-1	$\sqrt{2}$	$-\sqrt{2}$
240°	$-\frac{\sqrt{3}}{2}$	$-\frac{1}{2}$	$\sqrt{3}$	$\frac{\sqrt{3}}{3}$	-2	$-\frac{2\sqrt{3}}{3}$

2. **(a)** I. **(b)** III. **(c)** II. **(d)** III. **(e)** III. **(f)** II. **(g)** II. **(h)** III. **(i)** III. **(j)** IV. **3.** **(a)** $-\frac{1}{2}$. **(b)** 0. **(c)** $\frac{1}{2}$. **(d)** -1. **(e)** $\frac{-2\sqrt{3}}{3}$. **(f)** $\sqrt{2}$. **(g)** $\frac{\sqrt{2}}{2}$ **(h)** $\frac{\sqrt{3}}{2}$ **(i)** 0. **4.** **(a)** -1. **(b)** $\frac{3}{4}$. **(c)** 4. **5.** **(a)** sin 48°. **(b)** $-$cos 70°. **(c)** $-$tan 8°. **(d)** $-$sec 30°. **(e)** $-$cot 60°. **(f)** sin 60°. **(g)** $-$cos 65°. **(h)** $-$tan 42°. **(i)** $-$csc 63°.

6-7.

	$\frac{\pi}{2}$	$-\pi$	$-\frac{\pi}{2}$	2π	$\frac{5\pi}{2}$	-5π	$\frac{\pi}{3}+2n\pi$	$\frac{\pi}{6}\pm\frac{\pi}{2}$
sin	1	0	-1	0	1	0	$\frac{\sqrt{3}}{2}$	$\frac{\sqrt{3}}{2}, -\frac{\sqrt{3}}{2}$
cos	0	-1	0	1	0	-1	$\frac{1}{2}$	$-\frac{1}{2}, \frac{1}{2}$
tan	–	0	–	0	–	0	$\sqrt{3}$	$-\sqrt{3}, -\sqrt{3}$
cot	0	–	0	–	0	–	$\frac{\sqrt{3}}{3}$	$-\frac{\sqrt{3}}{3}, -\frac{\sqrt{3}}{3}$
sec	–	-1	–	1	–	-1	2	$-2, 2$
csc	1	–	-1	–	1	–	$\frac{2\sqrt{3}}{3}$	$\frac{2\sqrt{3}}{3}, -\frac{2\sqrt{3}}{3}$

8. (a) 2. (b) 9. (c) 5. (d) 3. (e) 11. (f) 5. **9.** (a) $\frac{1}{2}$. (b) $\frac{1}{4}$. (c) -1. (d) $\frac{1}{2}$. (e) $\frac{3\sqrt{3}}{4}$. **10.** (a) $f(n) = 0$. (b) $0, \pm 1$. (c) $\pm 2, 0$. (d) $0, 3\sqrt{3}, -3\sqrt{3}$. (e) $0, \sqrt{2}, -\sqrt{2}, 2, -2$. (f) $1, 1/2, -1/2, -1$.

11. (a) $\left\{\frac{\pi}{6} + 2n\pi, \frac{5\pi}{6} + 2n\pi\right\}$. (b) $\left\{\frac{\pi}{3} + n\pi\right\}$. (c) $\left\{\pm\frac{\pi}{6} + 2n\pi\right\}$.

(d) $\left\{\frac{\pi}{4} + 2n\pi, \frac{3\pi}{4} + 2n\pi\right\}$. (e) $\left\{\frac{2\pi}{3} + 2n\pi, \frac{4\pi}{3} + 2n\pi\right\}$.

(f) $\left\{\pm\frac{\pi}{3} + 2n\pi\right\}$. (g) $\left\{\frac{\pi}{4} + \pi\right\}$. (h) $\left\{\pm\frac{\pi}{3} + 2n\pi\right\}$.

(i) $\left\{\frac{\pi}{4} + 2n\pi, \frac{3\pi}{4} + 2n\pi\right\}$. **12.** (a) $\frac{\pi}{6}$. (b) $\frac{\pi}{4}$. (c) $-\frac{\pi}{3}$. (d) $\frac{\pi}{4}$. (e) $\frac{\pi}{3}$. (f) $\frac{\pi}{2}$. **13.** (a) $\frac{3}{5}$. (b) $\frac{4}{5}$. (c) $\frac{7}{24}$. (d) $(1 - y^2)^{-1/2}$. (e) $1/\sqrt{26}$. (f) $\frac{n}{\sqrt{1 - n^2}}$. **14.** (a) $\frac{\pi}{3}$. (b) $\frac{3\pi}{4}$. (c) $-30°$. (d) $45°$. **15.** (a) $\sin 114.59°$. (b) $\cos 90°$. (c) $\tan 180°$. (d) $\cot 57.3°$.

EXERCISE 3

2. See table on page 233.

EXERCISE 4

1.

	sin	cos	tan	cot	sec	csc
15°	$\frac{\sqrt{6} - \sqrt{2}}{4}$	$\frac{\sqrt{6} + \sqrt{2}}{4}$	$2 - \sqrt{3}$	$2 + \sqrt{3}$	$\sqrt{6} - \sqrt{2}$	$\sqrt{6} + \sqrt{2}$
75°	$\frac{\sqrt{6} + \sqrt{2}}{4}$	$\frac{\sqrt{6} - \sqrt{2}}{4}$	$2 + \sqrt{3}$	$2 - \sqrt{3}$	$\sqrt{6} + \sqrt{2}$	$\sqrt{6} - \sqrt{2}$
105°	$\frac{\sqrt{6} + \sqrt{2}}{4}$	$\frac{\sqrt{2} - \sqrt{6}}{4}$	$-2 - \sqrt{3}$	$\sqrt{3} - 2$	$-\sqrt{6} - \sqrt{2}$	$\sqrt{6} - \sqrt{2}$

3. (a) $-\frac{63}{65}, -\frac{63}{16}, \frac{56}{33}$. (b) $-\frac{171}{221}, \frac{171}{140}, \frac{220}{21}$. (c) $\frac{1508}{1517}, -\frac{1508}{165}, -\frac{795}{1292}$. **4.** (a) $\frac{416}{425}$. (b) $\frac{56}{65}$. (c) $-\frac{31}{480}$. (d) $\frac{4}{3}$. **6.** $\sin\alpha\cos\beta\cos\gamma + \cos\alpha\sin\beta\cos\gamma - \cos\alpha\cos\beta\sin\gamma - \sin\alpha\sin\beta\sin\gamma$.

sin t	$\sin t$	$\pm\sqrt{1-\cos^2 t}$	$\frac{\tan t}{\pm\sqrt{1+\tan^2 t}}$	$\frac{1}{\pm\sqrt{1+\cot^2 t}}$	$\frac{\pm\sqrt{\sec^2 t-1}}{\sec t}$	$\frac{1}{\csc t}$
cos t	$\pm\sqrt{1-\sin^2 t}$	$\cos t$	$\frac{1}{\pm\sqrt{1+\tan^2 t}}$	$\frac{\cot t}{\pm\sqrt{1+\cot^2 t}}$	$\frac{1}{\sec t}$	$\frac{\pm\sqrt{\csc^2 t-1}}{\csc t}$
tan t	$\frac{\sin t}{\pm\sqrt{1-\sin^2 t}}$	$\frac{\pm\sqrt{1-\cos^2 t}}{\cos t}$	$\tan t$	$\frac{1}{\cot t}$	$\pm\sqrt{\sec^2 t-1}$	$\frac{1}{\pm\sqrt{\csc^2 t-1}}$
cot t	$\frac{\pm\sqrt{1-\sin^2 t}}{\sin t}$	$\frac{\cos t}{\pm\sqrt{1-\cos^2 t}}$	$\frac{1}{\tan t}$	$\cot t$	$\frac{1}{\pm\sqrt{\sec^2 t-1}}$	$\pm\sqrt{\csc^2 t-1}$
sec t	$\frac{1}{\pm\sqrt{1-\sin^2 t}}$	$\frac{1}{\cos t}$	$\pm\sqrt{1+\tan^2 t}$	$\frac{\pm\sqrt{1+\cot^2 t}}{\cot t}$	$\sec t$	$\frac{\csc t}{\pm\sqrt{\csc^2 t-1}}$
csc t	$\frac{1}{\sin t}$	$\frac{1}{\pm\sqrt{1-\cos^2 t}}$	$\frac{\pm\sqrt{1+\tan^2 t}}{\tan t}$	$\pm\sqrt{1+\cot^2 t}$	$\frac{\sec t}{\pm\sqrt{\sec^2 t-1}}$	$\csc t$

EXERCISE 5

23. (a) $-\frac{119}{169}$. (b) $\frac{3}{2}$. (c) $\frac{\sqrt{2}}{10}$. (d) $-\frac{840}{1081}$. (e) $\frac{1}{3}$. (f) 5.

24. (a) $\frac{1}{4}$. (b) 0. (c) $\frac{\sqrt{3}+1}{4}$. (d) $\sqrt{4-2\sqrt{2}}$.

(e) $\frac{1}{4}(\sqrt{6}+\sqrt{2}+2)$. (f) $\sqrt{6}+\sqrt{3}-\sqrt{2}-2$.

(g) $\frac{1}{2}\sqrt{2+\sqrt{3}}$. **25.** (c) $\frac{1}{2}[\pm\sqrt{1+\sin x}\pm\sqrt{1-\sin x}]$.

The "±" depends on the size of x. Check for $x = 0, \frac{\pi}{2}, \pi, \frac{3\pi}{2}$.

EXERCISE 6

1. $5\cos^4 x \sin x - 10\cos^2 x \sin^3 x + \sin^5 x$,
$\cos^5 x - 10\cos^3 x \sin^2 x + 5\cos x \sin^4 x$,

$$\frac{5\tan x - 10\tan^3 x + \tan^5 x}{1 - \tan^2 x + 5\tan^4 x}.$$

2. $\sin 75° = 5\cos^4 15° \sin 15° - 10\cos^2 15° \sin^3 15° + \sin^5 15°$.

4.

	$7\frac{1}{2}°$	$9°$	$18°$	$22\frac{1}{2}°$	$27°$	$36°$
sin	$\frac{\sqrt{2-a}}{2}$	$\frac{\sqrt{8-2b}}{4}$	$\frac{\sqrt{5}-1}{4}$	$\frac{\sqrt{2-\sqrt{2}}}{2}$	$\frac{\sqrt{8-2c}}{4}$	$\frac{c}{4}$
cos	$\frac{\sqrt{2+a}}{2}$	$\frac{\sqrt{8+2b}}{4}$	$\frac{b}{4}$	$\frac{\sqrt{2+\sqrt{2}}}{2}$	$\frac{\sqrt{8+2c}}{4}$	$\frac{\sqrt{5}+1}{4}$

where $a = (2+\sqrt{3})^{1/2}$, $b = (10+2\sqrt{5})^{1/2}$, $c = (10-2\sqrt{5})^{1/2}$.

EXERCISE 7

3. (a) $1, \pi$. (b) $2, \pi$. (c) $5, \frac{2\pi}{3}$. (d) $10, \frac{\pi}{30}$.

10. (a) $r = 2(-\cos\theta + 2\sin\theta)$. (b) $r = \tan^2\theta\sec\theta$.

11. (a) $2x - 3y + 5 = 0$. (b) $4y^2 - 12x - 9 = 0$.
(c) $x^3 + xy^2 = 9y$. (d) $x^2 + y^2 = 25$. **12.** $\theta = .31 \doteq 18°$.

EXERCISE 8

1. (a) $\left\{\frac{\pi}{6} + 2n\pi, \frac{5\pi}{6} + 2n\pi\right\}$. (b) $\left\{2n\pi \pm \frac{\pi}{3}\right\}$. (c) $\left\{\frac{\pi}{4} + n\pi\right\}$.

(d) $\left\{\frac{7\pi}{6} + 2n\pi, -\frac{\pi}{6} + 2n\pi\right\}$. (e) $\left\{\frac{5\pi}{6} + 2n\pi, \frac{7\pi}{6} + 2n\pi\right\}$.

(f) $\left\{\frac{5\pi}{6} + 2n\pi, -\frac{\pi}{6} + 2n\pi\right\}$. **(g)** $\left\{\frac{3\pi}{2} + 2n\pi\right\}$. **(h)** ϕ. **(i)** $\{n\pi\}$.

(j) $\{n\pi\}$. **(k)** $\{(2n + 1)\pi\}$. **(l)** $\left\{\frac{\pi}{3} + n\pi\right\}$. **(m)** $\left\{\frac{\pi}{12} + \frac{2n\pi}{4}\right\}$.

(n) $\left\{\frac{\pi}{12} + \frac{n\pi}{3}, \frac{\pi}{4} + \frac{n\pi}{3}\right\}$. **(o)** $\left\{\frac{\pi}{12} + \frac{n\pi}{3}\right\}$. **2. (a)** $\frac{\pi}{2}$.

(b) $\frac{\pi}{4}, \frac{3\pi}{4}, \frac{5\pi}{4}, \frac{7\pi}{4}$. **(c)** $0, \frac{\pi}{3}, \frac{2\pi}{3}, \pi, \frac{4\pi}{3}, \frac{5\pi}{3}$. **(d)** $\left\{\frac{\pi}{30} + \frac{2n\pi}{5}, \frac{\pi}{6} + \frac{2n\pi}{5}, (n = 0, 1, 2, 3, 4)\right\}$. **(e)** π. **(f)** $\frac{\pi}{3}, \frac{5\pi}{6}, \frac{4\pi}{3}, \frac{11\pi}{6}$.

3. (a) $-180°, 0, 180°$. **(b)** $-60°, 60°$. **(c)** $-174°, -114°, -54°, 6°, 66°, 126°$. **(d)** $-150°, -30°, 90°$. **(e)** $\pm 18°, \pm 54°, \pm 90°, \pm 126°, \pm 162°$. **(f)** $15° + n45°$, $(n = 0, \pm 1, \pm 2, \pm 3, -4)$.

4. (a) $\{.322 + n\pi, 1.249 + n\pi\}$. **(b)** $\{\pm 1.231 + 2n\pi\}$.
(c) $\{1.107 + n\pi\}$. **(d)** $\{.294 + n\pi\}$. **(e)** ϕ. **(f)** $\left\{\pm .280 + \frac{2n\pi}{3}\right\}$.

EXERCISE 9

1. (a) $\frac{2n\pi}{5}, \frac{\pi}{8} + \frac{n\pi}{4}$. **(b)** $\frac{x}{2} + n\pi, \frac{n\pi}{4}$. **(c)** $\frac{2n\pi}{3}, \frac{2n\pi}{5}$.

(d) $\frac{(2n + 1)}{7}\pi, \frac{(2n + 1)}{3}\pi$. **(e)** $\frac{\pi}{14} + \frac{2n\pi}{7}, 2n\pi - \frac{\pi}{2}$.

(f) $-\frac{\pi}{10} + \frac{2n\pi}{5}, \frac{\pi}{2} + 2n\pi$. **(g)** $n\pi$. **(h)** $\frac{n + 1}{5}\pi$.

2. (a) $n\pi, \pm\frac{\pi}{3} + 2n\pi$. **(b)** $n\pi, \frac{\pi}{4} + \frac{n\pi}{2}$. **(c)** $n\pi$. **(d)** $2n\pi, \frac{3\pi}{2} + 2n\pi$. **(e)** $\frac{2\pi}{3} + 2n\pi, \frac{4\pi}{3} + 2n\pi, 2n\pi$. **(f)** $\pm\frac{\pi}{3} + 2n\pi, (2n + 1)\pi$. **(g)** $2n\pi, \frac{\pi}{2} + 2n\pi$. **(h)** $\frac{\pi}{6} + 2n\pi, \frac{5\pi}{6} + 2n\pi$.

(i) $\frac{\pi}{6} + 2n\pi, \frac{5\pi}{6} + 2n\pi, 2n\pi$. **(j)** $2n\pi$. **(k)** $\frac{n\pi}{2}$. **(l)** $\frac{\pi}{2} + 2n\pi$.

(m) $\frac{n\pi}{3} (n \neq 3k)$. **(n)** $\frac{7\pi}{6} + 2n\pi, -\frac{\pi}{6} + 2n\pi, \frac{\pi}{2} + 2n\pi$.

(o) Same as (n). **(p)** $n\pi, \frac{\pi}{6} + 2n\pi, \frac{5\pi}{6} + 2n\pi$. **(q)** $\frac{\pi}{6} + n\pi, \frac{5\pi}{6} + n\pi$.

(r) $\frac{2\pi}{3} + 2n\pi, \frac{4\pi}{3} + 2n\pi, (2n + 1)\pi$. **(s)** $\frac{\pi}{2} + n\pi, \frac{\pi}{3} + n\pi$.

(t) $\frac{\pi}{3} + n\pi$. **(u)** $\frac{\pi}{4} + 2n\pi, \frac{3\pi}{2} + 2n\pi$. **(v)** $\frac{\pi}{3} + n\pi$.

(w) $\frac{\pi}{6} + 2n\pi, \frac{5\pi}{6} + 2n\pi$. **(x)** $2n\pi, \frac{4\pi}{3} + 2n\pi$. **(y)** $\frac{7\pi}{6} + 2n\pi$,

$-\frac{\pi}{6} + 2n\pi$. **(z)** $\pm\frac{\pi}{3} + 2n\pi$. **3.** **(a)** $\frac{\pi}{6} + 2n\pi, \frac{5\pi}{6} + 2n\pi$, $\frac{3\pi}{2} + 2n\pi$. **(b)** $2n\pi, \frac{2\pi}{3} + \frac{4n\pi}{3}$. **(c)** $\frac{\pi}{12} + n\pi, \frac{5\pi}{12} + n\pi$. **(d)** $n\pi, \pm\frac{\pi}{3} + 2n\pi$. **(e)** $n\pi$. **(f)** $\pm\frac{\pi}{6} + n\pi$. **(g)** $2n\pi$, $\frac{\pi}{9} + \frac{4n\pi}{3}, \frac{5\pi}{9} + \frac{4n\pi}{3}$. **(h)** $\frac{\pi}{18} + \frac{2n\pi}{3}, \frac{5\pi}{18} + \frac{2n\pi}{3}$. **(i)** $\frac{\pi}{2} + 2n\pi$, $\frac{\pi}{6} + 2n\pi, \frac{5\pi}{6} + 2n\pi$. **(j)** $n\pi, \pm\frac{\pi}{3} + 2n\pi$. **4.** **(a)** $\frac{\sqrt{5}}{5}$. **(b)** $\pm\frac{1}{2}$. **(c)** ± 1. **(d)** $\frac{\sqrt{3}}{6}$. **(e)** $\frac{\sqrt{5}}{5}$. **(f)** $\pm\sqrt{\frac{5 + 2\sqrt{2}}{34}}$.

5. $x_1 = x_2$ if $c^2 = a^2 + b^2$. **6.** $C = \sqrt{a^2 + b^2}, \alpha = \text{Arcsin}\,\frac{a}{C}$.

EXERCISE 10

1. **(a)** 0.86. **(b)** $\frac{\pi}{2}$. **(c)** 0. **(d)** 1. **(e)** 0.88, 0. **2.** **(a)** $x = \frac{\pi}{2}$, $y = 0$. **(b)** $x = y = \frac{\pi}{6}, \frac{5\pi}{6}$. **(c)** $x \doteq 2.0027, y \doteq -2.7096$. **(d)** $r = \sqrt{2}, \theta = \frac{\pi}{8}$. **(e)** $r = \frac{2\sqrt{3}}{3}, \theta = \frac{\pi}{6}$. **(f)** $r = \sqrt{13}, \theta \doteq 0.98$.

EXERCISE 12

1. **(a)** 3.209. **(b)** 1.214. **2.** **(a)** 1.1568. **(b)** 2.2620. **(c)** 1.5479. **(d)** 3.6070. **(e)** 6.899 **(f)** 37.53. **4.** **(a)** 3.32. **(b)** −2. **(c)** 2.8768. **(d)** .7636. **(e)** $\frac{3 \pm \sqrt{5}}{2}$ **(f)** ± 2. **(g)** $\frac{25}{3}$. **(h)** 2.

EXERCISE 13

3. **(a)** $\beta = 60°, a = 5\sqrt{3}, c = 10\sqrt{3}$. **(b)** $\alpha = 45°, b = 11$, $c = 11\sqrt{2}$. **(c)** $\beta = 67°50', b \doteq 20.4, c \doteq 22.0$. **(d)** $c = \sqrt{113}$, $\alpha = 41°11', \beta = 48°49'$. **4.** **(a)** $\alpha = 43°16', \beta = 27°$, $\gamma = 109°44'$. **(b)** $\gamma = 80°, a = 12.69, b = 23.85$. **(c)** $\beta = 24°30'$, $b = 23.32, c = 40.00$. **(d)** $a = \sqrt{7}, \beta = 130°54', \gamma = 19°6'$. **(e)** $d_1 = 60°27', \beta_1 = 82°23', b_1 = 447.1, d_2 = 119°33'$, $\beta_2 = 23°17', b_2 = 178.4$. **5.** **(3a)** $\frac{75}{2}\sqrt{3}$. **(3b)** $\frac{121}{2}$. **(3c)** 84.66. **(3d)** 28. **(4a)** 2.149. **(4b)** 149.1 **(4c)** 437.9. **(4d)** $\sqrt{3}$. **(4e)** 5299; 2113. **6.** 9.198 yds. **7.** $V_x = 2263$ ft./sec., $V_y = 1750$ ft./sec. **8.** $\alpha \doteq 13°$, G.S. $\doteq$ 266.8 mi./hr., W 13° S. **9**: **(a)** 1.705. **(b)** 4.844. **(c)** 7.539. **(d)** 0.4135. **(e)** 95.28, 50.2.

10. N$\left(90° - \text{Arccos}\,\frac{25}{32}\right)$E.

EXERCISE 14

1. $-1, -i, 1, -i, -1, -i$. **2.** (a) $1 + 3i$. (b) $-\sqrt{15}$. (c) $14 + 2i$. (d) $4i$. **3.** (a) $2(\cos 45^\circ + i \sin 45^\circ)$. (b) $\sqrt{2} \operatorname{cis} 135^\circ$. (c) $5 \operatorname{cis} 90^\circ$. (d) $\operatorname{cis} \pi$. (e) $13(\cos 22^\circ 37.2' - i \sin 22^\circ 37.2')$. (f) $3\sqrt{2}(\cos 45^\circ - i \sin 45^\circ)$. **4.** (a) $\sqrt{3} + i$. (b) $\frac{3\sqrt{2}}{2} + \frac{3\sqrt{2}}{2} i$. (c) $5i$. (d) $-\frac{7\sqrt{2}}{2} + \frac{7\sqrt{2}}{2} i$. (e) $-3.9392 - .6946i$. (f) $-\frac{5}{2} + \frac{5\sqrt{3}}{2} i$. **5.** (a) $5 + 4i$. (b) $5 + i$. (c) $2i$. (d) 4. (e) $-4 + 4i$. (f) $-1 + i$. **6.** (a) $6i$. (b) $-5\sqrt{2} - 5\sqrt{2}i$. (c) $64i$. (d) $64 - 64\sqrt{3}i$. (e) $\sqrt{2} + \sqrt{2}i$. (f) $\frac{3\sqrt{3}}{4} + \frac{3}{4} i$. (g) $\frac{1}{2} + \frac{5}{2} i$. **7.** (a) $\frac{1}{2} + \frac{\sqrt{3}}{2} i, -1, \frac{1}{2} - \frac{\sqrt{3}}{2} i$. (b) $32^{1/10} \operatorname{cis}(45^\circ + n\, 72^\circ), n = 0, 1, 2, 3, 4$. (c) $\operatorname{cis}\left(\frac{\pi}{8} + n \frac{\pi}{2}\right), n = 0, 1, 2, 3$. (d) $\operatorname{cis}\left(\frac{\pi}{2} + \frac{2\pi}{3} n\right), n = 0, 1, 2$. (e) $28^{1/4} \operatorname{cis}(20^\circ 27' + n\, 180^\circ), 28^{1/4} \operatorname{cis}(159^\circ 33' + n\, 180^\circ)$, $n = 0, 1$. (f) $\operatorname{cis}\left(\frac{\pi}{4} + \frac{3\pi}{2} n\right), n = 0, 1$. **8.** $\operatorname{cis} n\, 72^\circ$, $n = 0, 1, 2, 3, 4$.

EXERCISE 15

5. The first and third quadrant.

EXERCISE 17

1. (a) $l = 42, s = 297$. (b) $l = \frac{10}{3}, s = \frac{77}{6}$. **2.** (a) $l = \frac{1}{256}$, $s = \frac{255}{256}$. (b) $l = 27\sqrt{2}, s = 13\sqrt{6} + 40\sqrt{2}$. **3.** $2, \frac{5}{2}, 3, \frac{7}{2}, 4, \frac{9}{2}, 5, \frac{11}{2}, 6$. **4.** $\frac{1}{2}, 2, 8, 32, 128, 512, 2048$. **5.** $\frac{19}{55}$.

6. (a) $(3x)^{1/3} - \frac{2}{3} \frac{y}{(3x)^{2/3}} - \frac{4}{9} \frac{y^2}{(3x)^{5/3}} - \frac{40}{81} \frac{y^3}{(3x)^{8/3}}$. (b) $x^{-1} + 4y^{1/3} x^{-3/2} + 12y^{2/3} x^{-2} + 32y\, x^{-5/2}$. (c) $a^5x^5 + 5a^4bx^4y + 10a^3b^2x^3y^2 + 10a^2b^3x^2y^3$.

7. $\left(\frac{77}{\sqrt{2}}\right)\left(\frac{3^7}{2^{16}}\right) \frac{y^3}{x^{13/3}}$. **8.** (a) $\cot x = x^{-1} - \frac{1}{3} x - \frac{1}{45} x^3 - \frac{2}{3^3 \cdot 5 \cdot 7} x^5 - \cdots$. (b) $\tanh x = x - \frac{x^3}{3} + \frac{2x^5}{15} - \frac{17x^7}{315} + \cdots$. (c) $\sin^2 x = x^2 - \frac{x^4}{3} + \frac{2}{45} x^6 - \frac{1}{315} x^8 + \cdots$. (d) $e^x \cos x = 1 + x - \frac{1}{3} x^3 - \frac{1}{6} x^4 + \cdots$.

(e) $e^x \sin x = x + x^2 + \frac{1}{3}x^3 - \frac{1}{30}x^5 + \cdots.$

(f) $\frac{1}{1 + e^x} = \frac{1}{2} - \frac{1}{4}x + \frac{1}{48}x^3 - \frac{1}{96}x^4 + \cdots.$

(g) $e = 1 + 1 + \frac{1}{2} + \frac{1}{6} + \cdots.$

(h) $\sin 1 = 1 - \frac{1}{3!} + \frac{1}{5!} - \frac{1}{7!} + \cdots.$

9. $x^2 - 2x^4 + 3x^6 - 4x^8 + \cdots.$

10. (a) $\sin iu = i\left[u + \frac{u^3}{3!} + \frac{u^5}{5!} + \frac{u^7}{7!} + \cdots\right] = i \sinh u.$

(b) $\cos iu = 1 + \frac{u^2}{2!} + \frac{u^4}{4!} + \frac{u^6}{6!} + \cdots = \cosh u.$

(c) $\sinh iu = i\left[u - \frac{u^3}{3!} + \frac{u^5}{5!} - \frac{u^7}{7!} + \cdots\right] = i \sin u.$

(d) $\cosh iu = 1 - \frac{u^2}{2!} + \frac{u^4}{4!} - \frac{u^6}{6!} + \cdots = \cos u.$

EXERCISE 18

3. $b = 53°39'$, $B = 58°15'$, $a = 57°14'$. **4.** See Kaj L. Nielsen and John H. Vanlonkhuyzen, *Plane and Spherical Trigonometry*, pp 124–125. **5 (a)** $b = 47°31'$, $a = 44°6'$, $A = 52°44'$. **(b)** $B = 110°47'$, $A = 37°46'$, $c = 119°21'$. **(c)** $A = 52°14'$, $c = 116°34'$, $b = 129°14'$. **(d)** $A_1 = 51°32'$, $c_1 = 59°37'$, $a_1 = 42°29'$, $A_2 = 128°28'$, $c_2 = 120°23'$, $a_2 = 137°31$. **6.** $A_1 = 51°$, $c_1 = 144°6'$, $C_1 = 152°23'$, $A_2 = 129°$, $c_2 = 35°54'$, $C_2 = 27°7'$.

EXERCISE 19

1(a) $B = 100°17'$, $C = 82°1'$, $a = 133°14'$. **(b)** $a = 64°29'$, $c = 134°5'$, $B = 75°40'$. **(c)** $B = 38°$, $C = 132°4'$, $c = 108°8'$. **(d)** $c = 76°6'$, $A = 32°20'$, $a = 33°48'$. **(e)** $A = 128°$, $B = 59°4'$, $C = 85°6'$. **(f)** $a = 91°46'$, $b = 112°54'$, $c = 125°26'$.

EXERCISE 20

1. $c = 37°2' = 2222$ nautical miles, $B = 99°39'$ or W 9°39′ S. **2.** $2^h\,50^m$. **3.** $6^h\,30.5^m$. **4.** 15°44′S.

PROBLEMS (A.8)

1. (a) .92164. **(b)** .8166. **(c)** .9424. **(d)** .3159. **(e)** .5412. **(f)** 6.5606. **(g)** .2399. **(h)** 1.7675. **(i)** .7135. **2. (a)** 32°40′. **(b)** 46°50′. **(c)** 37°40′. **(d)** 45°34′. **(e)** 46°45′. **(f)** 47°35′.

(g) 48°. (h) $53^\circ 8'$. (i) $59^\circ 53'$. **3.** (a) .9744. (b) $-.9205$. (c) 1.8940. (d) -3.1652. (e) $-.6900$. (f) $-.1650$. **4.** (a) .3146. (b) .5319. (c) .4831. (d) .9687. (e) 5.798. (f) .5403. **5.** 0.4702, 6.6345-10, 5.2362. **6.** .6673, 180.4. **7.** 9.4934-10, 9.9684-10. **8.** $57^\circ 11'$. **9.** (a) 6.012. (b) 25.54. (c) 1.706. **10.** (a) 1.22140, .74082, 4.48169, .13534. (b) $-1.20397, 0.33647$, 1.72277.

TEST 1

2. $\sqrt{65}$. **3.** $\overline{OP_1} = 5\sqrt{2}, \overline{OP_2} = 5$. **5.** $A(5\sqrt{2}, 0)$. **6.** $P_1\left(5\sqrt{2}, \frac{5\sqrt{2}}{4}\pi\right)$. **7.** $P_1\left(5\sqrt{2}, \frac{\pi}{4}\right)$. **8.** $x^2 + y^2 < 50$, $x > -3, y > 4$. **10.** No. **11.** $B(5,0), \frac{25}{2}$. **12.** $K = 50\pi$, $C = 10\sqrt{2}\pi$. **13.** $K = \frac{25}{4}\pi$. **14.** $v = \frac{5\sqrt{2}}{4}\pi$ units/sec., $\omega = \frac{\pi}{4}$ rad./sec. **15.** $x^2 + y^2 = 50$. **16.** CD = 5, $\overline{\text{BD}} = \sqrt{41}$. **17.** $D\left(4, \frac{\pi}{2}\right), D(4, 2\pi)$. **18.** $67\frac{1}{2}^\circ$. **19.** $-315^\circ, -\frac{7\pi}{4}$.

20. range of f: $0 \leq y \leq 5; f^{-1}: y = \sqrt{50 - x^2}, 0 \leq x \leq 5$.

TEST 2

1. (a) Let the argument be t of the circular coordinates of a point $P(1,t)$ on the unit circle, then $\cos t = x$, where x is the abscissa of P in a rectangular coordinate system. (b) $\cos\theta = \frac{x}{r}$ with θ in the standard position. (c) $\cos\theta = \frac{\text{side adjacent}}{\text{hypotenuse}}$ in a right triangle.

2.

	30°	$-\frac{\pi}{4}$	π	60°	$-\frac{3\pi}{2}$	-120°
sin	$\frac{1}{2}$	$-\frac{\sqrt{2}}{2}$	0	$\frac{\sqrt{3}}{2}$	1	$-\frac{\sqrt{3}}{2}$
cos	$\frac{\sqrt{3}}{2}$	$\frac{\sqrt{2}}{2}$	-1	$\frac{1}{2}$	0	$-\frac{1}{2}$
tan	$\frac{\sqrt{3}}{3}$	-1	0	$\sqrt{3}$	–	$\sqrt{3}$
cot	$\sqrt{3}$	-1	–	$\frac{\sqrt{3}}{3}$	0	$\frac{\sqrt{3}}{3}$

3. (a) $-\frac{\sqrt{2}}{4}$. (b) $\frac{\sqrt{2}}{4}$. (c) 0. **4.** $\frac{\sqrt{3}}{2}, -\frac{1}{2}, -\sqrt{3}, -\frac{\sqrt{3}}{3}, -2, \frac{2\sqrt{3}}{3}, -\frac{\sqrt{3}}{2}, \frac{1}{2}, -\sqrt{3}, -\frac{\sqrt{3}}{3}, 2, -\frac{2\sqrt{3}}{3}$. **5.** (a) $\frac{\pi}{6}$. (b) $-\frac{\pi}{3}$. (c) $\frac{4}{3}$. (d) $\frac{1}{y}$.

TEST 4

5. $\alpha = \frac{\pi}{3}$.

TEST 5

1. (a) $\frac{\pi}{9} + \frac{2n\pi}{3}$. (b) $\frac{\pi}{10} + \frac{2n\pi}{5}, -\frac{\pi}{2} + 2n\pi$. (c) $\frac{\pi}{3} + 2n\pi$. (d) $\frac{\pi}{8} + n\pi, \frac{5\pi}{8} + n\pi$. **2.** (a) $0, \pi, \frac{\pi}{6}, \frac{5\pi}{6}$. (b) $\frac{\pi}{24}, \frac{13\pi}{24}$. (c) $\frac{7\pi}{6}, \frac{11\pi}{6}, \frac{\pi}{2}$. (d) $\frac{\pi}{6}, \frac{5\pi}{6}, \frac{\pi}{2}$. (e) $x = 1$.

TEST 6

1. $f^{-1}: y = 2^x$. (See Figure 48.) **2.** (a) 1.90309. (b) 0.82660. **3.** (a) $8^{1/3} = 2, 5^{-2} = \frac{1}{25}$. (b) $\log_3 27 = 3, \log_{1/2} 8 = -3$. **4.** 1.4650. **5.** $x \doteq 13.9$.

TEST 7

3. $\frac{99}{2}$. **4.** $\angle BCD = 141°$, find BD and $\angle BDC$ given s.a.s. $\angle DEF = 83°$, find DF and $\angle FDE$ given s.a.s. $\angle BDF = 62° - \angle BDC + 52°30' - \angle FDE$, find BF given s.a.s. **5.** $h = 30 + 40\sqrt{3}$ ft.

TEST 8

1. (a) $-22 - 6i$. (b) $\frac{2}{13} + \frac{16}{13}i$. (c) $\sqrt{2}\operatorname{cis} 225°$. (d) $-3 - 4i$. **2.** $2\sqrt{3} + 2i$. **3.** $2^{1/10}\operatorname{cis}(9° + n\,72°), n = 0, 1, 2, 3, 4$. **4.** $\pm\frac{1}{4}(\sqrt{6} - \sqrt{2}) \pm \frac{1}{4}(\sqrt{6} + \sqrt{2})\,i$. **5.** $-1024\sqrt{2} - 1024\sqrt{2}\,i$.

TEST 9

2. $\ln(x + \sqrt{x^2 + 1})$. **5.** $(\cosh x + \sinh x)^3 = \cosh 3x + \sinh 3x$.

TEST 10

1. $x^{1/2} - x^{-1/2}y - \frac{1}{2}x^{-3/2}y^2 - \frac{1}{2}x^{-5/2}y^3 - \frac{5}{8}x^{-7/2}y^4 - \cdots$.

2. $\frac{7}{8} - \frac{1}{2}i$. **3.** 9.8994949375. **5.** $x + x^2 + \frac{2}{3!}x^3 + \frac{6}{5!}x^5 - \frac{14}{6}x^6 + \cdots$.

TEST 11

1. $\sin a = \sin A \sin c$, $\tan B = \cot A \sec c$, $\tan b = \cos A \tan c$.
2. $A = 121°13'$, $C = 114°17'$, $b = 158°44'$. **3.** See page 206.

EXAM 1

1. **(a)** $\sin x$. **(b)** $\tan x$. **(c)** $\cos x$. **(d)** $\sin x$. **3.** **(a)** $\frac{\pi}{2} + 2n\pi$, $\frac{3\pi}{2} + 2n\pi$, $\frac{\pi}{4} + n\pi$. **(b)** ϕ. **(c)** $\frac{\pi}{3} + 2n\pi$, $\frac{5\pi}{3} + 2n\pi$.

4. $\cos\alpha = \frac{b^2 + c^2 - a^2}{2bc}$, $\cos\beta = \frac{a^2 + c^2 - b^2}{2ac}$, $\cos\gamma = \frac{a^2 + b^2 - c^2}{2ab}$. **5.** $\text{cis}\,\frac{n\pi}{2}$, $n = 0, 1, 2, 3$. **6.** **(a)** $\frac{5}{13}$. **(b)** $\frac{3\pi}{4}$. **(c)** $\frac{63}{65}$. **7.** **(b)** $\left(\frac{1}{2} - \sqrt{2}\right) + \left(-\frac{\sqrt{3}}{2} + \sqrt{2}\right)i$.
(c) $-\frac{\sqrt{6} + \sqrt{2}}{8} + \frac{\sqrt{6} - \sqrt{2}}{8}i$. **8.** $y = \ln x$. **9.** 10.

EXAM 2

1. 7. **2.** $x = \frac{17}{8}$, $y = \frac{15}{8}$. **3.** 14π ft./sec. **4.** **(a)** 10. **(b)** $\sqrt{127}$. **(c)** 5. **7.** $\left(2, \frac{\pi}{3}\right)$, $(\sqrt{3}, 1)$. **8.** $f^{-1}: y = \sqrt{25 - x^2}$, $4 \le x \le 5$.

EXAM 3

2. **(a)** $0, \frac{\pi}{5}, \frac{3\pi}{5}$. **(b)** $0, \pi, \frac{\pi}{4}, \frac{5\pi}{4}$. **(c)** $0, \pi, \frac{\pi}{6}, \frac{5\pi}{6}$. **4.** $\pm\sqrt{3}$.
5. $1, \frac{1}{2} \pm \frac{\sqrt{3}}{2}i$. **7.** $\frac{\log 2}{\log 3 - \log 2}$. **8.** ± 1. **9.** $-128i$.
10. $1 + x - \frac{1}{3}x^3 - \frac{5}{24}x^4 + \cdots$.

EXAM 4

1. F. **2.** T. **3.** F. **4.** F. **5.** T. **6.** F. **7.** T.
8. T. **9.** T. **10.** T. **11.** F. **12.** T. **13.** T. **14.** F.
15. T. **16.** T. **17.** F. **18.** T. **19.** F. **20.** T. **21.** F.
22. F. **23.** F. **24.** T. **25.** F.

EXAM 5

1. $1 + x + x^2 + \frac{2}{3}x^3 + \frac{1}{2}x^4 + \cdots$. **2.** $b_1 = 41°11'$, $C_1 = 91°58'$, $c_1 = 48°14'$, $b_2 = 138°49'$, $C_2 = 149°16'$, $c_2 = 157°36'$. **3.** 2128 naut. mi. W 3°35′N.

CHALLENGING PROBLEMS

3. 12.31186. For a complete discussion of this problem, see Kaj L. Nielsen, *College Mathematics* (New York: Barnes and Noble, 1958), pp. 246–248. **4.** $a^2\left(1 + \frac{\pi}{3} - \sqrt{3}\right)$. **5.** 4. (In developing the formula $2R = a/\sin\alpha$, make good use of your knowledge of circles from plane geometry.)

Tables

DEGREES TO RADIANS 1 degree = 0.01745 32925 19943 radians

Degrees						Minutes		Seconds	
0	0.00000 00	60°	1.04719 76	120°	2.09439 51	0	0.00000 00	0	0.00000 00
1	0.01745 33	61	1.06465 08	121	2.11184 84	1	0.00029 09	1	0.00000 48
2	0.03490 66	62	1.08210 41	122	2.12930 17	2	0.00058 18	2	0.00000 97
3	0.05235 99	63	1.09955 74	123	2.14675 50	3	0.00087 27	3	0.00001 45
4	0.06981 32	64	1.11701 07	124	2.16420 83	4	0.00116 36	4	0.00001 94
5	0.08726 65	65	1.13446 40	125	2.18166 16	5	0.00145 44	5	0.00002 42
6	0.10471 98	66	1.15191 73	126	2.19911 49	6	0.00174 53	6	0.00002 91
7	0.12217 30	67	1.16937 06	127	2.21656 82	7	0.00203 62	7	0.00003 39
8	0.13962 63	68	1.18682 39	128	2.23402 14	8	0.00232 71	8	0.00003 88
9	0.15707 96	69	1.20427 72	129	2.25147 47	9	0.00261 80	9	0.00004 36
10	0.17453 29	70	1.22173 05	130	2.26892 80	10	0.00290 89	10	0.00004 85
11	0.19198 62	71	1.23918 38	131	2.28638 13	11	0.00319 98	11	0.00005 33
12	0.20943 95	72	1.25663 71	132	2.30383 46	12	0.00349 07	12	0.00005 82
13	0.22689 28	73	1.27409 04	133	2.32128 79	13	0.00378 15	13	0.00006 30
14	0.24434 61	74	1.29154 36	134	2.33874 12	14	0.00407 24	14	0.00006 79
15	0.26179 94	75	1.30899 69	135	2.35619 45	15	0.00436 33	15	0.00007 27
16	0.27925 27	76	1.32645 02	136	2.37364 78	16	0.00465 42	16	0.00007 76
17	0.29670 60	77	1.34390 35	137	2.39110 11	17	0.00494 51	17	0.00008 24
18	0.31415 93	78	1.36135 68	138	2.40855 44	18	0.00523 60	18	0.00008 73
19	0.33161 26	79	1.37881 01	139	2.42600 77	19	0.00552 69	19	0.00009 21
20	0.34906 59	80	1.39626 34	140	2.44346 10	20	0.00581 78	20	0.00009 70
21	0.36651 91	81	1.41371 67	141	2.46091 42	21	0.00610 87	21	0.00010 18
22	0.38397 24	82	1.43117 00	142	2.47836 75	22	0.00639 95	22	0.00010 67
23	0.40142 57	83	1.44862 33	143	2.49582 08	23	0.00669 04	23	0.00011 15
24	0.41887 90	84	1.46607 66	144	2.51327 41	24	0.00698 13	24	0.00011 64
25	0.43633 23	85	1.48352 99	145	2.53072 74	25	0.00727 22	25	0.00012 12
26	0.45378 56	86	1.50098 32	146	2.54818 07	26	0.00756 31	26	0.00012 61
27	0.47123 89	87	1.51843 64	147	2.56563 40	27	0.00785 40	27	0.00013 09
28	0.48869 22	88	1.53588 97	148	2.58308 73	28	0.00814 49	28	0.00013 57
29	0.50614 55	89	1.55334 30	149	2.60054 06	29	0.00843 58	29	0.00014 06
30	0.52359 88	90	1.57079 63	150	2.61799 39	30	0.00872 66	30	0.00014 54
31	0.54105 21	91	1.58824 96	151	2.63544 72	31	0.00901 75	31	0.00015 03
32	0.55850 54	92	1.60570 29	152	2.65290 05	32	0.00930 84	32	0.00015 51
33	0.57595 87	93	1.62315 62	153	2.67035 38	33	0.00959 93	33	0.00016 00
34	0.59341 19	94	1.64060 95	154	2.68780 70	34	0.00989 02	34	0.00016 48
35	0.61086 52	95	1.65806 28	155	2.70526 03	35	0.01018 11	35	0.00016 97
36	0.62831 85	96	1.67551 61	156	2.72271 36	36	0.01047 20	36	0.00017 45
37	0.64577 18	97	1.69296 94	157	2.74016 69	37	0.01076 29	37	0.00017 94
38	0.66322 51	98	1.71042 27	158	2.75762 02	38	0.01105 38	38	0.00018 42
39	0.68067 84	99	1.72787 60	159	2.77507 35	39	0.01134 46	39	0.00018 91
40	0.69813 17	100	1.74532 93	160	2.79252 68	40	0.01163 55	40	0.00019 39
41	0.71558 50	101	1.76278 25	161	2.80998 01	41	0.01192 64	41	0.00019 88
42	0.73303 83	102	1.78023 58	162	2.82743 34	42	0.01221 73	42	0.00020 36
43	0.75049 16	103	1.79768 91	163	2.84488 67	43	0.01250 82	43	0.00020 85
44	0.76794 49	104	1.81514 24	164	2.86234 00	44	0.01279 91	44	0.00021 33
45	0.78539 82	105	1.83259 57	165	2.87979 33	45	0.01309 00	45	0.00021 82
46	0.80285 15	106	1.85004 90	166	2.89724 66	46	0.01338 09	46	0.00022 30
47	0.82030 47	107	1.86750 23	167	2.91469 99	47	0.01367 17	47	0.00022 79
48	0.83775 80	108	1.88495 56	168	2.93215 31	48	0.01396 26	48	0.00023 27
49	0.85521 13	109	1.90240 89	169	2.94960 64	49	0.01425 35	49	0.00023 76
50	0.87266 46	110	1.91986 22	170	2.96705 97	50	0.01454 44	50	0.00024 24
51	0.89011 79	111	1.93731 55	171	2.98451 30	51	0.01483 53	51	0.00024 73
52	0.90757 12	112	1.95476 88	172	3.00196 63	52	0.01512 62	52	0.00025 21
53	0.92502 45	113	1.97222 21	173	3.01941 96	53	0.01541 71	53	0.00025 70
54	0.94247 78	114	1.98967 53	174	3.03687 29	54	0.01570 80	54	0.00026 18
55	0.95993 11	115	2.00712 86	175	3.05432 62	55	0.01599 89	55	0.00026 66
56	.097738 44	116	2.02458 19	176	3.07177 95	56	0.01628 97	56	0.00027 15
57	0.99483 77	117	2.04203 52	177	3.08923 28	57	0.01658 06	57	0.00027 63
58	1.01229 10	118	2.05948 85	178	3.10668 61	58	0.01687 15	58	0.00028 12
59	1.02974 43	119	2.07694 18	179	3.12413 94	59	0.01716 24	59	0.00028 60
60	1.04719 76	120	2.09439 51	180	3.14159 27	60	0.01745 33	60	0.00029 09

RADIANS TO DEGREES

1 radian = 57.29577 95130 82321 degrees

	Radians	Tenths	Hundredths	Thousandths	Ten-thousandths
1	57°17′44″.8	5°43′46″.5	0°34′22″.6	0° 3′26″.3	0°0′20″.6
2	114°35′29″.6	11°27′33″.0	1° 8′45″.3	0° 6′52″.5	0°0′41″.3
3	171°53′14″.4	17°11′19″.4	1°43′07″.9	0°10′18″.8	0°1′01″.9
4	229°10′59″.2	22°55′05″.9	2°17′30″.6	0°13′45″.1	0°1′22″.5
5	286°28′44″.0	28°38′52″.4	2°51′53″.2	0°17′11″.3	0°1′43″.1
6	343°46′28″.8	34°22′38″.9	3°26′15″.9	0°20′37″.6	0°2′03″.8
7	401° 4′13″.6	40° 6′25″.4	4° 0′38″.5	0°24′03″.9	0°2′24″.4
8	458°21′58″.4	45°50′11″.8	4°35′01″.2	0°27′30″.1	0°2′45″.0
9	515°39′43″.3	51°33′58″.3	5° 9′23″.8	0°30′56″.4	0°3′05″.6

TABLE OF CONSTANTS

Values				Reciprocals			
π	3.14159	26535	89793	$\frac{1}{\pi}$	0.31830	98861	83791
$\frac{\pi}{2}$	1.57079	63267	94897	$\frac{2}{\pi}$	0.63661	97723	67582
2π	6.28318	53071	79586	$\frac{1}{2\pi}$	0.15915	49430	91895
π^2	9.86960	44010	89359	$\frac{1}{\pi^2}$	0.10132	11836	42338
$\sqrt{\pi}$	1.77245	38509	05516	$\frac{1}{\sqrt{\pi}}$	0.56418	95835	47756
$\sqrt{\frac{\pi}{2}}$	1.25331	41373	15500	$\sqrt{\frac{2}{\pi}}$	0.79788	45608	02865
$\sqrt{2\pi}$	2.50622	82746	31001	$\frac{1}{\sqrt{2\pi}}$	0.39894	22804	01433
e	2.71828	18284	59045	$\frac{1}{e}$	0.36787	94411	71442
e^2	7.38905	60989	30650	$\frac{1}{e^2}$	0.13533	52832	36613
$\sqrt{e}$	1.64872	12707	00128	$\frac{1}{\sqrt{e}}$	0.60653	06597	12633
$\log_{10} e$	0.43429	44819	03252	$\log_e 10$	2.30258	50929	94046

$\log_{10} \pi$ =	0.49714	98726	94133		
1″ =	0.00000	48481	36811	095	radians
sin 1″ =	0.00000	48481	36811	076	
tan 1″ =	0.00000	48481	36811	133	

Table 3. FOUR-PLACE COMMON LOGARITHMS

N	0	1	2	3	4	5	6	7	8	9
10	0000	0043	0086	0128	0170	0212	0253	0294	0334	0374
11	0414	0453	0492	0531	0569	0607	0645	0682	0719	0755
12	0792	0828	0864	0899	0934	0969	1004	1038	1072	1106
13	1139	1173	1206	1239	1271	1303	1335	1367	1399	1430
14	1461	1492	1523	1553	1584	1614	1644	1673	1703	1732
15	1761	1790	1818	1847	1875	1903	1931	1959	1987	2014
16	2041	2068	2095	2122	2148	2175	2201	2227	2253	2279
17	2304	2330	2355	2380	2405	2430	2455	2480	2504	2529
18	2553	2577	2601	2625	2648	2672	2695	2718	2742	2765
19	2788	2810	2833	2856	2878	2900	2923	2945	2967	2989
20	3010	3032	3054	3075	3096	3118	3139	3160	3181	3201
21	3222	3243	3263	3284	3304	3324	3345	3365	3385	3404
22	3424	3444	3464	3483	3502	3522	3541	3560	3579	3598
23	3617	3636	3655	3674	3692	3711	3729	3747	3766	3784
24	3802	3820	3838	3856	3874	3892	3909	3927	3945	3962
25	3979	3997	4014	4031	4048	4065	4082	4099	4116	4133
26	4150	4166	4183	4200	4216	4232	4249	4265	4281	4298
27	4314	4330	4346	4362	4378	4393	4409	4425	4440	4456
28	4472	4487	4502	4518	4533	4548	4564	4579	4594	4609
29	4624	4639	4654	4669	4683	4698	4713	4728	4742	4757
30	4771	4786	4800	4814	4829	4843	4857	4871	4886	4900
31	4914	4928	4942	4955	4969	4983	4997	5011	5024	5038
32	5051	5065	5079	5092	5105	5119	5132	5145	5159	5172
33	5185	5198	5211	5224	5237	5250	5263	5276	5289	5302
34	5315	5328	5340	5353	5366	5378	5391	5403	5416	5428
35	5441	5453	5465	5478	5490	5502	5514	5527	5539	5551
36	5563	5575	5587	5599	5611	5623	5635	5647	5658	5670
37	5682	5694	5705	5717	5729	5740	5752	5763	5775	5786
38	5798	5809	5821	5832	5843	5855	5866	5877	5888	5899
39	5911	5922	5933	5944	5955	5966	5977	5988	5999	6010
40	6021	6031	6042	6053	6064	6075	6085	6096	6107	6117
41	6128	6138	6149	6160	6170	6180	6191	6201	6212	6222
42	6232	6243	6253	6263	6274	6284	6294	6304	6314	6325
43	6335	6345	6355	6365	6375	6385	6395	6405	6415	6425
44	6435	6444	6454	6464	6474	6484	6493	6503	6513	6522
45	6532	6542	6551	6561	6571	6580	6590	6599	6609	6618
46	6628	6637	6646	6656	6665	6675	6684	6693	6702	6712
47	6721	6730	6739	6749	6758	6767	6776	6785	6794	6803
48	6812	6821	6830	6839	6848	6857	6866	6875	6884	6893
49	6902	6911	6920	6928	6937	6946	6955	6964	6972	6981
50	6990	6998	7007	7016	7024	7033	7042	7050	7059	7067
51	7076	7084	7093	7101	7110	7118	7126	7135	7143	7152
52	7160	7168	7177	7185	7193	7202	7210	7218	7226	7235
53	7243	7251	7259	7267	7275	7284	7292	7300	7308	7316
54	7324	7332	7340	7348	7356	7364	7372	7380	7388	7396

Table 3. CONTINUED

N	0	1	2	3	4	5	6	7	8	9
55	7404	7412	7419	7427	7435	7443	7451	7459	7466	7474
56	7482	7490	7497	7505	7513	7520	7528	7536	7543	7551
57	7559	7566	7574	7582	7589	7597	7604	7612	7619	7627
58	7634	7642	7649	7657	7664	7672	7679	7686	7694	7701
59	7709	7716	7723	7731	7738	7745	7752	7760	7767	7774
60	7782	7789	7796	7803	7810	7818	7825	7832	7839	7846
61	7853	7860	7868	7875	7882	7889	7896	7903	7910	7917
62	7924	7931	7938	7945	7952	7959	7966	7973	7980	7987
63	7993	8000	8007	8014	8021	8028	8035	8041	8048	8055
64	8062	8069	8075	8082	8089	8096	8102	8109	8116	8122
65	8129	8136	8142	8149	8156	8162	8169	8176	8182	8189
66	8195	8202	8209	8215	8222	8228	8235	8241	8248	8254
67	8261	8267	8274	8280	8287	8293	8299	8306	8312	8319
68	8325	8331	8338	8344	8351	8357	8363	8370	8376	8382
69	8388	8395	8401	8407	8414	8420	8426	8432	8439	8445
70	8451	8457	8463	8470	8476	8482	8488	8494	8500	8506
71	8513	8519	8525	8531	8537	8543	8549	8555	8561	8567
72	8573	8579	8585	8591	8597	8603	8609	8615	8621	8627
73	8633	8639	8645	8651	8657	8663	8669	8675	8681	8686
74	8692	8698	8704	8710	8716	8722	8727	8733	8739	8745
75	8751	8756	8762	8768	8774	8779	8785	8791	8797	8802
76	8808	8814	8820	8825	8831	8837	8842	8848	8854	8859
77	8865	8871	8876	8882	8887	8893	8899	8904	8910	8915
78	8921	8927	8932	8938	8943	8949	8954	8960	8965	8971
79	8976	8982	8987	8993	8998	9004	9009	9015	9020	9025
80	9031	9036	9042	9047	9053	9058	9063	9069	9074	9079
81	9085	9090	9096	9101	9106	9112	9117	9122	9128	9133
82	9138	9143	9149	9154	9159	9165	9170	9175	9180	9186
83	9191	9196	9201	9206	9212	9217	9222	9227	9232	9238
84	9243	9248	9253	9258	9263	9269	9274	9279	9284	9289
85	9294	9299	9304	9309	9315	9320	9325	9330	9335	9340
86	9345	9350	9355	9360	9365	9370	9375	9380	9385	9390
87	9395	9400	9405	9410	9415	9420	9425	9430	9435	9440
88	9445	9450	9455	9460	9465	9469	9474	9479	9484	9489
89	9494	9499	9504	9509	9513	9518	9523	9528	9533	9538
90	9542	9547	9552	9557	9562	9566	9571	9576	9581	9586
91	9590	9595	9600	9605	9609	9614	9619	9624	9628	9633
92	9638	9643	9647	9652	9657	9661	9666	9671	9675	9680
93	9685	9689	9694	9699	9703	9708	9713	9717	9722	9727
94	9731	9736	9741	9745	9750	9754	9759	9763	9768	9773
95	9777	9782	9786	9791	9795	9800	9805	9809	9814	9818
96	9823	9827	9832	9836	9841	9845	9850	9854	9859	9863
97	9868	9872	9877	9881	9886	9890	9894	9899	9903	9908
98	9912	9917	9921	9926	9930	9934	9939	9943	9948	9952
99	9956	9961	9965	9969	9974	9978	9983	9987	9991	9996

Table 4. LOGARITHMS OF TRIGONOMETRIC FUNCTIONS

Angle	L sin	d	L cos	d	L tan	cd	L cot	Angle
0° 0'	- ∞		0.0000	0	- ∞		∞	90° 0'
10'	7.4637	3011	0.0000	0	7.4637	3011	2.5363	50'
20'	7.7648	1760	0.0000	0	7.7648	1761	2.2352	40'
30'	7.9408	1250	0.0000	0	7.9409	1249	2.0591	30'
40'	8.0658	969	0.0000	0	8.0658	969	1.9342	20'
50'	8.1627	792	0.0000	1	8.1627	792	1.8373	10'
1° 0'	8.2419	669	9.9999	0	8.2419	670	1.7581	89° 0'
10'	8.3088	580	9.9999	0	8.3089	580	1.6911	50'
20'	8.3668	511	9.9999	0	8.3669	512	1.6331	40'
30'	8.4179	458	9.9999	1	8.4181	457	1.5819	30'
40'	8.4637	413	9.9998	0	8.4638	415	1.5362	20'
50'	8.5050	378	9.9998	1	8.5053	378	1.4947	10'
2° 0'	8.5428	348	9.9997	0	8.5431	348	1.4569	88° 0'
10'	8.5776	321	9.9997	1	8.5779	322	1.4221	50'
20'	8.6097	300	9.9996	0	8.6101	300	1.3899	40'
30'	8.6397	280	9.9996	1	8.6401	281	1.3599	30'
40'	8.6677	263	9.9995	0	8.6682	263	1.3318	20'
50'	8.6940	248	9.9995	1	8.6945	249	1.3055	10'
3° 0'	8.7188	235	9.9994	1	8.7194	235	1.2806	87° 0'
10'	8.7423	222	9.9993	0	8.7429	223	1.2571	50'
20'	8.7645	212	9.9993	1	8.7652	213	1.2348	40'
30'	8.7857	202	9.9992	1	8.7865	202	1.2135	30'
40'	8.8059	192	9.9991	1	8.8067	194	1.1933	20'
50'	8.8251	185	9.9990	1	8.8261	185	1.1739	10'
4° 0'	8.8436	177	9.9989	1	8.8446	178	1.1554	86° 0'
10'	8.8613	170	9.9988	0	8.8624	171	1.1376	50'
20'	8.8783	163	9.9988	1	8.8795	165	1.1205	40'
30'	8.8946	158	9.9987	1	8.8960	158	1.1040	30'
40'	8.9104	152	9.9986	2	8.9118	154	1.0882	20'
50'	8.9256	147	9.9984	1	8.9272	148	1.0728	10'
5° 0'	8.9403	142	9.9983	1	8.9420	143	1.0580	85° 0'
10'	8.9545	137	9.9982	1	8.9563	138	1.0437	50'
20'	8.9682	134	9.9981	1	8.9701	135	1.0299	40'
30'	8.9816	129	9.9980	1	8.9836	130	1.0164	30'
40'	8.9945	125	9.9979	2	8.9966	127	1.0034	20'
50'	9.0070	122	9.9977	1	9.0093	123	0.9907	10'
6° 0'	9.0192	119	9.9976	1	9.0216	120	0.9784	84° 0'
10'	9.0311	115	9.9975	2	9.0336	117	0.9664	50'
20'	9.0426	113	9.9973	1	9.0453	114	0.9547	40'
30'	9.0539	109	9.9972	2	9.0567	111	0.9433	30'
40'	9.0648	107	9.9970	1	9.0678	108	0.9322	20'
50'	9.0755	104	9.9969	1	9.0786	105	0.9214	10'
7° 0'	9.0859	102	9.9968	2	9.0891	104	0.9109	83° 0'
10'	9.0961	99	9.9966	2	9.0995	101	0.9005	50'
20'	9.1060	97	9.9964	1	9.1096	98	0.8904	40'
30'	9.1157	95	9.9963	2	9.1194	97	0.8806	30'
Angle	L cos	d	L sin	d	L cot	cd	L tan	Angle

Table 4. CONTINUED

Angle	L sin	d	L cos	d	L tan	cd	L cot	Angle
7° 30'	9.1157	95	9.9963	2	9.1194	97	0.8806	83° 30'
40'	9.1252	93	9.9961	2	9.1291	94	0.8709	20'
50'	9.1345	91	9.9959	1	9.1385	93	0.8615	10'
8° 0'	9.1436	89	9.9958	2	9.1478	91	0.8522	82° 0'
10'	9.1525	87	9.9956	2	9.1569	89	0.8431	50'
20'	9.1612	85	9.9954	2	9.1658	87	0.8342	40'
30'	9.1697	84	9.9952	2	9.1745	86	0.8255	30'
40'	9.1781	82	9.9950	2	9.1831	84	0.8169	20'
50'	9.1863	80	9.9948	2	9.1915	82	0.8085	10'
9° 0'	9.1943	79	9.9946	2	9.1997	81	0.8003	81° 0'
10'	9.2022	78	9.9944	2	9.2078	80	0.7922	50'
20'	9.2100	76	9.9942	2	9.2158	78	0.7842	40'
30'	9.2176	75	9.9940	2	9.2236	77	0.7764	30'
40'	9.2251	73	9.9938	2	9.2313	76	0.7687	20'
50'	9.2324	73	9.9936	2	9.2389	74	0.7611	10'
10° 0'	9.2397	71	9.9934	3	9.2463	73	0.7537	80° 0'
10'	9.2468	70	9.9931	2	9.2536	73	0.7464	50'
20'	9.2538	68	9.9929	2	9.2609	71	0.7391	40'
30'	9.2606	68	9.9927	3	9.2680	70	0.7320	30'
40'	9.2674	66	9.9924	2	9.2750	69	0.7250	20'
50'	9.2740	66	9.9922	2	9.2819	68	0.7181	10'
11° 0'	9.2806	64	9.9919	2	9.2887	66	0.7113	79° 0'
10'	9.2870	64	9.9917	3	9.2953	67	0.7047	50'
20'	9.2934	63	9.9914	2	9.3020	65	0.6980	40'
30'	9.2997	61	9.9912	3	9.3085	64	0.6915	30'
40'	9.3058	61	9.9909	2	9.3149	63	0.6851	20'
50'	9.3119	60	9.9907	3	9.3212	63	0.6788	10'
12° 0'	9.3179	59	9.9904	3	9.3275	61	0.6725	78° 0'
10'	9.3238	58	9.9901	2	9.3336	61	0.6664	50'
20'	9.3296	57	9.9899	3	9.3397	61	0.6603	40'
30'	9.3353	57	9.9896	3	9.3458	59	0.6542	30'
40'	9.3410	56	9.9893	3	9.3517	59	0.6483	20'
50'	9.3466	55	9.9890	3	9.3576	58	0.6424	10'
13° 0'	9.3521	54	9.9887	3	9.3634	57	0.6366	77° 0'
10'	9.3575	54	9.9884	3	9.3691	57	0.6309	50'
20'	9.3629	53	9.9881	3	9.3748	56	0.6252	40'
30'	9.3682	52	9.9878	3	9.3804	55	0.6196	30'
40'	9.3734	52	9.9875	3	9.3859	55	0.6141	20'
50'	9.3786	51	9.9872	3	9.3914	54	0.6086	10'
14° 0'	9.3837	50	9.9869	3	9.3968	53	0.6032	76° 0'
10'	9.3887	50	9.9866	3	9.4021	53	0.5979	50'
20'	9.3937	49	9.9863	4	9.4074	53	0.5926	40'
30'	9.3986	49	9.9859	3	9.4127	51	0.5873	30'
40'	9.4035	48	9.9856	3	9.4178	52	0.5822	20'
50'	9.4083	47	9.9853	4	9.4230	50	0.5770	10'
Angle	L cos	d	L sin	d	L cot	cd	L tan	Angle

Table 4. CONTINUED

Angle		L sin	d	L cos	d	L tan	cd	L cot	Angle	
15°	0'	9.4130	47	9.9849	3	9.4280	51	0.5720	75°	0'
	10'	9.4177	46	9.9846	3	9.4331	50	0.5669		50'
	20'	9.4223	46	9.9843	4	9.4381	49	0.5619		40'
	30'	9.4269	45	9.9839	3	9.4430	49	0.5570		30'
	40'	9.4314	45	9.9836	4	9.4479	48	0.5521		20'
	50'	9.4359	44	9.9832	4	9.4527	48	0.5473		10'
16°	0'	9.4403	44	9.9828	3	9.4575	47	0.5425	74°	0'
	10'	9.4447	44	9.9825	4	9.4622	47	0.5378		50'
	20'	9.4491	42	9.9821	4	9.4669	47	0.5331		40'
	30'	9.4533	43	9.9817	3	9.4716	46	0.5284		30'
	40'	9.4576	42	9.9814	4	9.4762	46	0.5238		20'
	50'	9.4618	41	9.9810	4	9.4808	45	0.5192		10'
17°	0'	9.4659	41	9.9806	4	9.4853	45	0.5147	73°	0'
	10'	9.4700	41	9.9802	4	9.4898	45	0.5102		50'
	20'	9.4741	40	9.9798	4	9.4943	44	0.5057		40'
	30'	9.4781	40	9.9794	4	9.4987	44	0.5013		30'
	40'	9.4821	40	9.9790	4	9.5031	44	0.4969		20'
	50'	9.4861	39	9.9786	4	9.5075	43	0.4925		10'
18°	0'	9.4900	39	9.9782	4	9.5118	43	0.4882	72°	0'
	10'	9.4939	38	9.9778	4	9.5161	42	0.4839		50'
	20'	9.4977	38	9.9774	4	9.5203	42	0.4797		40'
	30'	9.5015	37	9.9770	5	9.5245	42	0.4755		30'
	40'	9.5052	38	9.9765	4	9.5287	42	0.4713		20'
	50'	9.5090	36	9.9761	4	9.5329	41	0.4671		10'
19°	0'	9.5126	37	9.9757	5	9.5370	41	0.4630	71°	0'
	10'	9.5163	36	9.9752	4	9.5411	40	0.4589		50'
	20'	9.5199	36	9.9748	5	9.5451	40	0.4549		40'
	30'	9.5235	35	9.9743	4	9.5491	40	0.4509		30'
	40'	9.5270	36	9.9739	5	9.5531	40	0.4469		20'
	50'	9.5306	34	9.9734	4	9.5571	40	0.4429		10'
20°	0'	9.5340	35	9.9730	5	9.5611	39	0.4389	70°	0'
	10'	9.5375	34	9.9725	4	9.5650	39	0.4350		50'
	20'	9.5409	34	9.9721	5	9.5689	38	0.4311		40'
	30'	9.5443	34	9.9716	5	9.5727	39	0.4273		30'
	40'	9.5477	33	9.9711	5	9.5766	38	0.4234		20'
	50'	9.5510	33	9.9706	4	9.5804	38	0.4196		10'
21°	0'	9.5543	33	9.9702	5	9.5842	37	0.4158	69°	0'
	10'	9.5576	33	9.9697	5	9.5879	38	0.4121		50'
	20'	9.5609	32	9.9692	5	9.5917	37	0.4083		40'
	30'	9.5641	32	9.9687	5	9.5954	37	0.4046		30'
	40'	9.5673	31	9.9682	5	9.5991	37	0.4009		20'
	50'	9.5704	32	9.9677	5	9.6028	36	0.3972		10'
22°	0'	9.5736	31	9.9672	5	9.6064	36	0.3936	68°	0'
	10'	9.5767	31	9.9667	6	9.6100	36	0.3900		50'
	20'	9.5798	30	9.9661	5	9.6136	36	0.3864		40'
	30'	9.5828	31	9.9656	5	9.6172	36	0.3828		30'
Angle		L cos	d	L sin	d	L cot	cd	L tan	Angle	

Table 4. CONTINUED

Angle		L sin	d	L cos	d	L tan	cd	L cot	Angle	
22°	30'	9.5828	31	9.9656	5	9.6172	36	0.3828	68°	30'
	40'	9.5859	30	9.9651	5	9.6208	35	0.3792		20'
	50'	9.5889	30	9.9646	6	9.6243	36	0.3757		10'
23°	0'	9.5919	29	9.9640	5	9.6279	35	0.3721	67°	0'
	10'	9.5948	30	9.9635	6	9.6314	34	0.3686		50'
	20'	9.5978	29	9.9629	5	9.6348	35	0.3652		40'
	30'	9.6007	29	9.9624	6	9.6383	34	0.3617		30'
	40'	9.6036	29	9.9618	5	9.6417	35	0.3583		20'
	50'	9.6065	28	9.9613	6	9.6452	34	0.3548		10'
24°	0'	9.6093	28	9.9607	5	9.6486	34	0.3514	66°	0'
	10'	9.6121	28	9.9602	6	9.6520	33	0.3480		50'
	20'	9.6149	28	9.9596	6	9.6553	34	0.3447		40'
	30'	9.6177	28	9.9590	6	9.6587	33	0.3413		30'
	40'	9.6205	27	9.9584	5	9.6620	34	0.3380		20'
	50'	9.6232	27	9.9579	6	9.6654	33	0.3346		10'
25°	0'	9.6259	27	9.9573	6	9.6687	33	0.3313	65°	0'
	10'	9.6286	27	9.9567	6	9.6720	32	0.3280		50'
	20'	9.6313	27	9.9561	6	9.6752	33	0.3248		40'
	30'	9.6340	26	9.9555	6	9.6785	32	0.3215		30'
	40'	9.6366	26	9.9549	6	9.6817	33	0.3183		20'
	50'	9.6392	26	9.9543	6	9.6850	32	0.3150		10'
26°	0'	9.6418	26	9.9537	7	9.6882	32	0.3118	64°	0'
	10'	9.6444	26	9.9530	6	9.6914	32	0.3086		50'
	20'	9.6470	25	9.9524	6	9.6946	31	0.3054		40'
	30'	9.6495	26	9.9518	6	9.6977	32	0.3023		30'
	40'	9.6521	25	9.9512	7	9.7009	31	0.2991		20'
	50'	9.6546	24	9.9505	6	9.7040	32	0.2960		10'
27°	0'	9.6570	25	9.9499	7	9.7072	31	0.2928	63°	0'
	10'	9.6595	25	9.9492	6	9.7103	31	0.2897		50'
	20'	9.6620	24	9.9486	7	9.7134	31	0.2866		40'
	30'	9.6644	24	9.9479	6	9.7165	31	0.2835		30'
	40'	9.6668	24	9.9473	7	9.7196	30	0.2804		20'
	50'	9.6692	24	9.9466	7	9.7226	31	0.2774		10'
28°	0'	9.6716	24	9.9459	6	9.7257	30	0.2743	62°	0'
	10'	9.6740	23	9.9453	7	9.7287	30	0.2713		50'
	20'	9.6763	24	9.9446	7	9.7317	31	0.2683		40'
	30'	9.6787	23	9.9439	7	9.7348	30	0.2652		30'
	40'	9.6810	23	9.9432	7	9.7378	30	0.2622		20'
	50'	9.6833	23	9.9425	7	9.7408	30	0.2592		10'
29°	0'	9.6856	22	9.9418	7	9.7438	29	0.2562	61°	0'
	10'	9.6878	23	9.9411	7	9.7467	30	0.2533		50'
	20'	9.6901	22	9.9404	7	9.7497	29	0.2503		40'
	30'	9.6923	23	9.9397	7	9.7526	30	0.2474		30'
	40'	9.6946	22	9.9390	7	9.7556	29	0.2444		20'
	50'	9.6968	22	9.9383	8	9.7585	29	0.2415		10'
Angle		L cos	d	L sin	d	L cot	cd	L tan	Angle	

Table 4. CONTINUED

Angle	L sin	d	L cos	d	L tan	cd	L cot	Angle
30° 0'	9.6990	22	9.9375	7	9.7614	30	0.2386	60° 0'
10'	9.7012	21	9.9368	7	9.7644	29	0.2356	50'
20'	9.7033	22	9.9361	8	9.7673	28	0.2327	40'
30'	9.7055	21	9.9353	7	9.7701	29	0.2299	30'
40'	9.7076	21	9.9346	8	9.7730	29	0.2270	20'
50'	9.7097	21	9.9338	7	9.7759	29	0.2241	10'
31° 0'	9.7118	21	9.9331	8	9.7788	28	0.2212	59° 0'
10'	9.7139	21	9.9323	8	9.7816	29	0.2184	50'
20'	9.7160	21	9.9315	7	9.7845	28	0.2155	40'
30'	9.7181	20	9.9308	8	9.7873	29	0.2127	30'
40'	9.7201	21	9.9300	8	9.7902	28	0.2098	20'
50'	9.7222	20	9.9292	8	9.7930	28	0.2070	10'
32° 0'	9.7242	20	9.9284	8	9.7958	28	0.2042	58° 0'
10'	9.7262	20	9.9276	8	9.7986	28	0.2014	50'
20'	9.7282	20	9.9268	8	9.8014	28	0.1986	40'
30'	9.7302	20	9.9260	8	9.8042	28	0.1958	30'
40'	9.7322	20	9.9252	8	9.8070	27	0.1930	20'
50'	9.7342	19	9.9244	8	9.8097	28	0.1903	10'
33° 0'	9.7361	19	9.9236	8	9.8125	28	0.1875	57° 0'
10'	9.7380	20	9.9228	9	9.8153	27	0.1847	50'
20'	9.7400	19	9.9219	8	9.8180	28	0.1820	40'
30'	9.7419	19	9.9211	8	9.8208	27	0.1792	30'
40'	9.7438	19	9.9203	9	9.8235	28	0.1765	20'
50'	9.7457	19	9.9194	8	9.8263	27	0.1737	10'
34° 0'	9.7476	18	9.9186	9	9.8290	27	0.1710	56° 0'
10'	9.7494	19	9.9177	8	9.8317	27	0.1683	50'
20'	9.7513	18	9.9169	9	9.8344	27	0.1656	40'
30'	9.7531	19	9.9160	9	9.8371	27	0.1629	30'
40'	9.7550	18	9.9151	9	9.8398	27	0.1602	20'
50'	9.7568	18	9.9142	8	9.8425	27	0.1575	10'
35° 0'	9.7586	18	9.9134	9	9.8452	27	0.1548	55° 0'
10'	9.7604	18	9.9125	9	9.8479	27	0.1521	50'
20'	9.7622	18	9.9116	9	9.8506	27	0.1494	40'
30'	9.7640	17	9.9107	9	9.8533	26	0.1467	30'
40'	9.7657	18	9.9098	9	9.8559	27	0.1441	20'
50'	9.7675	17	9.9089	9	9.8586	27	0.1414	10'
36° 0'	9.7692	18	9.9080	10	9.8613	26	0.1387	54° 0'
10'	9.7710	17	9.9070	9	9.8639	27	0.1361	50'
20'	9.7727	17	9.9061	9	9.8666	26	0.1334	40'
30'	9.7744	17	9.9052	10	9.8692	26	0.1308	30'
40'	9.7761	17	9.9042	9	9.8718	27	0.1282	20'
50'	9.7778	17	9.9033	10	9.8745	26	0.1255	10'
37° 0'	9.7795	16	9.9023	9	9.8771	26	0.1229	53° 0'
10'	9.7811	17	9.9014	10	9.8797	27	0.1203	50'
20'	9.7828	16	9.9004	9	9.8824	26	0.1176	40'
30'	9.7844	17	9.8995	10	9.8850	26	0.1150	30'
Angle	L cos	d	L sin	d	L cot	cd	L tan	Angle

Table 4. CONTINUED

Angle	L sin	d	L cos	d	L tan	cd	L cot	Angle
37° 40'	9.7861		9.8985		9.8876		0.1124	53° 20'
		16		10		26		
50'	9.7877		9.8975		9.8902		0.1098	10'
		16		10		26		
38° 0'	9.7893		9.8965		9.8928		0.1072	52° 0'
		17		10		26		
10'	9.7910		9.8955		9.8954		0.1046	50'
		16		10		26		
20'	9.7926		9.8945		9.8980		0.1020	40'
		15		10		26		
30'	9.7941		9.8935		9.9006		0.0994	30'
		16		10		26		
40'	9.7957		9.8925		9.9032		0.0968	20'
		16		10		26		
50'	9.7973		9.8915		9.9058		0.0942	10'
		16		10		26		
39° 0'	9.7989		9.8905		9.9084		0.0916	51° 0'
		15		10		26		
10'	9.8004		9.8895		9.9110		0.0890	50'
		16		11		25		
20'	9.8020		9.8884		9.9135		0.0865	40'
		15		10		26		
30'	9.8035		9.8874		9.9161		0.0839	30'
		15		10		26		
40'	9.8050		9.8864		9.9187		0.0813	20'
		16		11		25		
50'	9.8066		9.8853		9.9212		0.0788	10'
		15		10		26		
40° 0'	9.8081		9.8843		9.9238		0.0762	50° 0'
		15		11		26		
10'	9.8096		9.8832		9.9264		0.0736	50'
		15		11		25		
20'	9.8111		9.8821		9.9289		0.0711	40'
		14		11		26		
30'	9.8125		9.8810		9.9315		0.0685	30'
		15		10		26		
40'	9.8140		9.8800		9.9341		0.0659	20'
		15		11		25		
50'	9.8155		9.8789		9.9366		0.0634	10'
		14		11		26		
41° 0'	9.8169		9.8778		9.9392		0.0608	49° 0'
		15		11		25		
10'	9.8184		9.8767		9.9417		0.0583	50'
		14		11		26		
20'	9.8198		9.8756		9.9443		0.0557	40'
		15		11		25		
30'	9.8213		9.8745		9.9468		0.0532	30'
		14		12		26		
40'	9.8227		9.8733		9.9494		0.0506	20'
		14		11		25		
50'	9.8241		9.8722		9.9519		0.0481	10'
		14		11		25		
42° 0'	9.8255		9.8711		9.9544		0.0456	48° 0'
		14		12		26		
10'	9.8269		9.8699		9.9570		0.0430	50'
		14		11		25		
20'	9.8283		9.8688		9.9595		0.0405	40'
		14		12		26		
30'	9.8297		9.8676		9.9621		0.0379	30'
		14		11		25		
40'	9.8311		9.8665		9.9646		0.0354	20'
		13		12		25		
50'	9.8324		9.8653		9.9671		0.0329	10'
		14		12		26		
43° 0'	9.8338		9.8641		9.9697		0.0303	47° 0'
		13		12		25		
10'	9.8351		9.8629		9.9722		0.0278	50'
		14		11		25		
20'	9.8365		9.8618		9.9747		0.0253	40'
		13		12		25		
30'	9.8378		9.8606		9.9772		0.0228	30'
		13		12		26		
40'	9.8391		9.8594		9.9798		0.0202	20'
		14		12		25		
50'	9.8405		9.8582		9.9823		0.0177	10'
		13		13		25		
44° 0'	9.8418		9.8569		9.9848		0.0152	46° 0'
		13		12		26		
10'	9.8431		9.8557		9.9874		0.0126	50'
		13		12		25		
20'	9.8444		9.8545		9.9899		0.0101	40'
		13		13		25		
30'	9.8457		9.8532		9.9924		0.0076	30'
		12		12		25		
40'	9.8469		9.8520		9.9949		0.0051	20'
		13		13		26		
50'	9.8482		9.8507		9.9975		0.0025	10'
		13		12		25		
45° 0'	9.8495		9.8495		0.0000		0.0000	45° 0'
Angle	L cos	d	L sin	d	L cot	cd	L tan	Angle

Table 5. NATURAL TRIGONOMETRIC FUNCTIONS

Angle	sin	cos	tan	cot	Angle
0° 0'	.0000	1.0000	.0000	∞	90° 0'
10'	.0029	1.0000	.0029	343.774	50'
20'	.0058	1.0000	.0058	171.885	40'
30'	.0087	1.0000	.0087	114.589	30'
40'	.0116	.9999	.0116	85.9398	20'
50'	.0145	.9999	.0145	68.7501	10'
1° 0'	.0175	.9998	.0175	57.2900	89° 0'
10'	.0204	.9998	.0204	49.1039	50'
20'	.0233	.9997	.0233	42.9641	40'
30'	.0262	.9997	.0262	38.1885	30'
40'	.0291	.9996	.0291	34.3678	20'
50'	.0320	.9995	.0320	31.2416	10'
2° 0'	.0349	.9994	.0349	28.6363	88° 0'
10'	.0378	.9993	.0378	26.4316	50'
20'	.0407	.9992	.0407	24.5418	40'
30'	.0436	.9990	.0437	22.9038	30'
40'	.0465	.9989	.0466	21.4704	20'
50'	.0494	.9988	.0495	20.2056	10'
3° 0'	.0523	.9986	.0524	19.0811	87° 0'
10'	.0552	.9985	.0553	18.0750	50'
20'	.0581	.9983	.0582	17.1693	40'
30'	.0610	.9981	.0612	16.3499	30'
40'	.0640	.9980	.0641	15.6048	20'
50'	.0669	.9978	.0670	14.9244	10'
4° 0'	.0698	.9976	.0699	14.3007	86° 0'
10'	.0727	.9974	.0729	13.7267	50'
20'	.0756	.9971	.0758	13.1969	40'
30'	.0785	.9969	.0787	12.7062	30'
40'	.0814	.9967	.0816	12.2505	20'
50'	.0843	.9964	.0846	11.8262	10'
5° 0'	.0872	.9962	.0875	11.4301	85° 0'
10'	.0901	.9959	.0904	11.0594	50'
20'	.0929	.9957	.0934	10.7119	40'
30'	.0958	.9954	.0963	10.3854	30'
40'	.0987	.9951	.0992	10.0780	20'
50'	.1016	.9948	.1022	9.7882	10'
6° 0'	.1045	.9945	.1051	9.5144	84° 0'
10'	.1074	.9942	.1080	9.2553	50'
20'	.1103	.9939	.1110	9.0098	40'
30'	.1132	.9936	.1139	8.7769	30'
40'	.1161	.9932	.1169	8.5555	20'
50'	.1190	.9929	.1198	8.3450	10'
7° 0'	.1219	.9925	.1228	8.1443	83° 0'
10'	.1248	.9922	.1257	7.9530	50'
20'	.1276	.9918	.1287	7.7704	40'
30'	.1305	.9914	.1317	7.5958	30'
Angle	cos	sin	cot	tan	Angle

Table 5. CONTINUED

Angle		sin	cos	tan	cot	Angle	
	30'	.1305	.9914	.1317	7.5958		30'
	40'	.1334	.9911	.1346	7.4287		20'
	50'	.1363	.9907	.1376	7.2687		10'
8°	0'	.1392	.9903	.1405	7.1154	82°	0'
	10'	.1421	.9899	.1435	6.9682		50'
	20'	.1449	.9894	.1465	6.8269		40'
	30'	.1478	.9890	.1495	6.6912		30'
	40'	.1507	.9886	.1524	6.5606		20'
	50'	.1536	.9881	.1554	6.4348		10'
9°	0'	.1564	.9877	.1584	6.3138	81°	0'
	10'	.1593	.9872	.1614	6.1970		50'
	20'	.1622	.9868	.1644	6.0844		40'
	30'	.1650	.9863	.1673	5.9758		30'
	40'	.1679	.9858	.1703	5.8708		20'
	50'	.1708	.9853	.1733	5.7694		10'
10°	0'	.1736	.9848	.1763	5.6713	80°	0'
	10'	.1765	.9843	.1793	5.5764		50'
	20'	.1794	.9838	.1823	5.4845		40'
	30'	.1822	.9833	.1853	5.3955		30'
	40'	.1851	.9827	.1883	5.3093		20'
	50'	.1880	.9822	.1914	5.2257		10'
11°	0'	.1908	.9816	.1944	5.1446	79°	0'
	10'	.1937	.9811	.1974	5.0658		50'
	20'	.1965	.9805	.2004	4.9894		40'
	30'	.1994	.9799	.2035	4.9152		30'
	40'	.2022	.9793	.2065	4.8430		20'
	50'	.2051	.9787	.2095	4.7729		10'
12°	0'	.2079	.9781	.2126	4.7046	78°	0'
	10'	.2108	.9775	.2156	4.6382		50'
	20'	.2136	.9769	.2186	4.5736		40'
	30'	.2164	.9763	.2217	4.5107		30'
	40'	.2193	.9757	.2247	4.4494		20'
	50'	.2221	.9750	.2278	4.3897		10'
13°	0'	.2250	.9744	.2309	4.3315	77°	0'
	10'	.2278	.9737	.2339	4.2747		50'
	20'	.2306	.9730	.2370	4.2193		40'
	30'	.2334	.9724	.2401	4.1653		30'
	40'	.2363	.9717	.2432	4.1126		20'
	50'	.2391	.9710	.2462	4.0611		10'
14°	0'	.2419	.9703	.2493	4.0108	76°	0'
	10'	.2447	.9696	.2524	3.9617		50'
	20'	.2476	.9689	.2555	3.9136		40'
	30'	.2504	.9681	.2586	3.8667		30'
	40'	.2532	.9674	.2617	3.8208		20'
	50'	.2560	.9667	.2648	3.7760		10'
15°	0'	.2588	.9659	.2679	3.7321	75°	0'
Angle		cos	sin	cot	tan	Angle	

Table 5. CONTINUED

Angle	sin	cos	tan	cot	Angle
15° 0'	.2588	.9659	.2679	3.7321	75° 0'
10'	.2616	.9652	.2711	3.6891	50'
20'	.2644	.9644	.2742	3.6470	40'
30'	.2672	.9636	.2773	3.6059	30'
40'	.2700	.9628	.2805	3.5656	20'
50'	.2728	.9621	.2836	3.5261	10'
16° 0'	.2756	.9613	.2867	3.4874	74° 0'
10'	.2784	.9605	.2899	3.4495	50'
20'	.2812	.9596	.2931	3.4124	40'
30'	.2840	.9588	.2962	3.3759	30'
40'	.2868	.9580	.2994	3.3402	20'
50'	.2896	.9572	.3026	3.3052	10'
17° 0'	.2924	.9563	.3057	3.2709	73° 0'
10'	.2952	.9555	.3089	3.2371	50'
20'	.2979	.9546	.3121	3.2041	40'
30'	.3007	.9537	.3153	3.1716	30'
40'	.3035	.9528	.3185	3.1397	20'
50'	.3062	.9520	.3217	3.1084	10'
18° 0'	.3090	.9511	.3249	3.0777	72° 0'
10'	.3118	.9502	.3281	3.0475	50'
20'	.3145	.9492	.3314	3.0178	40'
30'	.3173	.9483	.3346	2.9887	30'
40'	.3201	.9474	.3378	2.9600	20'
50'	.3228	.9465	.3411	2.9319	10'
19° 0'	.3256	.9455	.3443	2.9042	71° 0'
10'	.3283	.9446	.3476	2.8770	50'
20'	.3311	.9436	.3508	2.8502	40'
30'	.3338	.9426	.3541	2.8239	30'
40'	.3365	.9417	.3574	2.7980	20'
50'	.3393	.9407	.3607	2.7725	10'
20° 0'	.3420	.9397	.3640	2.7475	70° 0'
10'	.3448	.9387	.3673	2.7228	50'
20'	.3475	.9377	.3706	2.6985	40'
30'	.3502	.9367	.3739	2.6746	30'
40'	.3529	.9356	.3772	2.6511	20'
50'	.3557	.9346	.3805	2.6279	10'
21° 0'	.3584	.9336	.3839	2.6051	69° 0'
10'	.3611	.9325	.3872	2.5826	50'
20'	.3638	.9315	.3906	2.5605	40'
30'	.3665	.9304	.3939	2.5386	30'
40'	.3692	.9293	.3973	2.5172	20'
50'	.3719	.9283	.4006	2.4960	10'
22° 0'	.3746	.9272	.4040	2.4751	68° 0'
10'	.3773	.9261	.4074	2.4545	50'
20'	.3800	.9250	.4108	2.4342	40'
30'	.3827	.9239	.4142	2.4142	30'
Angle	cos	sin	cot	tan	Angle

Table 5. CONTINUED

Angle		sin	cos	tan	cot	Angle	
	30'	.3827	.9239	.4142	2.4142		30'
	40'	.3854	.9228	.4176	2.3945		20'
	50'	.3881	.9216	.4210	2.3750		10'
23°	0'	.3907	.9205	.4245	2.3559	67°	0'
	10'	.3934	.9194	.4279	2.3369		50'
	20'	.3961	.9182	.4314	2.3183		40'
	30'	.3987	.9171	.4348	2.2998		30'
	40'	.4014	.9159	.4383	2.2817		20'
	50'	.4041	.9147	.4417	2.2637		10'
24°	0'	.4067	.9135	.4452	2.2460	66°	0'
	10'	.4094	.9124	.4487	2.2286		50'
	20'	.4120	.9112	.4522	2.2113		40'
	30'	.4147	.9100	.4557	2.1943		30'
	40'	.4173	.9088	.4592	2.1775		20'
	50'	.4200	.9075	.4628	2.1609		10'
25°	0'	.4226	.9063	.4663	2.1445	65°	0'
	10'	.4253	.9051	.4699	2.1283		50'
	20'	.4279	.9038	.4734	2.1123		40'
	30'	.4305	.9026	.4770	2.0965		30'
	40'	.4331	.9013	.4806	2.0809		20'
	50'	.4358	.9001	.4841	2.0655		10'
26°	0'	.4384	.8988	.4877	2.0503	64°	0'
	10'	.4410	.8975	.4913	2.0353		50'
	20'	.4436	.8962	.4950	2.0204		40'
	30'	.4462	.8949	.4986	2.0057		30'
	40'	.4488	.8936	.5022	1.9912		20'
	50'	.4514	.8923	.5059	1.9768		10'
27°	0'	.4540	.8910	.5095	1.9626	63°	0'
	10'	.4566	.8897	.5132	1.9486		50'
	20'	.4592	.8884	.5169	1.9347		40'
	30'	.4617	.8870	.5206	1.9210		30'
	40'	.4643	.8857	.5243	1.9074		20'
	50'	.4669	.8843	.5280	1.8940		10'
28°	0'	.4695	.8829	.5317	1.8807	62°	0'
	10'	.4720	.8816	.5354	1.8676		50'
	20'	.4746	.8802	.5392	1.8546		40'
	30'	.4772	.8788	.5430	1.8418		30'
	40'	.4797	.8774	.5467	1.8291		20'
	50'	.4823	.8760	.5505	1.8165		10'
29°	0'	.4848	.8746	.5543	1.8040	61°	0'
	10'	.4874	.8732	.5581	1.7917		50'
	20'	.4899	.8718	.5619	1.7796		40'
	30'	.4924	.8704	.5658	1.7675		30'
	40'	.4950	.8689	.5696	1.7556		20'
	50'	.4975	.8675	.5735	1.7437		10'
30°	0'	.5000	.8660	.5774	1.7321	60°	0'
Angle		cos	sin	cot	tan	Angle	

Table 5. CONTINUED

Angle	sin	cos	tan	cot	Angle
30° 0'	.5000	.8660	.5774	1.7321	60° 0'
10'	.5025	.8646	.5812	1.7205	50'
20'	.5050	.8631	.5851	1.7090	40'
30'	.5075	.8616	.5890	1.6977	30'
40'	.5100	.8601	.5930	1.6864	20'
50'	.5125	.8587	.5969	1.6753	10'
31° 0'	.5150	.8572	.6009	1.6643	59° 0'
10'	.5175	.8557	.6048	1.6534	50'
20'	.5200	.8542	.6088	1.6426	40'
30'	.5225	.8526	.6128	1.6319	30'
40'	.5250	.8511	.6168	1.6212	20'
50'	.5275	.8496	.6208	1.6107	10'
32° 0'	.5299	.8480	.6249	1.6003	58° 0'
10'	.5324	.8465	.6289	1.5900	50'
20'	.5348	.8450	.6330	1.5798	40'
30'	.5373	.8434	.6371	1.5697	30'
40'	.5398	.8418	.6412	1.5597	20'
50'	.5422	.8403	.6453	1.5497	10'
33° 0'	.5446	.8387	.6494	1.5399	57° 0'
10'	.5471	.8371	.6536	1.5301	50'
20'	.5495	.8355	.6577	1.5204	40'
30'	.5519	.8339	.6619	1.5108	30'
40'	.5544	.8323	.6661	1.5013	20'
50'	.5568	.8307	.6703	1.4919	10'
34° 0'	.5592	.8290	.6745	1.4826	56° 0'
10'	.5616	.8274	.6787	1.4733	50'
20'	.5640	.8258	.6830	1.4641	40'
30'	.5664	.8241	.6873	1.4550	30'
40'	.5688	.8225	.6916	1.4460	20'
50'	.5712	.8208	.6959	1.4370	10'
35° 0'	.5736	.8192	.7002	1.4281	55° 0'
10'	.5760	.8175	.7046	1.4193	50'
20'	.5783	.8158	.7089	1.4106	40'
30'	.5807	.8141	.7133	1.4019	30'
40'	.5831	.8124	.7177	1.3934	20'
50'	.5854	.8107	.7221	1.3848	10'
36° 0'	.5878	.8090	.7265	1.3764	54° 0'
10'	.5901	.8073	.7310	1.3680	50'
20'	.5925	.8056	.7355	1.3597	40'
30'	.5948	.8039	.7400	1.3514	30'
40'	.5972	.8021	.7445	1.3432	20'
50'	.5995	.8004	.7490	1.3351	10'
37° 0'	.6018	.7986	.7536	1.3270	53° 0'
10'	.6041	.7969	.7581	1.3190	50'
20'	.6065	.7951	.7627	1.3111	40'
30'	.6088	.7934	.7673	1.3032	30'
Angle	cos	sin	cot	tan	Angle

Table 5. CONTINUED

Angle	sin	cos	tan	cot	Angle
30'	.6088	.7934	.7673	1.3032	30'
40'	.6111	.7916	.7720	1.2954	20'
50'	.6134	.7898	.7766	1.2876	10'
38° 0'	.6157	.7880	.7813	1.2799	52° 0'
10'	.6180	.7862	.7860	1.2723	50'
20'	.6202	.7844	.7907	1.2647	40'
30'	.6225	.7826	.7954	1.2572	30'
40'	.6248	.7808	.8002	1.2497	20'
50'	.6271	.7790	.8050	1.2423	10'
39° 0'	.6293	.7771	.8098	1.2349	51° 0'
10'	.6316	.7753	.8146	1.2276	50'
20'	.6338	.7735	.8195	1.2203	40'
30'	.6361	.7716	.8243	1.2131	30'
40'	.6383	.7698	.8292	1.2059	20'
50'	.6406	.7679	.8342	1.1988	10'
40° 0'	.6428	.7660	.8391	1.1918	50° 0'
10'	.6450	.7642	.8441	1.1847	50'
20'	.6472	.7623	.8491	1.1778	40'
30'	.6494	.7604	.8541	1.1708	30'
40'	.6517	.7585	.8591	1.1640	20'
50'	.6539	.7566	.8642	1.1571	10'
41° 0'	.6561	.7547	.8693	1.1504	49° 0'
10'	.6583	.7528	.8744	1.1436	50'
20'	.6604	.7509	.8796	1.1369	40'
30'	.6626	.7490	.8847	1.1303	30'
40'	.6648	.7470	.8899	1.1237	20'
50'	.6670	.7451	.8952	1.1171	10'
42° 0'	.6691	.7431	.9004	1.1106	48° 0'
10'	.6713	.7412	.9057	1.1041	50'
20'	.6734	.7392	.9110	1.0977	40'
30'	.6756	.7373	.9163	1.0913	30'
40'	.6777	.7353	.9217	1.0850	20'
50'	.6799	.7333	.9271	1.0786	10'
43° 0'	.6820	.7314	.9325	1.0724	47° 0'
10'	.6841	.7294	.9380	1.0661	50'
20'	.6862	.7274	.9435	1.0599	40'
30'	.6884	.7254	.9490	1.0538	30'
40'	.6905	.7234	.9545	1.0477	20'
50'	.6926	.7214	.9601	1.0416	10'
44° 0'	.6947	.7193	.9657	1.0355	46° 0'
10'	.6967	.7173	.9713	1.0295	50'
20'	.6988	.7153	.9770	1.0235	40'
30'	.7009	.7133	.9827	1.0176	30'
40'	.7030	.7112	.9884	1.0117	20'
50'	.7050	.7092	.9942	1.0058	10'
45° 0'	.7071	.7071	1.0000	1.0000	45° 0'
Angle	cos	sin	cot	tan	Angle

Table 6. RADIAN ARGUMENT

rad	sin	tan	cot	cos
.00	.0000	.0000	......	1.0000
.01	.0100	.0100	99.997	1.0000
.02	.0200	.0200	49.993	.9998
.03	.0300	.0300	33.323	.9996
.04	.0400	.0400	24.987	.9992
.05	.0500	.0500	19.983	.9988
.06	.0600	.0601	16.647	.9982
.07	.0699	.0701	14.262	.9976
.08	.0799	.0802	12.473	.9968
.09	.0899	.0902	11.081	.9960
.10	.0998	.1003	9.967	.9950
.11	.1098	.1104	9.054	.9940
.12	.1197	.1206	8.293	.9928
.13	.1296	.1307	7.649	.9916
.14	.1395	.1409	7.096	.9902
.15	.1494	.1511	6.617	.9888
.16	.1593	.1614	6.197	.9872
.17	.1692	.1717	5.826	.9856
.18	.1790	.1820	5.495	.9838
.19	.1889	.1923	5.200	.9820
.20	.1987	.2027	4.933	.9801
.21	.2085	.2131	4.692	.9780
.22	.2182	.2236	4.472	.9759
.23	.2280	.2341	4.271	.9737
.24	.2377	.2447	4.086	.9713
.25	.2474	.2553	3.916	.9689
.26	.2571	.2660	3.759	.9664
.27	.2667	.2768	3.613	.9638
.28	.2764	.2876	3.478	.9611
.29	.2860	.2984	3.351	.9582
.30	.2955	.3093	3.233	.9553
.31	.3051	.3203	3.122	.9523
.32	.3146	.3314	3.018	.9492
.33	.3240	.3425	2.920	.9460
.34	.3335	.3537	2.827	.9428
.35	.3429	.3650	2.740	.9394
.36	.3523	.3764	2.657	.9359
.37	.3616	.3879	2.578	.9323
.38	.3709	.3994	2.504	.9287
.39	.3802	.4111	2.433	.9249
.40	.3894	.4228	2.365	.9211

Table 6. CONTINUED

rad	sin	tan	cot	cos
.40	.3894	.4228	2.365	.9211
.41	.3986	.4346	2.301	.9171
.42	.4078	.4466	2.239	.9131
.43	.4169	.4586	2.180	.9090
.44	.4259	.4708	2.124	.9048
.45	.4350	.4831	2.070	.9004
.46	.4439	.4954	2.018	.8961
.47	.4529	.5080	1.969	.8916
.48	.4618	.5206	1.921	.8870
.49	.4706	.5334	1.875	.8823
.50	.4794	.5463	1.830	.8776
.51	.4882	.5594	1.788	.8727
.52	.4969	.5726	1.747	.8678
.53	.5055	.5859	1.707	.8628
.54	.5141	.5994	1.668	.8577
.55	.5227	.6131	1.631	.8525
.56	.5312	.6269	1.595	.8473
.57	.5396	.6410	1.560	.8419
.58	.5480	.6552	1.526	.8365
.59	.5564	.6696	1.494	.8309
.60	.5646	.6841	1.462	.8253
.61	.5729	.6989	1.431	.8196
.62	.5810	.7139	1.401	.8139
.63	.5891	.7291	1.372	.8080
.64	.5972	.7445	1.343	.8021
.65	.6052	.7602	1.315	.7961
.66	.6131	.7761	1.288	.7900
.67	.6210	.7923	1.262	.7838
.68	.6288	.8087	1.237	.7776
.69	.6365	.8253	1.212	.7712
.70	.6442	.8423	1.187	.7648
.71	.6518	.8595	1.163	.7584
.72	.6594	.8771	1.140	.7518
.73	.6669	.8949	1.117	.7452
.74	.6743	.9131	1.095	.7385
.75	.6816	.9316	1.073	.7317
.76	.6889	.9505	1.052	.7248
.77	.6961	.9897	1.031	.7179
.78	.7033	.9893	1.011	.7109
.79	.7104	1.0093	.9908	.7038
.80	.7174	1.0296	.9712	.6967

Table 6. CONTINUED

rad	sin	tan	cot	cos
.80	.7174	1.030	.9712	.6967
.81	.7243	1.050	.9520	.6895
.82	.7311	1.072	.9331	.6822
.83	.7379	1.093	.9146	.6749
.84	.7446	1.116	.8964	.6675
.85	.7513	1.138	.8785	.6600
.86	.7578	1.162	.8609	.6524
.87	.7643	1.185	.8437	.6448
.88	.7707	1.210	.8267	.6372
.89	.7771	1.235	.8100	.6294
.90	.7833	1.260	.7936	.6216
.91	.7895	1.286	.7774	.6137
.92	.7956	1.313	.7615	.6058
.93	.8016	1.341	.7450	.5978
.94	.8076	1.369	.7303	.5898
.95	.8134	1.398	.7151	.5817
.96	.8192	1.428	.7001	.5735
.97	.8249	1.459	.6853	.5653
.98	.8305	1.491	.6707	.5570
.99	.8360	1.524	.6563	.5487
1.00	.8415	1.557	.6421	.5403
1.01	.8468	1.592	.6281	.5319
1.02	.8521	1.628	.6142	.5234
1.03	.8573	1.665	.6005	.5148
1.04	.8624	1.704	.5870	.5062
1.05	.8674	1.743	.5736	.4976
1.06	.8724	1.784	.5604	.4889
1.07	.8772	1.827	.5473	.4801
1.08	.8820	1.871	.5344	.4713
1.09	.8866	1.917	.5216	.4625
1.10	.8912	1.965	.5090	.4536
1.11	.8957	2.014	.4964	.4447
1.12	.9001	2.066	.4840	.4357
1.13	.9044	2.120	.4718	.4267
1.14	.9086	2.176	.4596	.4176
1.15	.9128	2.234	.4475	.4085
1.16	.9168	2.296	.4356	.3993
1.17	.9208	2.360	.4237	.3902
1.18	.9246	2.427	.4120	.3809
1.19	.9284	2.498	.4003	.3717
1.20	.9320	2.572	.3888	.3624

Table 6. CONTINUED

rad	sin	tan	cot	cos
1.20	.9320	2.572	.3888	.3624
1.21	.9356	2.650	.3773	.3530
1.22	.9391	2.733	.3659	.3436
1.23	.9425	2.820	.3546	.3342
1.24	.9458	2.912	.3434	.3248
1.25	.9490	3.010	.3323	.3153
1.26	.9521	3.113	.3212	.3058
1.27	.9551	3.224	.3102	.2963
1.28	.9580	3.341	.2993	.2867
1.29	.9608	3.467	.2884	.2771
1.30	.9636	3.602	.2776	.2675
1.31	.9662	3.747	.2669	.2579
1.32	.9687	3.903	.2562	.2482
1.33	.9711	4.072	.2456	.2385
1.34	.9735	4.256	.2350	.2288
1.35	.9757	4.455	.2245	.2190
1.36	.9779	4.673	.2140	.2092
1.37	.9799	4.913	.2035	.1994
1.38	.9819	5.177	.1931	.1896
1.39	.9837	5.471	.1828	.1798
1.40	.9854	5.798	.1725	.1700
1.41	.9871	6.165	.1622	.1601
1.42	.9887	6.581	.1519	.1502
1.43	.9901	7.055	.1417	.1403
1.44	.9915	7.602	.1315	.1304
1.45	.9927	8.238	.1214	.1205
1.46	.9939	8.989	.1113	.1106
1.47	.9949	9.887	.1011	.1006
1.48	.9959	10.983	.0910	.0907
1.49	.9967	12.350	.0810	.0807
1.50	.9975	14.101	.0709	.0707
1.51	.9982	16.428	.0609	.0608
1.52	.9987	19.670	.0508	.0508
1.53	.9992	24.498	.0408	.0408
1.54	.9995	32.461	.0308	.0308
1.55	.9998	48.078	.0208	.0208
1.56	.9999	92.620	.0108	.0108
1.57	1.0000	1255.8	.0008	.0008
1.58	1.0000	−108.65	−.0092	−.0092

Table 7. ln x, e^x, e^{-x}, sinh x, cosh x

x	$\ln x$	e^x	e^{-x}	$\sinh x$	$\cosh x$
0.0	....	1.00000	1.00000	0.0000	1.0000
0.1	−2.30258	1.10517	9.04837 − 1	0.1002	1.0050
0.2	−1.60943	1.22140	8.18731 − 1	0.2013	1.0201
0.3	−1.20397	1.34986	7.40818 − 1	0.3045	1.0453
0.4	−0.91629	1.49182	6.70320 − 1	0.4108	1.0811
0.5	−0.69314	1.64872	6.06531 − 1	0.5211	1.1276
0.6	−0.51082	1.82212	5.48812 − 1	0.6367	1.1855
0.7	−0.35667	2.01375	4.96585 − 1	0.7586	1.2552
0.8	−0.22314	2.22554	4.49329 − 1	0.8881	1.3374
0.9	−0.10536	2.45960	4.06570 − 1	1.0265	1.4331
1.0	0.00000	2.71828	3.67879 − 1	1.1752	1.5431
1.1	0.09531	3.00417	3.32871 − 1	1.3356	1.6685
1.2	0.18232	3.32012	3.01194 − 1	1.5095	1.8107
1.3	0.26236	3.66930	2.72532 − 1	1.6984	1.9709
1.4	0.33647	4.05520	2.46597 − 1	1.9043	2.1509
1.5	0.40547	4.48169	2.23130 − 1	2.1293	2.3524
1.6	0.47000	4.95303	2.01897 − 1	2.3756	2.5775
1.7	0.53063	5.47395	1.82684 − 1	2.6456	2.8283
1.8	0.58779	6.04965	1.65299 − 1	2.9422	3.1075
1.9	0.64185	6.68589	1.49569 − 1	3.2682	3.4177
2.0	0.69315	7.38906	1.35335 − 1	3.6269	3.7622
2.1	0.74194	8.16617	1.22456 − 1	4.0219	4.1443
2.2	0.78846	9.02501	1.10803 − 1	4.4571	4.5679
2.3	0.83291	9.97418	1.00259 − 1	4.9370	5.0372
2.4	0.87547	11.0232	9.07180 − 2	5.4662	5.5569
2.5	0.91629	12.1825	8.20850 − 2	6.0502	6.1323
2.6	0.95551	13.4637	7.42736 − 2	6.6947	6.7690
2.7	0.99325	14.8797	6.72055 − 2	7.4063	7.4735
2.8	1.02962	16.4446	6.08101 − 2	8.1919	8.2527
2.9	1.06471	18.1741	5.50232 − 2	9.0596	9.1146
3.0	1.09861	20.0855	4.97871 − 2	10.018	10.068
3.1	1.13140	22.1980	4.50492 − 2	11.076	11.122
3.2	1.16315	24.5325	4.07622 − 2	12.246	12.287
3.3	1.19392	27.1126	3.68832 − 2	13.538	13.575
3.4	1.22378	29.9641	3.33733 − 2	14.965	14.999
3.5	1.25276	33.1155	3.01974 − 2	16.543	16.573
3.6	1.28093	36.5982	2.73237 − 2	18.286	18.313
3.7	1.30833	40.4473	2.47235 − 2	20.211	20.236
3.8	1.33500	44.7012	2.23708 − 2	22.339	22.362
3.9	1.36098	49.4024	2.02419 − 2	24.691	24.711
4.0	1.38629	54.5982	1.83156 − 2	27.290	27.308
4.1	1.41099	60.3403	1.65727 − 2	30.162	30.178
4.2	1.43508	66.6863	1.49956 − 2	33.336	33.351
4.3	1.45861	73.6998	1.35686 − 2	36.843	36.857
4.4	1.48160	81.4509	1.22773 − 2	40.719	40.732
4.5	1.50408	90.0171	1.11090 − 2	45.003	45.014
4.6	1.52606	99.4843	1.00518 − 2	49.737	49.747
4.7	1.54756	109.947	9.09528 − 3	54.969	54.978
4.8	1.56862	121.510	8.22975 − 3	60.751	60.759
4.9	1.58924	134.290	7.44658 − 3	67.141	67.149
x	$\ln x$	e^x	e^{-x}	$\sinh x$	$\cosh x$

Table 7. CONTINUED

x	$\ln x$	e^x	e^{-x}	$\sinh x$	$\cosh x$
5.0	1.60944	148.413	6.73795 − 3	74.203	74.210
5.1	1.62924	164.022	6.09675 − 3	82.008	82.014
5.2	1.64866	181.272	5.51656 − 3	90.633	90.639
5.3	1.66771	200.337	4.99159 − 3	100.17	100.17
5.4	1.68640	221.406	4.51658 − 3	110.70	110.71
5.5	1.70475	244.692	4.08677 − 3	122.34	122.35
5.6	1.72277	270.426	3.69786 − 3	135.21	135.22
5.7	1.74047	298.867	3.34597 − 3	149.43	149.44
5.8	1.75786	330.300	3.02755 − 3	165.15	165.15
5.9	1.77495	365.037	2.73944 − 3	182.52	182.52
6.0	1.79176	403.429	2.47875 − 3	201.71	201.72
6.1	1.80829	445.858	2.24287 − 3	222.93	222.93
6.2	1.82455	492.749	2.02943 − 3	246.37	246.38
6.3	1.84055	544.572	1.83630 − 3	272.29	272.29
6.4	1.85630	601.845	1.66156 − 3	300.92	300.92
6.5	1.87180	665.142	1.50344 − 3	332.57	332.57
6.6	1.88707	735.093	1.36037 − 3	367.55	367.55
6.7	1.90211	812.406	1.23091 − 3	406.20	406.20
6.8	1.91692	897.847	1.11378 − 3	448.92	448.92
6.9	1.93152	992.275	1.00779 − 3	496.14	496.14
7.0	1.94591	1096.63	9.11882 − 4	548.32	548.32
7.1	1.96009	1211.97	8.25105 − 4	605.98	605.98
7.2	1.97408	1339.43	7.46586 − 4	669.72	669.72
7.3	1.98787	1480.30	6.75539 − 4	740.15	740.15
7.4	2.00148	1635.98	6.11253 − 4	817.99	817.99
7.5	2.01490	1808.04	5.53084 − 4	904.02	904.02
7.6	2.02815	1998.20	5.00451 − 4	999.10	999.10
7.7	2.04122	2208.35	4.52827 − 4	1104.2	1104.2
7.8	2.05412	2440.60	4.09735 − 4	1220.3	1220.3
7.9	2.06686	2697.28	3.70744 − 4	1348.6	1348.6
8.0	2.07944	2980.96	3.35463 − 4	1490.5	1490.5
8.1	2.09186	3294.47	3.03539 − 4	1647.2	1647.2
8.2	2.10413	3640.95	2.74654 − 4	1820.5	1820.5
8.3	2.11626	4023.87	2.48517 − 4	2011.9	2011.9
8.4	2.12823	4447.07	2.24867 − 4	2223.5	2223.5
8.5	2.14007	4914.77	2.03468 − 4	2457.4	2457.4
8.6	2.15176	5431.66	1.84106 − 4	2715.8	2715.8
8.7	2.16332	6002.91	1.66586 − 4	3001.5	3001.5
8.8	2.17475	6634.24	1.50733 − 4	3317.1	3317.1
8.9	2.18605	7331.97	1.36389 − 4	3666.0	3666.0
9.0	2.19722	8103.08	1.23410 − 4	4051.5	4051.5
9.1	2.20827	8955.29	1.11666 − 4	4477.6	4477.6
9.2	2.21920	9897.13	1.01039 − 4	4948.6	4948.6
9.3	2.23001	10938.0	9.14242 − 5	5469.0	5469.0
9.4	2.24071	12088.4	8.27241 − 5	6044.2	6044.2
9.5	2.25129	13359.7	7.48518 − 5	6679.9	6679.9
9.6	2.26176	14764.8	6.77287 − 5	7382.4	7382.4
9.7	2.27213	16317.6	6.12835 − 5	8158.8	8158.8
9.8	2.28238	18033.7	5.54516 − 5	9016.9	9016.9
9.9	2.29253	19930.4	5.01747 − 5	9965.2	9965.2
x	$\ln x$	e^x	e^{-x}	$\sinh x$	$\cosh x$

Table 8. POWERS AND ROOTS

No.	Square	Cube	Square Root	Cube Root	No.	Square	Cube	Square Root	Cube Root
1	1	1	1.000	1.000	51	2601	132651	7.141	3.708
2	4	8	1.414	1.260	52	2704	140608	7.211	3.733
3	9	27	1.732	1.442	53	2809	148877	7.280	3.756
4	16	64	2.000	1.587	54	2916	157464	7.348	3.780
5	25	125	2.236	1.710	55	3025	166375	7.416	3.803
6	36	216	2.449	1.817	56	3136	175616	7.483	3.826
7	49	343	2.646	1.913	57	3249	185193	7.550	3.849
8	64	512	2.828	2.000	58	3364	195112	7.616	3.871
9	81	729	3.000	2.080	59	3481	205379	7.681	3.893
10	100	1000	3.162	2.154	60	3600	216000	7.746	3.915
11	121	1331	3.317	2.224	61	3721	226981	7.810	3.936
12	144	1728	3.464	2.289	62	3844	238328	7.874	3.958
13	169	2197	3.606	2.351	63	3969	250047	7.937	3.979
14	196	2744	3.742	2.410	64	4096	262144	8.000	4.000
15	225	3375	3.873	2.466	65	4225	274625	8.062	4.021
16	256	4096	4.000	2.520	66	4356	287496	8.124	4.041
17	289	4913	4.123	2.571	67	4489	300763	8.185	4.062
18	324	5832	4.243	2.621	68	4624	314432	8.246	4.082
19	361	6859	4.359	2.668	69	4761	328509	8.307	4.102
20	400	8000	4.472	2.714	70	4900	343000	8.367	4.121
21	441	9261	4.583	2.759	71	5041	357911	8.426	4.141
22	484	10648	4.690	2.802	72	5184	373248	8.485	4.160
23	529	12167	4.796	2.844	73	5329	389017	8.544	4.179
24	576	13824	4.899	2.884	74	5476	405224	8.602	4.198
25	625	15625	5.000	2.924	75	5625	421875	8.660	4.217
26	676	17576	5.099	2.962	76	5776	438976	8.718	4.236
27	729	19683	5.196	3.000	77	5929	456533	8.775	4.254
28	784	21952	5.292	3.037	78	6084	474552	8.832	4.273
29	841	24389	5.385	3.072	79	6241	493039	8.888	4.291
30	900	27000	5.477	3.107	80	6400	512000	8.944	4.309
31	961	29791	5.568	3.141	81	6561	531441	9.000	4.327
32	1024	32768	5.657	3.175	82	6724	551368	9.055	4.344
33	1089	35937	5.745	3.208	83	6889	571787	9.110	4.362
34	1156	39304	5.831	3.240	84	7056	592704	9.165	4.380
35	1225	42875	5.916	3.271	85	7225	614125	9.220	4.397
36	1296	46656	6.000	3.302	86	7396	636056	9.274	4.414
37	1369	50653	6.083	3.332	87	7569	658503	9.327	4.431
38	1444	54872	6.164	3.362	88	7744	681472	9.381	4.448
39	1521	59319	6.245	3.391	89	7921	704969	9.434	4.465
40	1600	64000	6.325	3.420	90	8100	729000	9.487	4.481
41	1681	68921	6.403	3.448	91	8281	753571	9.539	4.498
42	1764	74088	6.481	3.476	92	8464	778688	9.592	4.514
43	1849	79507	6.557	3.503	93	8649	804357	9.644	4.531
44	1936	85184	6.633	3.530	94	8836	830584	9.695	4.547
45	2025	91125	6.708	3.557	95	9025	857375	9.747	4.563
46	2116	97336	6.782	3.583	96	9216	884736	9.798	4.579
47	2209	103823	6.856	3.609	97	9409	912673	9.849	4.595
48	2304	110592	6.928	3.634	98	9604	941192	9.899	4.610
49	2401	117649	7.000	3.659	99	9801	970299	9.950	4.626
50	2500	125000	7.071	3.684	100	10000	1000000	10.000	4.642

Index